Springer INdAM Series

Volume 51

This series will publish textbooks, multi-authors books, thesis and monographs in English language resulting from workshops, conferences, courses, schools, seminars, doctoral thesis, and research activities carried out at INDAM - Istituto Nazionale di Alta Matematica, http://www.altamatematica.it/en. The books in the series will discuss recent results and analyze new trends in mathematics and its applications.

THE SERIES IS INDEXED IN SCOPUS

Paolo Barbante • Francesco D. Belgiorno •
Silvia Lorenzani • Lorenzo Valdettaro
Editors

From Kinetic Theory to Turbulence Modeling

The Legacy of Carlo Cercignani

Springer

Editors
Paolo Barbante
Dipartimento di Matematica
Politecnico di Milano
Milano, Italy

Francesco D. Belgiorno
Dipartimento di Matematica
Politecnico di Milano
Milano, Italy

Silvia Lorenzani
Dipartimento di Matematica
Politecnico di Milano
MIlano, Italy

Lorenzo Valdettaro
Dipartimento di Matematica
Politecnico di Milano
Milano, Italy

ISSN 2281-518X ISSN 2281-5198 (electronic)
Springer INdAM Series
ISBN 978-981-19-6461-9 ISBN 978-981-19-6462-6 (eBook)
https://doi.org/10.1007/978-981-19-6462-6

This Springer imprint is published by the registered company Springer Nature Singapore Pte Ltd.
The registered company address is: 152 Beach Road, #21-01/04 Gateway East, Singapore 189721, Singapore

Preface

In this volume are collected some of the contributions presented at a conference organized in memory of Carlo Cercignani, which took place at Politecnico di Milano on May 24–28, 2021, 11 years after his death. A delay of 1 year was imposed by the difficulties due to the COVID-19 pandemic. Carlo Cercignani was unanimously considered one of the world's leading experts of the Boltzmann equation, and he carried out his research for decades in the Department of Mathematics at Politecnico di Milano. He obtained significant results in many different areas of kinetic theory relevant for applications, related, in particular, to the derivation of a gas-surface interaction kernel (the so-called Cercignani-Lampis boundary conditions), to the propagation of shock waves, to the modeling of polyatomic gases and mixtures, and to the evaporation and condensation phenomena. As a mathematician, Carlo Cercignani established important theoretical findings, including a conjecture on the behavior of the solutions to the Boltzmann equation for large times, along with the derivation of this equation from microscopic models. He also devoted particular care to analytical and semianalytical methods of solution, formulating, in particular, a variational principle for the integro-differential form of the linearized Boltzmann equation applied successfully, in the last years of his scientific career, to the study of microelectromechanical systems (MEMS) devices. This versatility reflects the feeling, which Carlo Cercignani had throughout his life, of trying to reconcile the theoretical investigation with its practical application. In addition, it is perhaps thanks to this intuition that he has become a giant in the field of rarefied gas dynamics. However, the scientific interests of Carlo Cercignani were not confined to kinetic theory. He was also very active in the turbulence research area, obtaining, among other results, a new class of subgrid-scale models for large eddy simulations (LESs). Furthermore, in the 1970s, Carlo Cercignani was among the first ones in Italy to understand the relevance of dynamical systems theory and its developments in mechanics such as KAM theory. Finally, it is worth mentioning his deep interest in the foundations of physics, in the spirit of the so-called "Einstein Classical Program" of proving that quantum mechanics can be recovered from a theory presenting some "realistic character."

Carlo Cercignani published 10 research monographs and more than 300 scientific papers. His most famous book, *The Boltzmann Equation and its Applications* (Springer-Verlag, New York, 1988), is still nowadays considered the standard reference in kinetic theory. To pay homage to the extreme versatility of this prominent scientist, recent developments in all the themes summarized above have been included in the present book.

In the framework of kinetic theory, simplified models for polyatomic gases and mixtures have been presented, along with applications of the Boltzmann equation to electron transport, social phenomena, and epidemic spread. Regarding turbulence modeling, a description of turbulence based on generalized central moments instead of fluctuations has been presented. This unconventional point of view has provided fruitful enhancements to turbulence modeling. Additionally, a class of anisotropic LES models, inspired by the work of Carlo Cercignani, has been studied and applied to different kinds of turbulent flows. Lattice Boltzmann and gas-kinetic methods are also addressed to describe complex flows. Some contributions concern the Einstein Classical Program where the recovery of quantum mechanics from classical electrodynamics is discussed, as well as Cercignani's conjecture on the possibility of having a classical zero-point energy. A few papers are also related to dynamical systems theory.

Milano, Italy

Paolo Barbante
Francesco D. Belgiorno
Silvia Lorenzani
Lorenzo Valdettaro

Contents

About the Editors

Paolo Barbante is a researcher in mathematical physics at Politecnico di Milano, where he teaches theoretical mechanics and ordinary differential equations.

He graduated in aerospace engineering from Politecnico di Milano (Italy) and obtained his PhD degree from Université Libre de Bruxelles (Belgium).

He is author of research articles mainly in the fields of reacting and multiphase flows and phase transition phenomena.

He is a member of the Italian National Group for Mathematical Physics (GNFM-INdAM).

Francesco Domenico Belgiorno is Associate Professor of Mathematical Physics at Politecnico di Milano, where he teaches theoretical mechanics. He graduated in physics and obtained his PhD degree in physics from the University of Milano.

He is author of many research articles in the field of analogue gravity and black hole thermodynamics, thermodynamics formalism, and quantum field theory in curved spacetime and in external fields.

He is a member of the National Group for Mathematical Physics (GNFM-INdAM) since 2012.

Silvia Lorenzani is Associate Professor of Mathematical Physics at Politecnico di Milano, where she teaches theoretical mechanics, and methods and models for statistical mechanics.

She graduated from the University of Bologna in both physics and mathematics, and obtained her PhD degree from the University of Goettingen.

She is author of several research articles mainly in the field of rarefied gas dynamics and microfluidics, but also on turbulence and hydrodynamic instabilities.

Since 2003, she is a member of the National Group for Mathematical Physics (GNFM-INdAM), and since 2020, she is an editorial board member of the scientific journal *Fluids* (published by MDPI).

Lorenzo Valdettaro is Associate Professor of Mathematical Physics at Politecnico di Milano, where he teaches theoretical mechanics and computational fluid mechanics.

He graduated in physics from the University of Pisa (Italy), and obtained the PhD degree from the University of Toulouse (France).

He is author of several research articles mainly in the field of turbulence modeling, numerical simulation of turbulent flows, hydrodynamic instabilities, and wave propagation.

The *"Divertissements"* of Carlo Cercignani

Luigi Galgani

Abstract A short description is given of Carlo Cercignani's intellectual activities that he performed as a kind of refreshment from his scientific research. Some samples of his poems are also included. Being not really gifted for sport activities, Carlo, so much busy with his kinetic theory, had essentially two refreshing activities, still of an intellectual character, i.e., literature and foundations of physics, that are briefly illustrated below.

Literature

The most regenerating activity for Carlo, so much busy with his work on Kinetic Theory, was *literature*, in the creative sense of actually writing himself poems and novels, or of performing translations. He liked poetry immensely, and, gifted with a prodigious memory as he was (for instance, he could at any moment recite all of Dante's *Divine Comedy* by heart), he enjoyed composing rhymes in the style of jokes that imitated celebrated Italian poems, popular songs, or ballads. These *poems* were collected in an unpublished booklet entitled *"Scherzi in versi"* (*"Jokes in verses"*) that was prepared by his wife Silvana as a surprise gift at a conference for his 60th birthday.

His activity as a *translator* was huge. Indeed, he not only translated, for example, all of Shakespeare's *Sonnets*, and the Queneau's poem *"Petite cosmogonie portative"*. But he also translated the Iliad and the Odyssey from Greek, and eventually the Aeneid from Latin, using verses suitably chosen to best reproduce the rhythm of the hexameter.

Novels

He also wrote three *novels*. I only have a vague memory of the *second* one, related to the atmosphere of the "student revolution" of 1968. I was instead fascinated by his *first novel*, written perhaps in the Seventies, by the title *"Morte di un professore"*

L. Galgani (✉)
Department of Mathematics, Università degli Studi di Milano, Milano, Italy
e-mail: luigi.galgani@unimi.it

P. Barbante et al. (eds.), *From Kinetic Theory to Turbulence Modeling*, Springer INdAM Series 51, https://doi.org/10.1007/978-981-19-6462-6_1

(i.e., Death of a professor), that is really amusing. It is mainly a parody, in which one may recognize real life professors behind certain characters. However, Chapter 6 is somehow more serious. A character named Veroviro (i.e., *true man*, alluding to a well known American professor) gives a talk at the Milan Politecnico, about kinetic theory. In fact, the talk deals rather deeply with the problem of whether the existence of a conscience (or of a soul, or of freedom) may be compatible with a deterministic vision of the world. It is clear that Carlo himself is speaking here through Veroviro. However, the only rather modest result obtained is the suggestion that the emergence of complexity (the so-called *butterfly effect*) might help in this connection.

Such a hard theme is instead, implicitly, the main objective of his *third novel* (32 short chapters, for 136 pages) entitled *La creazione secondo Michele* (i.e., *The creation according to Michael*), on which he worked in the final part of his life. In the last pages he also introduced an explicit autobiographical reference to his illness, albeit, as usual for him, in an extremely reserved way.

Apparently, the novel tells the story of how God created the world through quantum fluctuations, letting it evolve according to the laws of physics, that also account for biology and evolution. This is the first part, mainly of a didascalic character. Then, human beings come in, and Lucifer is precipitated on the Earth, bringing evil in the world. Here, the second part of the novel starts up. As the human beings seem unable to understand how the world was created, God sends the Archangel Michael on Earth in order to instruct them. To this end Michael goes in search of a suited writer, or a "typist," as he says. This is indeed the opportunity for the most beautiful part of the novel, a fascinating round up of human history, the Egyptians, the Hebrews, the Greeks, the Italian renaissance, Michelangelo, Luther, Bruno, Galileo and Newton, Faraday and Maxwell, and finally Einstein, with a mention of his love for realism, against the orthodox interpretation of Quantum Mechanics. But according to Michael all those people were so genial that they could not conceive of just acting as typists. Eventually Michael finds the good, sufficiently humble, person, Carlo himself, while he was, ill, in Paris. Here, a few pages of a very touching type (but still with irony) follow. In particular, Carlo reveals himself as the true author (and not just the typist) of the book.

Then he says: *"Man seems to go ahead towards the nothing, with a life that to most people appears devoid of any aim"*. So he refers to God, and actually fights with Him, saying: *"Men loved Him insanely all the time the world about them brought the signs of mystery and enchantment. But when they entered the luminous shadow of the tree of science, and they believed they were mastering ... the secrets of the universe, they loved God less and less. Perhaps knowledge and power made the human beings so superb that they dumped the idea of God, asking themselves, in the night, at the dim light of the stars, or in the moments of discouragement, what were they doing in the world. They no more needed His blessing."* Then he asks himself whether this is progress, but he finds no clear answer, just saying: *"Life is an adventure that is worth living."*

About Soul Within Physics

In conclusion, about the main theme of the novel, i.e., the problem of soul within physics, Carlo seems to have been unable to even imagine a clear answer. My impression, however, is that he dares to witness he feels they coexist. For example, in the novel he says the following touching words: *"the human being, so easily dominated by tenderness and wonder. A being made up of the stuff of the stars, born from a womb, but able to understand, as by enchantement, the symbols of the spirit. Living as if he were bewitched . . . "*

Foundations of Physics as a Further Divertissement

I close this short contribution by recalling how, along all his life, devoted to Kinetic Theory somehow as a work, Carlo had another divertissement in addition to literature, namely the foundations of physics in the spirit of the Einstein Classical Program. Namely, to understand whether Quantum Mechanics might be "explained" in some "realistic" way. Indeed, already in the year 1972 (see [1]) he had advanced the idea that zero-point energy might be "explained" classically within dynamical systems theory, in terms of a transition from ordered to chaotic motions, whereas in 1998 (see [2]) he discussed how Planck's distribution law may be derived in a nonquantum frame. A review of such works is given in a paper of mine (see [3]).

In particular, in an Appendix to that review are reproduced two poems by Carlo: my preferred one, by the title "1921," a kind of a lyrical recollection of the themes and characters of Mathematical Physics, and the one by the title *"Beethoven in Cielo (i.e., in Heaven)"*. In the review, erroneously (by a the reason explained there) I attributed the latter poem to Carlo. It is instead his translation of a Boltzmann's rather dramatic poem, an English version of which is available in Carlo's book on Boltzmann. Here such poems are again reproduced in an Appendix, together with another one, inspired to Virgil's Georgics.

Appendix

1921 (Year 1987)

Il pendolo semplice e il verso degli angoli, coppie e momenti,
e l'orientazion dei segmenti, che sembrano avere un po' perso
quel ruolo di bei caposaldi, che noi studiavamo convinti
sui tomi, a caratteri stinti, del buon Levi Civita-Amaldi;
i solidi rigidi e i fili poggiati e sospesi per aria,
disposti per far catenaria, e, ancor, coniugati, i profili;
le formule dell'ellissoide d'inerzia, le tre rotazioni,
il calcolo delle reazioni, il grafico della cicloide;
e dopo? anche il moto centrale e la geometria delle masse,
e come se ciò non bastasse le formule del potenziale;
la velocità areolare, insieme col noto rapporto

tra assi e periodi e uno storto poligono funicolare;
l'epicicloide ordinaria, le verghe, i vettori, i versori.
la legge d'inerzia, i cursori, l'odografo, la legge oraria;
e quei giroscopi, che ognuno ricorda, che sembrano armille;
rinasco, rinasco nel mille, eh sì, novecentoventuno!
Il Finzi è un ragazzo aitante, che svolge, sicuro e veloce,
il calcolo, pure il più atroce, e sgomina . . . un determinante;
e ancor la Pastori, serafica, in certe serate d'inverno,
risolve, su un lindo quaderno, problemi di statica grafica.
Sul tavolo la lucernetta fa un orbe di luce conchiuso,
nel quale discutere è d'uso problemi di base e rulletta.
Max Abraham, presso alla morte, ancor polemizza ma è stanco;
la teppa, rizzata sul banco, ne aveva deciso la sorte
da anni: doveva lontano andare la Kultur germanica
che quel professor di meccanica voleva illustrare a Milano.
I giorni si fanni piu' foschi; che importa se poi il suo tensore
potrà risultare migliore di quello che ha dato Minkowski?
Ma è giunta notizia che oggi la vecchia teoria newtoniana,
ed è una notizia ben strana, su basi sicure non poggi.
Oltr'alpe, se pur tra i fragori di guerra, son stati trovati
dei nuovi concetti, e, applicati, dei vecchi si trovan migliori.
Qualcuno scuotendo la testa, davanti ai propositi empi,
ripete: Che tempi! Che tempi! Dovevo sentire anche questa!
Tal altro si sente già certo di quella scoperta recente
e dice: Che mente! Che mente quel tipo, quell'Einstein . . . Alberto!
Oh sere, passare a studiare le formule dell'avvenire,
passate a cercar di capire, sapendo pur sol balbettare
ja, bitte, Forelle e Kartoffel quei fogli di stampa ancor fresca,
riempiti di lingua tedesca, e i simboli di E.B. Christoffel!
Ci sono certuni che stiman che sia una pura follia
studiar quella nuova teoria con dentro il tensore di Riemann:
che cuore partire all'assalto di pagine, quindici a quindici,
ripiene di formule e indici che stanno un po' in basso e un po' in alto!
E Weyl, che negli ultimi mesi, seguendo i precetti di Mie,
produce ancor nuove teorie, chiamate di gauge dagli inglesi!
Misteri grandissimi ancor ha in serbo la quantizzazione,
descritta con la condizione che fu escogitata da Bohr.
Quel mondo si è rotto ed ormai i giovani studiano a caso
le cose più strane e col vaso Pandora ancor semina guai!
Sì certi equilibri son rotti e circolan libri un po' strani
che srivono, qui, il Cercignani, e, un poco piú in là, il Gallavotti.
Perfino ai congressi si sente parlare di cose un po' strane:
scompaiono le lagrangiane, emerge il fibrato tangente.
O tempo vicino e lontano, sei sempre presente nel cuore!
quand'era un versore un versore e non un simbolo strano,

ch'è simile a una derivata parziale, che invece non è;
nessuno mi spiega il perché di questa Babele sfrenata.
Eh sı, non lo spiega nessuno ed io vorrei che tornasse
quel tempo, che ci si trovasse nel milnovecentoventuno.
Tornare nel tempo che fu, poter imparare i tensori
col Finzi, amar la Pastori! quei giorni non tornano più !

Beethoven in cielo (About 2000)

L'anima mia dal corpo si è staccata
con una lotta che sarà obliata.
Ma dopo tanta angoscia e sofferenza,
gioia è librarsi come pura essenza!
Il brulichio del mondo è come un velo
e salgo verso un'alta meta: il cielo!
Lunga e svelta qual freccia è la mia via.
Sento lungi una splendida armonia.
D'angeli un coro dolcemente aspetta
per accogliere chi lassù si affretta.
Tra poco saro giunto; quale incanto!
Che monotono sembri però il canto
a nascondere agli angeli non riesco.
Ridon felici: –E' un animo tedesco!
La musica da voi sale fin qua!
Dio glorifica, allor, *l'eternità*
cantiamo; e veda che ce ne intendiamo!
Ma di andare all'unì sono cerchiamo!–
E cantano un corale grande; e presto
penso: – E' lo stile di Beethoven questo;
quel pezzo lì però mi è sconosciuto."
Chiedo allora: – Cos'è?– –Ordine ha avuto
di scriverlo – mi dicon – dal Signore
l'anima di Beethoven. Con fervore
lo eseguiamo ogni volta che c'è festa;
musica qui non c'è miglior di questa!–
–Lo credo! Ma vorrei che a me mostrata
fosse l'anima sua. Non sarà stata
vana così la gita.– Divertiti,
mi conducon per bei prati fioriti
e m'indican lo spirito divino
che solitario va, in lento cammino,
sotto le palme. – Da lui, benché indegno,
da lui or voglio andare in questo regno,
colui che onora più l'ingegno umano!–
Mi vede allora e mi porge la mano:
– Benvenuto, o terreno ospite, prono

al poter della musica e al suo suono!
Per te fu il coro angelico eseguito,
da me composto in cielo, ed ho gradito
che gli angeli lo affrontin con impegno:
del mio corale, pure in questo regno,
son le quarte eccedenti assai temute!
Ma le mie note, di', ti son piaciute? –
Confuso non rispondo. Ed egli lesto
e cortese prosegue: – Animo onesto,
tu sei sincero! E' giusta l'opinione:
fuggivi in terra pur l'adulazione. –
Allora del suo dir colgo il vantaggio;
dico: – O mio eroe –, facendomi coraggio, –
o mio maestro! Ora ho ascoltato il canto
con entusiasmo. Devo dir soltanto
che, tra gli angeli, io, su queste cime,
musica mi aspettavo più sublime!–
Egli risponde sorridendo: – Senti,
la penso come te se ti lamenti.
Tutto in ciel mi vien male, che iattura!
Ho smesso di comporre addirittura.
Solo per il giudizio universale
ho dovuto impegnarmi, bene o male,
per non imbarazzare assai il buon Dio,
di scriver per gli ottoni un pezzo mio:
lo devo far, ma non ne sono lieto.
Ma sai perché, altrimenti, ormai mi vieto
di comporre? Mi manca la scintilla
più creativa, la nota che più brilla:
questa nota è il dolore! Sì, il dolore
che ti afferra e ti fa stringere il cuore;
come un metallo forte suona e vibra
e ti fa risuonare in ogni fibra.
E' un vero amico e ti fa superiore:
solo chi piange e geme dal dolore
avrà l'umanità, dono divino.
Cosa lega alla madre ogni bambino?
La grandi pene della notte in cui
c'è Dio soltanto in veglia con lei e lui.
Non hai mai pianto insieme con tua moglie?
Chi non lo fa l'amor vero non coglie,
un dolore profondo e condiviso:
il suo ricordo è come un paradiso.
Sopporta il santo pena ed afflizione:
brilla in lui il raggio della perfezione.

Fama di eroe ottener sol ti è concesso,
se fermamente domini te stesso.
Tremi il tuo cuore nella sofferenza!
Vivrai nel canto della discendenza.
Dio stesso quando qui fra noi si scorse,
fu forse un re, si volle ricco forse?
Fu figlio d'uomo, pieno del dolore,
che ognor s'incontra in qualsisia maggiore
cosa; è la nota mia fondamentale.
Ma qui tutto è beato e senza male;
la cetra mi e caduta allor di mano.–
Lo guardo ora atterrito: – Come 'e strano
lo scorrere del mondo. Poche ore
fa, chiedevo alla morte che il dolore
al cuore mio venisse risparmiato.
Ora qui, in questo mondo alto e beato,
si rimpiange il dolore! Oh, cuore umano,
veramente insondabile e ben strano!–

Leggendo le georgiche (Year 1994)
Che dir delle stelle, del cielo d'autunno, dell'ansia che prende
se il sole ogni giorno discende piu' presto e si copre d'un velo;
o se primavera finisce piovosa e nei campi matura
la messe di spighe e Natura le steli d'umori arricchisce?
E quando nei campi dorati falciare vuoi il fragile orzo
e i venti, rigonfi di sforzo, tu vedi scontrarsi adirati
e, come guerrieri nemici, avvolti di nuvole scure,
strappare le spighe mature, svellendo perfin le radici,
scagliarle nell'aria; e le nere tempeste avvolgere in spire,
facendoli in alto salire, gli steli e le stoppie leggere.
Il cielo non e' piu' celeste e senti continua scrosciare
la pioggia; salita dal mare, la nube si addensa in tempeste
oscure d'orribile pioggia; il cielo precipita in terra
e scende feroce a far guerra. Si gonfiano il fiume e la roggia,
con strepito il mare ribolle; si allagano i campi ridenti
dell'opra di giorni pazienti e sul seminato e sul colle.
Furente con noi Giove Pluvio, brandendo la folgore iroso
ci manda col buio nuvoloso, o sembra, un suo nuovo Diluvio...

References

1. Cercignani, C., Galgani, L., Scotti, A.: Zero-point energy in classical nonlinear mechanics. Phys. Lett. **38A**, 403 (1972)
2. Cercignani, C.: On a nonquantum derivation of Planck's distribution law. Found. Phys. Lett. **11**, 189 (1998)
3. Galgani, L.: Carlo Cercignani's interests for the foundations of physics. Meccanica **47**, 1723 (2012)

Tensorial Turbulent Viscosity Model for LES: Properties and Applications

Antonella Abbà and Lorenzo Valdettaro

Abstract A review of the dynamic anisotropic eddy viscosity model and its applications is presented here. Aim of the model is to overcome the limit of classical eddy viscosity models where the subgrid stress tensor is forced to be aligned to the velocity strain rate tensor. Based on the original intuition of Carlo Cercignani to model the eddy viscosity by means of a fourth order tensor contracting locally to a second order tensor, and joined to the Germano dynamic procedure, the model demonstrated to take into account the anisotropicity of turbulent structures and at the same time to well reproduce the energy transfer.

1 Anisotropic LES Models

For an incompressible, constant density flow, the filtered Navier–Stokes equations read:

$$\rho \frac{\partial \overline{v}_i}{\partial t} + \rho \frac{\partial}{\partial x_j}(\overline{v}_i \overline{v}_j) = -\frac{\partial \overline{p}}{\partial x_i} + \frac{\partial \overline{\tau}_{ij}}{\partial x_j} - \rho \frac{\partial \tau_{ij}^{\mathrm{SGS}}}{\partial x_j} + \overline{f}_i$$

where τ_{ij}^{SGS} is the *subgrid-scale* stress tensor:

$$\tau_{ij}^{SGS} = \overline{v_i v_j} - \overline{v}_i \overline{v}_j$$

A. Abbà (✉)
Department of Aerospace Science and Technology, Politecnico di Milano, Milan, Italy
e-mail: antonella.abba@polimi.it

L. Valdettaro
MOX, Department of Mathematics, Politecnico di Milano, Milano, Italy
e-mail: lorenzo.valdettaro@polimi.it

P. Barbante et al. (eds.), *From Kinetic Theory to Turbulence Modeling*, Springer INdAM Series 51, https://doi.org/10.1007/978-981-19-6462-6_2

and overbar denotes filtered quantities

$$\overline{f}(\boldsymbol{x},t) = \int_D f(\boldsymbol{x}-\boldsymbol{r},t)G_\Delta(\boldsymbol{r},\boldsymbol{x})\,\mathrm{d}^3\boldsymbol{r}$$

with G the filter function and Δ its amplitude. Frequently used eddy viscosity models assume proportionality between the anisotropic part of the SGS stress tensor $\tau_{ij}^a = \tau_{ij}^{SGS} - \frac{1}{3}\delta_{ij}\tau_{kk}^{SGS}$ and strain rate tensor $\overline{S}_{ij} = \frac{1}{2}\left(\frac{\partial \overline{v}_i}{\partial x_j} + \frac{\partial \overline{v}_j}{\partial x_i}\right)$:

$$\tau_{ij}^a = -\nu_{SGS}\overline{S}_{ij}$$

It usually contains a model constant to be provided. For Smagorinsky model, for example:

$$\nu_{SGS} = C_S^2\overline{\Delta}^2|\overline{S}|$$

here $\overline{\Delta}$ is the filter width and $|\overline{S}|^2 = 2\overline{S}_{ij}\overline{S}_{ij}$. If $\nu_{SGS} > 0$, this term has a dissipative character and it cannot account for backscatter. The dynamic procedure [7] aims to determine the model constant during the numerical simulation. The idea is to use two filters, the *grid filter* ($\bar{}$) of amplitude $\overline{\Delta}$ and the *test filter* ($\widehat{}$) of amplitude $\widehat{\Delta} > \overline{\Delta}$. It is assumed that both amplitudes are within the inertial band, and the same turbulence model is assumed to be valid for equations filtered with both filters.

To illustrate in general the dynamic procedure, let us consider the following set of differential equations (u could be a vector field):

$$\frac{\partial u(\boldsymbol{x},t)}{\partial t} = f(u(\boldsymbol{x},t))$$

where $f(u(\boldsymbol{x},t))$ is a nonlinear differential operator. Let us apply the filter of amplitude $\overline{\Delta}$:

$$\frac{\partial \overline{u}}{\partial t} = \overline{f(u)} = f(\overline{u}) + \left(\overline{f(u)} - f(\overline{u})\right)$$

and suppose that the term $\overline{f(u)} - f(\overline{u})$ can be modelled through a function $g(\boldsymbol{c},\overline{u},\overline{\Delta})$ where $\boldsymbol{c} = (c_1,\ldots,c_n)$ is a set of n model coefficients whose scale of variation is larger than the filters used, so that the same set of coefficients $\boldsymbol{c}$ can be applied to both filters:

$$\begin{aligned}\overline{f(u)} - f(\overline{u}) &= g(\boldsymbol{c},\overline{u},\overline{\Delta}),\\ \widehat{f(u)} - f(\widehat{u}) &= g(\boldsymbol{c},\widehat{u},\widehat{\Delta})\end{aligned} \tag{1}$$

By applying subsequently the two filters we obtain:

$$\widehat{\overline{f(u)}} - \widehat{f(\overline{u})} = \widehat{g(\boldsymbol{c},\overline{u},\overline{\Delta})} \tag{2}$$

Now, (2) – (1) provides:

$$\widehat{\overline{f(u)}} - \widehat{f(\overline{u})} - \widehat{f(u)} + f(\widehat{u}) = g(\widehat{c, \overline{u}, \overline{\Delta}}) - g(c, \widehat{u}, \widehat{\Delta})$$

Under the following assumptions (strictly true only when the sharp cut-off filter is used):

$$\widehat{\overline{f(u)}} = \widehat{f(u)}, \quad \widehat{\overline{u}} = \widehat{u}$$

an equation for the coefficients c is obtained:

$$\widehat{f(\overline{u})} - f(\widehat{\overline{u}}) = g(c, \widehat{\overline{u}}, \widehat{\Delta}) - g(\widehat{c, \overline{u}, \overline{\Delta}}) \tag{3}$$

By solving this equation numerically at each node of the computational grid, the values of the coefficients c are obtained. Of course if the term to be modelled had not been a scalar but a vector or a tensor, we would have obtained more than one equation for the coefficients c_i.

We note that these coefficients depend on space and time. As a matter of fact, for consistency reasons it will be necessary to check a posteriori that their variations are small for distances of the order of the test filter.

In Germano's model [7] the dynamic procedure is applied to the Smagorinsky model for the subgrid stress tensor. Let us define:

$$\begin{cases} \tau_{ij} = \overline{v_i v_j} - \overline{v}_i \overline{v}_j & \textit{\textbf{grid} filter stress tensor} \\ T_{ij} = \widehat{\overline{v_i v_j}} - \widehat{\overline{v}}_i \widehat{\overline{v}}_j & \textit{\textbf{test} filter stress tensor} \\ L_{ij} = \widehat{\overline{v}_i \overline{v}_j} - \widehat{\overline{v}}_i \widehat{\overline{v}}_j & \textit{Leonard \textbf{resolved} scales stress tensor} \end{cases}$$

With $f(v_i) = \tau_{ij}^a$, eq. (3) provides the so-called *Germano's identity*:

$$L_{ij}^{(a)} = T_{ij}^{(a)} - \widehat{\tau}_{ij}^{(a)} \tag{4}$$

using Smagorinsky model we set $g(C, \overline{\Delta}, \overline{v}_i) = -2C\overline{\Delta}^2|\overline{S}|\overline{S}_{ij}$. Germano's identity becomes:

$$L_{ij}^a = -2CM_{ij} \tag{5}$$

where

$$M_{ij} = \widehat{\Delta}^2|\widehat{\overline{S}}|\widehat{\overline{S}}_{ij} - \overline{\Delta}^2\widehat{|\overline{S}|\overline{S}_{ij}}$$

This set of five equations (since the tensors that appear are symmetric and traceless) for one unknown (the constant C of the model) is solved by applying a least squares procedure [10] and it reads:

$$C(\boldsymbol{x},t) = -\frac{L^a_{ij} M_{ij}}{2 M_{kl} M_{kl}}$$

Smagorinsky model is based on an isotropy hypothesis of the unresolved scales. To partially overcome the limits of the above dynamic model, following Cercignani idea [2], a fourth order symmetric tensor B_{ijrs} is used:

$$\tau^a_{ij} = -2 B_{ijrs} \overline{\Delta}^2 |\overline{S}| \overline{S}_{rs} + \frac{2}{3} \delta_{ij} B_{kkrs} \overline{\Delta}^2 |\overline{S}| \overline{S}_{rs}$$

The number of unknowns of the tensor B_{ijrs} is too large and one cannot determine all of them via the dynamic procedure. The following simplified form was assumed:

$$B_{ijrs} = \sum_{\alpha,\beta} C_{\alpha\beta} a_{i\alpha} a_{j\beta} a_{r\alpha} a_{s\beta}$$

Note that summation over r and s is implied by the Einstein notation, while Greek letters appear more than two times and an explicit sum sign is required.

- $C_{\alpha\beta} = C_{\beta\alpha}$ are the elements of a 3×3 symmetric matrix which replaces the coefficient C of the isotropic model. In this case too the $C_{\alpha\beta}$ coefficients depend on space and time.
- $a_{i\alpha}$ are the components of a rotation matrix to be determined, or alternatively they can be viewed as the unit vectors $\boldsymbol{a}_\alpha\,(\alpha = 1, 2, 3)$ of an orthonormal triad. Thus the transformation matrix $\boldsymbol{a}$ defines a local reference system in which the tensor B_{ijrs} is reduced diagonally with respect to two indices.
- Smagorinsky formulation is retrieved by setting $B_{ijrs} = C\delta_{ir}\delta_{js}$, and this is obtained when $C_{\alpha\beta} = C\ \forall \alpha, \beta$.

The anisotropic model does not prescribe how to choose the matrix $a_{i\alpha}$, which in principle can be any rotation matrix, possibly varying in space and time. Of course, the values of the components $C_{\alpha\beta}$ computed with the dynamic procedure depend on the chosen tensor. One possible Galilean invariant triple is the set of eigenvectors of symmetric anisotropic Leonard tensor L^a_{ij}.

Coefficients $C_{\alpha\beta}$ can be determined by applying the dynamic procedure. Germano identity (4) becomes:

$$L^a_{ij} = g(C, \widehat{\overline{S}}, \widehat{\Delta}) - \widehat{g(C, \overline{S}, \overline{\Delta})} = -2 B_{ijrs} M_{rs}$$

with

$$g(C, \overline{\Delta}, \overline{v}_i) = -2\overline{\Delta}^2 |\overline{S}| B_{ijrs} \overline{S}_{rs} = -2\overline{\Delta}^2 |\overline{S}| \sum_{\alpha,\beta} C_{\alpha\beta} a_{i\alpha} a_{j\beta} a_{r\alpha} a_{s\beta} \overline{S}_{rs}$$

This equation can be easily solved for the unknowns $C_{\alpha\beta}$:

$$C_{\alpha\beta} = -\frac{\boldsymbol{a}^T \boldsymbol{L}^a \boldsymbol{a}\big|_{\alpha\beta}}{2\,\boldsymbol{a}^T \boldsymbol{M}\boldsymbol{a}\big|_{\alpha\beta}}$$

Mixed models combine eddy viscosity models and scale similarity models. Liu et al. [11] proposed the following model:

$$\tau^a_{ij} = K\,L^a_{ij} - 2C\overline{\Delta}^2|\overline{S}|\overline{S_{ij}}$$

with model parameters K and C to be determined using the dynamic procedure (five equations in two unknowns, least squares solution). An anisotropic mixed model has also been introduced [2]:

$$\tau^a_{ij} = L^a_{ij} - 2B_{ijrs}\overline{\Delta}^2|\overline{S}|\overline{S}_{rs} + \frac{2}{3}\delta_{ij}B_{kkrs}\overline{\Delta}^2|\overline{S}|\overline{S}_{rs}$$

2 A Priori Tests for Homogeneous Isotropic Turbulence

In Fig. 1 we show the correlation between exact and modelled stress tensor for various LES models. The figures have been obtained using results of a DNS of a homogeneous isotropic turbulence flow at $Re = 1000$ which was obtained by a pseudo-spectral method on a $512 \times 512 \times 512$ grid (courtesy M. Meneguzzi). The exact filtered stress tensor is compared to the one predicted by the model on the

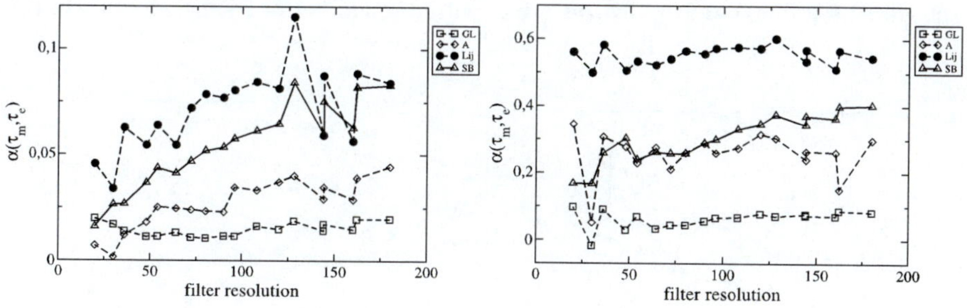

Fig. 1 Correlations between exact and modelled stress tensor. Left: the average is performed over the whole domain; right: only the high vorticity regions (*worms*) are considered. GL: isotropic dynamic model from [7, 10]. A: anisotropic model. The triple $a_{i\alpha} = \delta_{i\alpha}$ (δ is the Kronecker delta) has been used. L_{ij}: Leonard resolved stress tensor. *SB*: the dynamic mixed model from [13]

filtered flow field. The correlation coefficient for component τ_{12} is defined as

$$\text{corr. for } \tau_{12} = \frac{\sum_{ijk} \tau^{(e)}_{12,ijk} \tau^{(m)}_{12,ijk}}{\sqrt{\sum_{ijk} \tau^{2(e)}_{12,ijk}} \sqrt{\sum_{ijk} \tau^{2(m)}_{12,ijk}}}$$

where subscripts $^{(e)}$ and $^{(m)}$ denote, respectively, filtered DNS and modelled quantities, and i, j, k are the mesh points on the filtered grid. The filtered DNS data are obtained using a sharp cut-off filter in spectral space. The filter resolution is the number of grid points in each direction, or equivalently the number of harmonics in each direction in Fourier space.

The figure shows that correlations are generally low, but larger inside the high vorticity regions. In descending order, The Leonard resolved stress tensor is the one with the larges correlation(L_{ij}), followed by the dynamic mixed model (SB), the anisotropic model (A) and finally the isotropic dynamic model (GL) from [7, 10].

We next define

$$\alpha = \arccos\left\langle \frac{Tr(A * B)}{\sqrt{Tr(A * A) Tr(B * B)}} \right\rangle \tag{6}$$

If the two tensors A and B are proportional to each other, this angle is zero, while if they were completely uncorrelated and random, the average angle would be about 90°. In Fig. 2 we show the calculated mean angle between the modelled stress tensor and the exact one using formula (6). We observe that the average angle is about 90°, demonstrating that the modelled tensor is practically not related to the exact one. We also notice that the angle does not change so much with the resolution of the filter. When restricted to the regions that contain most of the vorticity, the angle becomes instead much smaller than 90° (see Fig. 2 right). This confirms that the coherent structures that make up the *worms* are captured much better by the models. In this

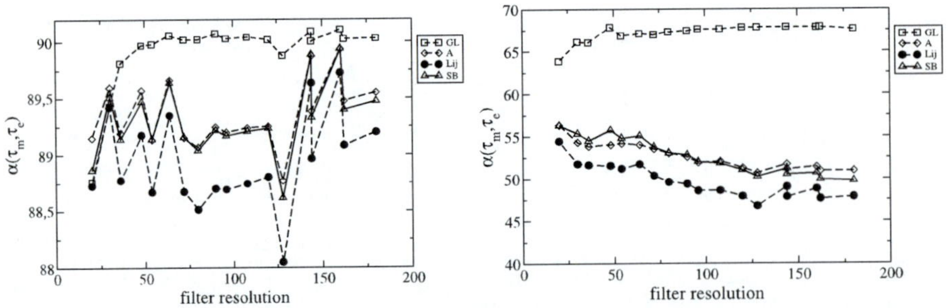

Fig. 2 Left: computed average angle between the modelled stress tensor and the exact one, using formula (6). Right: Same as left figure but restricted to *worms*. The abbreviations are the same as in Fig. 1

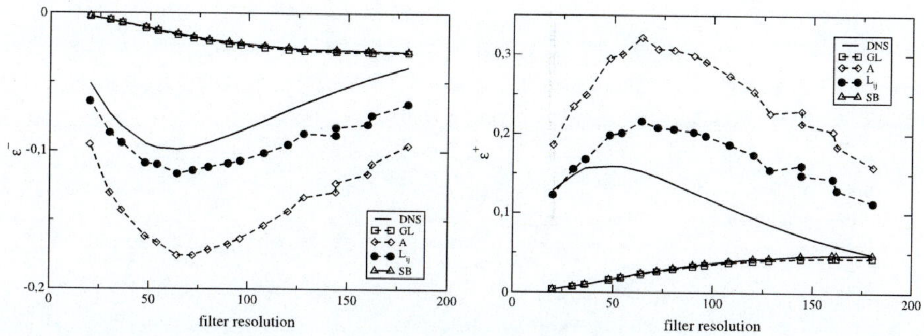

Fig. 3 Left: backward scatter $\epsilon^- = \left\langle \frac{\epsilon_{SGS} - |\epsilon_{SGS}|}{2} \right\rangle$. Right: forward scatter $\epsilon^+ = \left\langle \frac{\epsilon_{SGS} + |\epsilon_{SGS}|}{2} \right\rangle$. The abbreviations are the same as in Fig. 1

respect the anisotropic model behaves much better than the Germano isotropic one, and almost as well as the resolved Leonard tensor $\mathrm{L_{ij}}$.

Finally in Fig. 3 we consider the energetics of the flow. The forward scatter ϵ^+ and backward scatter ϵ^-, defined as:

$$\epsilon^+ = \left\langle \frac{\epsilon_{\mathrm{SGS}} + |\epsilon_{\mathrm{SGS}}|}{2} \right\rangle, \quad \epsilon^- = \left\langle \frac{\epsilon_{\mathrm{SGS}} - |\epsilon_{\mathrm{SGS}}|}{2} \right\rangle, \quad \epsilon_{\mathrm{SGS}} = \left\langle \tau_{ij} \overline{S}_{ij} \right\rangle,$$

are shown for different models and compared to DNS data. All the models except $\mathrm{L_{ij}}$ and A underestimate the amount of backward and forward scatter.

3 Rotating Boundary Layer

In order to highlight the properties of the SGS models in an anisotropic flow, the results of an *a priori* test on the velocity field of a three dimensional boundary layer generated on an infinite plate by a freestream rotating velocity [15] are presented in this section. The external velocity vector has constant magnitude U_0 and rotates with angular frequency f in planes parallel to the wall as represented in Fig. 4.

This flow presents the advantage to be three dimensional and very simple at the same time. Indeed the turbulence statistics, adimensionalized in a reference frame rotating with the angular frequency f, depend only on the distance from the wall and on the Reynolds number allowing the use of periodic boundary conditions in the homogeneous parallel to the wall directions. Although its simplicity, this flow is strongly anisotropic, making this rotating boundary layer a very convenient test for the anisotropic turbulence models. A numerical code based on mixed finite difference-pseudospectral discretization, kindly supplied by Passoni et al. [12], has been used for the DNS. The simulation has been performed at a Reynolds number $Re_{\delta_l} = \frac{U_0 \delta_l}{\nu} = 767$ based on the freestream velocity U_0 and on the viscous thickness

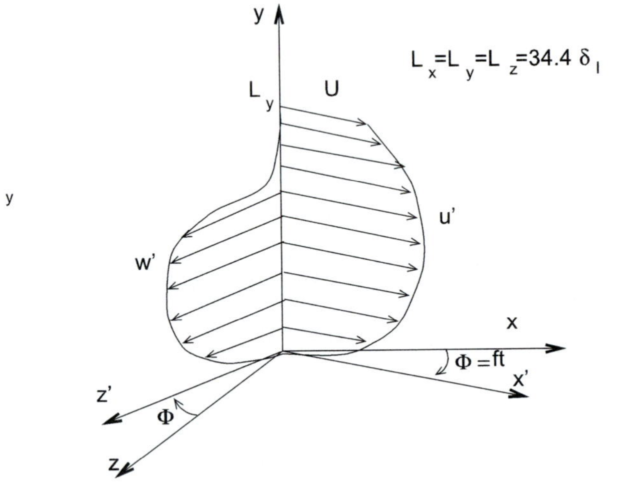

Fig. 4 Scheme of the boundary layer generated by a rotating freestream velocity over an infinite flat plate

$\delta_l = \sqrt{2\nu/f}$ corresponding to a Reynolds number based on the friction velocity $Re_\tau = 914$. A grid with $192 \times 180 \times 192$ cells has been adopted, corresponding to a resolution in parallel to the wall directions $\delta_x^+ = \delta_z^+ = 7$ while the distance of the first grid point from the wall is $\delta_y^+ = 0.8$ in wall unit. In order to eliminate numerical errors, a sharp cut-off filter is applied in the homogeneous directions to the DNS velocity field, while in the normal to the wall direction, where the mesh is stretched, a high accuracy compact filter, characterised by null commutation error with spatial derivatives, is applied in the physical space. The filtered quantities have then been located on the DNS mesh instead of on the LES grid in order to avoid the commutation errors occurring in finite difference discretization of derivatives [8].

In Fig. 5 (left) the SGS dissipation associated with the modelled stresses, averaged in time and in directions parallel to the wall, is compared to the exact one. The plot confirms that the Bardina model is poorly dissipative. The peak close to the wall is well captured by the two anisotropic models and by the dynamic mixed one. Far from the wall the anisotropic mixed model is a bit more dissipative while the dynamic mixed model and the anisotropic one are a bit less, compared to the exact stresses.

In Figs. 5 (right) and 6 (left) the mean forward and backward scatter are represented. For these quantities the anisotropic model presents too low values except for the too high peak close to the wall. Globally, from the energetic point of view, the anisotropic mixed model performs better than the other models.

Observing Fig. 6 (right), where the probability density function of the energy transfer of the models is compared to the exact one, it is evident that the anisotropic models, in particular, the anisotropic mixed one, present a distribution similar to that of the DNS. The large symmetric queues are representative of almost equally

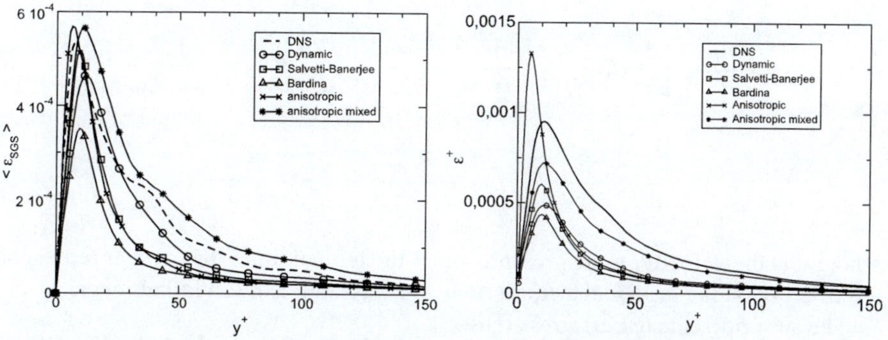

Fig. 5 Left: mean SGS dissipation $\langle \epsilon_{SGS} \rangle = \langle \tau_{ij} \overline{S}_{ij} \rangle$, where the brackets denote averaging over horizontal planes and time. versus the distance from the wall in the rotating boundary layer [1]. Right: mean SGS forward scatter ϵ^+ versus the distance from the wall [1]. Inside the graphics, Bardina refers to the original dynamic model from [5], while Salvetti-Banerjee is the dynamic mixed model from [13] and dynamic is the Germano dynamic model [7]

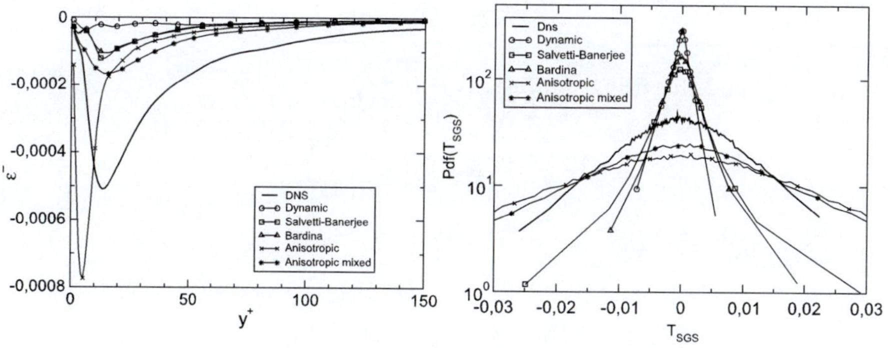

Fig. 6 Left: Mean SGS backward scatter ϵ^- versus the distance from the wall [1]. Right: Probability distribution function for the energy transfer $T_{SGS} = -\frac{\partial}{\partial x_j}(\overline{u}_i \tau_{ij}) - \epsilon_{SGS}$

distributed positive and negative energy transfer. On the contrary, all the other models show a too high peak for values around zero.

From this results it can be assumed that the anisotropic mixed model well represents the energetic behaviour of unresolved scales of motion.

4 Application to Rayleigh–Bénard Turbulent Convection

The nondimensional filtered Boussinesq system of equations reads

$$\frac{\partial \overline{u}_i}{\partial x_i} = 0$$

$$\frac{\partial \overline{u}_i}{\partial t} + \frac{\partial}{\partial x_j}(\overline{u}_i \overline{u}_j) = \frac{\partial \overline{p}}{\partial x_i} - \frac{\partial \tau_{ij}^{SGS}}{\partial x_j} + \frac{Ra}{Pr}\overline{\theta}\delta_{i3} + \frac{\partial^2 \overline{u}_i}{\partial x_j \partial x_j}$$

$$\frac{\partial \overline{\theta}}{\partial t} + \frac{\partial}{\partial x_j}(\overline{\theta}\overline{u}_j) = -\frac{\partial q_j^{SGS}}{\partial x_j} + \frac{1}{Pr}\frac{\partial^2 \overline{\theta}}{\partial x_j \partial x_j}$$

where **u** is the velocity, p the pressure and θ the temperature; the overbar represents the filtering on the grid-scale. Ra and Pr are Rayleigh and Prandtl number.

The subgrid-scale (SGS) stress tensor

$$\tau_{ij}^{SGS} = \overline{u_i u_j} - \overline{u}_i \overline{u}_j$$

is modelled using the anisotropic approach presented above.

The SGS heat flux

$$q_i^{SGS} = \overline{\theta u_i} - \overline{\theta}\overline{u}_i$$

is similarly modelled using a diffusivity tensor ν_{ir}^{ϑ}:

$$q_i^{SGS} = -\sum_r \nu_{ir}^{\vartheta} \frac{\partial \overline{\vartheta}}{\partial x_r}$$

and prescribing

$$\nu_{ir}^{\vartheta} = \overline{\Delta}^{4/3} \sum_\alpha C_\alpha^\theta a_{i\alpha} a_{r\alpha}$$

we obtain:

$$C_\alpha^\theta = -\frac{a^T S\big|_\alpha}{\left(\widehat{\Delta}^{4/3} - \overline{\Delta}^{4/3}\right) \sum_r a_{r\alpha} \frac{\partial \widehat{\overline{\theta}}}{\partial x_r}}$$

In Fig. 7 we evaluate the performance of the above model compared to dynamic isotropic model [7] and Smagorinsky [14]. In left panel the vertical fluctuations obtained using LES models at resolution $24 \times 24 \times 31$ are compared to DNS data by R.M. Kerr [9] using a numerical grid $288 \times 288 \times 96$. Velocity is in units of $Ra\ \nu\ k/L^2$ (L is the height of the layer). On the right panel an a priori test is performed on LES models using numerical results of a DNS at $Ra = 4 \times 10^6$.

In both panels the results show that the anisotropic model behaves far better than the isotropic ones in the boundary layer and its vicinity, that is, in the regions where the flow is more anisotropic.

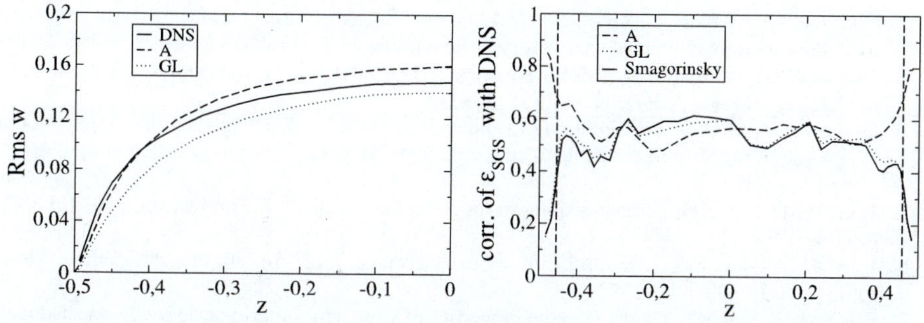

Fig. 7 Left: Normal fluctuating velocity component (RMS values) at $Ra = 2 \times 10^7$. DNS: Results of DNS by Kerr [9] A: anisotropic model; GL: isotropic dynamic model by Germano et al. [7]; Right: Correlation of ε_{SGS} between model and DNS data, as a function of height z. Smagorinsky: original Smagorinsky model; A and GL are the same as in left panel. The long-dashed vertical lines delimit the thermal boundary layer

5 Conclusions

A review of the tensorial dynamic eddy viscosity model and his applications to different kinds of flows is presented here. The a priori test and the numerical simulation of different flows such as 3D turbulent boundary layer and Rayleigh–Bénard convection demonstrated good performances of the model in the reproduction of anisotropicity of the turbulence and energetic aspects. The tensorial dynamic eddy viscosity represents a general approach applicable to model any filtered moments. Indeed the extension to model SGS terms in the energy equations for compressible flows [3] presented very good performances too. The versatility and robustness of the model are also demonstrated by its use in different numerical approach with polynomial adaptivity [4]. Finally another variation of tensorial eddy viscosity model recently proposed [6] proves the fruitfulness of this approach.

References

1. Abbà, A., Cercignani, C., Picarella, G., Valdettaro, L.: A 3D turbulent boundary layer test for LES models. In: Satofuka, N. (ed.) Computational Fluid Dynamics 2000. First International Conference on Computational Fluid Dynamics (2001)
2. Abbà, A., Cercignani, C., Valdettaro, L.: Analysis of subgrid scale models. Comput. Math. Appl. **46**, 521–535 (2003)
3. Abbà, A., Campaniello, D., Nini, M.: Filter size definition in anisotropic subgrid models for large eddy simulation on irregular grids. J. Turbul. **18**(6), 589–610 (2017)
4. Abbà, A., Recanati, A., Tugnoli, M., Bonaventura, L.: Dynamical p–adaptivity for LES of compressible flows in a high order DG framework. J. Comput. Phys. **420**, 109720 (2020)
5. Bardina, J., Ferziger, J.H., Reynolds, W.C.: Improved turbulence models based on large eddy simulation of homogeneous, incompressible, turbulent flows. PhD thesis. Department of Mechanical Engineering, Stanford University (1983)

6. Cimarelli, A., Abbà, A., Germano, M.: General formalism for a reduced description and modelling of momentum and energy transfer in turbulence. J. Fluid Mech. **866**, 865–896 (2019)
7. Germano, M., Piomelli, U., Moin, P., Cabot, H.: A dynamic subgrid-scale eddy viscosity model. Phys. Fluids A **3**(7), 1760–1765 (1991)
8. Gonze, M.-A.: Revue, amélioration et validation de modélisations sous-mailles. application à la simulation numérique des écoulements turbulents. Rapport de stage post-doctoral, CERCA (1994)
9. Kerr, R.M.: Rayleigh number scaling in numerical convection. J. Fluid Mech. **310**, 139–179 (1996)
10. Lilly, D.K.: A proposed modification of the Germano subgrid-scale closure method. Phys. Fluids A **4**(3), 633–635 (1992)
11. Liu, S., Menevau, C., Katz, J.: On the properties of similarity subgrid-scale models as deduced from measurements in a turbulent jet. J. Fluid Mech. **275**, 83–119 (1994)
12. Passoni, G., Alfonsi, G., Tula, G., Cardu, U.: A wavenumber parallel computational code for the numerical integration of the Navier-Stokes equations. Parallel Comput. **25**, 593–611 (1999)
13. Salvetti, M.V., Banerjee, S.: A priori tests of a new dynamic subgrid-scale model for finite difference large-eddy simulations. Phys. Fluids **7**(11), 2831–2847 (1995)
14. Smagorinsky, J.: General circulation experiments with the primitive equations, I. The basic experiment. Mon. Weather Rev. **91**(3), 99–164 (1963)
15. Spalart, P.R.: Theoretical and numerical study of a three-dimensional turbulent boundary layer. J. Fluid Mech. **205**, 319–340 (1989)

FPU Model and Toda Model: A Survey, a View

Giancarlo Benettin and Antonio Ponno

Abstract A huge amount of papers investigated, over more than 65 years, the Fermi–Pasta–Ulam problem. One of the leading ideas, present already in the early literature, is that the unexpected regular behavior observed by the authors, quite different from the expected ergodicity, could be explained by the presence of a close hidden nonlinear integrable dynamics. This was initially searched among nonlinear wave equations, but rather soon, after the discovery of the integrability of the Toda model, it was progressively understood that Toda provides the natural integrable approximation to FPU. The aim of this paper is to provide a short updated review of the relation between the FPU dynamics and the Toda dynamics. Updated means it also includes new results, see Sect. 3. The paper is ideally addressed to the wide—very wide!—community of people who feel *The legacy of Carlo Cercignani*, so it includes a few introductory comments.

Foreword *It is a great honor for me to contribute to this volume in honor and memory of Carlo Cercignani. Carlo has been an important person in my life. In my scientific life of course, where Carlo, as for all of us, has been a light and a model. But beyond science, I wish to say I always felt from him warm friendship. In conferences, in particular, in Ravello, I remember his lectures, but even more, I remember staying together: with him, his wife Silvana, Luigi, a few friends. Conversating, about so many different topics; walking together, when it was still possible. It was a great opportunity for me to meet Carlo, and I feel deep sincere gratitude.* [G.B.]

G. Benettin (✉) · A. Ponno
Università di Padova, Dipartimento di Matematica "Tullio Levi-Civita", Padova, Italy
e-mail: benettin@math.unipd.it; ponno@math.unipd.it

P. Barbante et al. (eds.), *From Kinetic Theory to Turbulence Modeling*, Springer INdAM Series 51, https://doi.org/10.1007/978-981-19-6462-6_3

1 Introduction

In 1955 Fermi, Pasta, and Ulam wrote a paper [1] which was destined to have a deep influence in different branches of research.

- It started Molecular Dynamics (more generally, numerical experiments on dynamical systems), namely investigating the statistical properties of a system by numerically solving its microscopic equations of motion.
- It raised "elementary" questions in the dynamical foundations of Statistical Mechanics, which still are not clearly answered.
- It motivated a relevant branch of the theory of nonlinear oscillations, namely modern theory of nonlinear wave equations (Boussinesq, KdV...), and more generally the study of nonlinear integrability for systems with many degrees of freedom.

Hundreds of papers have been devoted to the subject, with a great variety of theoretical and numerical approaches, still far from merging in a unitary clear view.[1]

The aim of this paper is to focus on one of the main ideas, namely that the reference integrable dynamics for FPU is Toda dynamics [4], nonlinear and highly nontrivial. This view was suggested already in 1974 in [5], one of the three simultaneous papers where integrability of the Toda model was proved [5–7]. The perspective was reconsidered and widely developed in 1982 [8], but nevertheless not much exploited in the later literature, and emphasized again only recently, in a few papers; among them [9–14].

We aim to show that viewing FPU as a perturbed Toda model provides a unitary perspective, which can possibly give order to the complex phenomenology of FPU. We refer here to the standard, generic case of the so-called α (also called $\alpha + \beta$) FPU model. We shall not discuss instead the β-model, which is not at all close to Toda nor we shall enter extensions to dimension two and three, although physically very important (see in particular the papers by Carati and Galgani and by Gangemi in this volume).

The paper is addressed, ideally, to the wide community of researchers feeling *The legacy of Carlo Cercignani*, joined together in his memory and contributing to this Conference in his honor. Carlo was indeed very interested in FPU (see [15, 16] and the comments in the paper by Carati and Galgani in this volume). Likely, however, some of us are not very familiar with FPU, and so, in the remaining part of this Introduction, a very short tentative introduction to FPU is provided.

After that, Sects. 2 and 3 are devoted, respectively, to the role of Toda dynamics in the long-time approach to statistical equilibrium and to the mechanism of formation of the so-called FPU state, that is the first, crucial part of the FPU dynamics, where the underlying integrable Toda dynamics is particularly transparent. Section 3 also includes new results.

[1] See, for example, the collections of papers [2, 3], appeared in occasion of the 50th anniversary of FPU.

1.1 FPU in a Nutshell

The specific problem Fermi, Pasta, and Ulam confronted with, is the problem of energy sharing in weakly nonlinear chains of oscillators. The Hamiltonian has the form

$$H(p,q) = \frac{1}{2}\sum_{i=1}^{N} p_i^2 + \sum_{i=0}^{N} V(q_{i+1} - q_i)\,, \tag{1}$$

V being some nearest-neighbors potential with a minimum in zero,

$$V(r) = \frac{r^2}{2} + \alpha\frac{r^3}{3} + \beta\frac{r^4}{4} + \cdots\,, \qquad \beta > 0\,. \tag{2}$$

In (1) the boundary conditions are still not specified; they are generally either fixed ends, i.e., $q_0 = q_{N+1} = 0$, like in the original FPU paper, or periodic, $q_N = q_0$, with not much difference.

The value of α, if different from zero, is irrelevant, since a trivial rescaling reports it to any prefixed value; the effective parameters determining the dynamics are indeed $|\alpha|\sqrt{\varepsilon}$, $\varepsilon = E/N$ being the specific energy, and then $\beta/\alpha^2, \ldots$ The choice of α actually fixes the energy scale, as well as the scale of β and of possibly further coefficients in (2). Throughout the paper we shall use $\alpha = 1$.

Fermi, Pasta, and Ulam aimed to investigate how the system reaches the statistical equilibrium, identified with the equipartition of energy among normal modes, if started very far from equilibrium, the whole energy being given to only one or two long-wavelength normal modes. This is indeed part of the general problem of the energy flowing from macroscopic to microscopic scale, so crucial in quite different fields of physics. With great surprise they found—within the time scale accessible to their computer—no equilibrium at all: the system apparently reached a stationary state, different both from the initial state and from equipartition, in which only a few normal modes significantly share energy, and dynamics looks quasi-periodic with long time almost exact recurrencies.

Consider, to be definite, the case of fixed ends. Normal modes are then

$$Q_k = \sqrt{\tfrac{2}{N+1}}\sum_{i=1}^{N} q_i \sin\frac{\pi k i}{N+1}\,, \qquad P_k = \sqrt{\tfrac{2}{N+1}}\sum_{i=1}^{N} p_i \sin\frac{\pi k i}{N+1}\,;$$

their energies E_k and frequencies ω_k are

$$E_k(P_k, Q_k) = \tfrac{1}{2}(P_k^2 + \omega_k^2 Q_k^2)\,, \qquad \omega_k = 2\sin\frac{\pi k}{2(N+1)}\,,$$

and the Hamiltonian in such coordinates assumes the form

$$\tilde{H}(P, Q) = \sum_{k=1}^{N} E_k(P_k, Q_k) + \alpha U_3(Q) + \beta U_4(Q) + \cdots ,$$

U_j being a homogeneous polynomial of degree j. For given initial data, let

$$\overline{E}_k(T) = \frac{1}{T} \int_0^T E_k(P(t), Q(t))\, \mathrm{d}t \ ;$$

statistical mechanics is based on the *ergodic hypothesis*, which implies

$$\overline{E}_k(T) \overset{T\to\infty}{\longrightarrow} \langle E_k \rangle \quad \simeq \varepsilon ,$$

$\langle E_k \rangle$ denoting the microcanonical phase average. Figures 1 and 2 summarize the heart of the FPU results. They both refer to a model with $N = 32$ and an initial datum in which only mode $k = 1$ is excited, at small energy $\varepsilon = 4 \times 10^{-5}$. Figure 1 shows $\overline{E}_k(T)$ as function of T, for the first few modes. Quite evidently, there is no indication at all of any tendency to energy equipartition: on the contrary, an asymptotic state is apparently reached, in which only a few modes, and not at the same extent, are involved in energy sharing. Figure 2 reports, for the same dynamics, the instantaneous values $E_k(t)$ as functions of t: panel a (left) shows, on a short time scale, the behavior of modes $k = 1, 2, 3$, while panel b (right) reports only mode $k = 1$, on a longer time scale. The presence of long time recurrencies is quite impressive, the dynamics appearing quasi-periodic and thus possibly integrable. Such longer time recurrencies have been observed a few years after FPU, in [17].

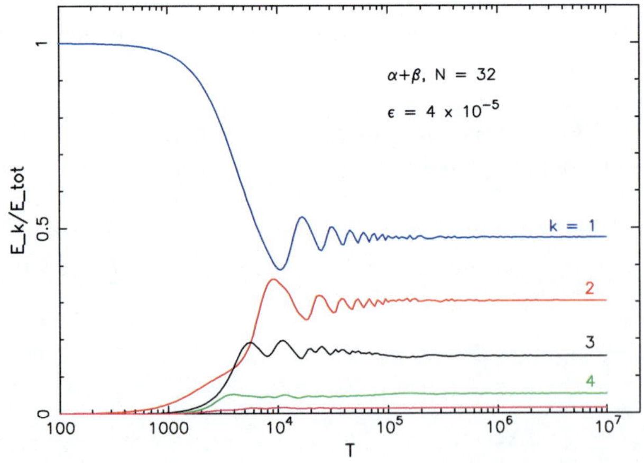

Fig. 1 The time average $\overline{E}_k(T)$ as a function of T, for the first few modes. FPU model with $N = 32$, $\alpha = 1$, $\beta = 2$; specific energy $\varepsilon = 4 \times 10^{-5}$, initial excitation of mode $k = 1$

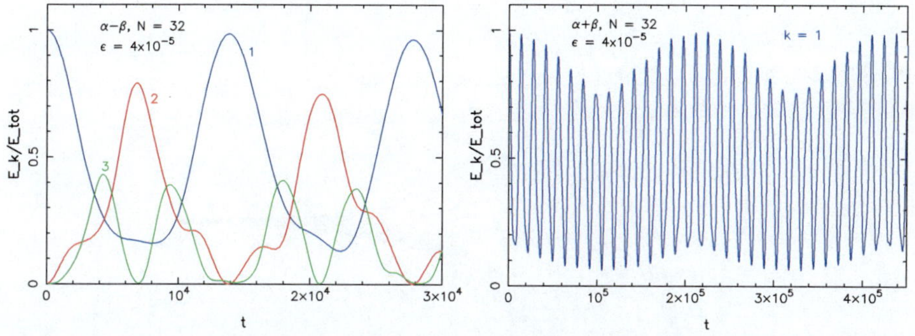

Fig. 2 The instantaneous values $E_k(t)$ as functions of t, in the same conditions as Fig. 1. Left: modes $k = 1, 2, 3$; right: only mode $k = 1$, for a longer time scale

1.2 The Search for an Underlying Integrable Dynamics

The suspect the dynamics is close to integrable, in the conditions studied by the authors—that is when only long-wave modes are excited, and so the discrete chain appears almost continuous—prompted the idea to approximate the FPU model with a convenient nonlinear wave equation.

The first attempt in this direction, going back to 1965, is [18], a fundamental paper which is at the basis of modern theory of nonlinear integrable wave equations. FPU appears there as a main motivation to study again, after years, the KdV equation

$$u_t = \frac{\alpha}{\sqrt{2}} u u_x + \frac{1}{24} u_{xxx} \,.$$

The progress in the field is then rapid: a couple of years later the method of inverse scattering [19] and Lax pairs [20] are introduced, and the presence of infinitely many constants of motion is established [21]. Finally, in 1971 [22], KdV is shown to be an infinite dimensional completely integrable Hamiltonian system.[2]

In parallel with the research on the integrability, numerical work clearly established that FPU is not integrable: it is enough to raise the energy, to recover the expected normal statistical behavior [24]. This is clear, for example, in Fig. 3, which differs from Fig. 1 only for the larger energy $\varepsilon = 10^{-2}$. It is not easy to reconcile the two views; the suggestion in [23], spontaneous in that moment, was that the lack of integrability of FPU is possibly due to the discretization.

[2] In fact, the nonlinear wave equation which is more immediately related to FPU, if one searches for a continuum limit in which the first nonlinear and the first dispersive terms beyond the wave equation are kept, is the Boussinesq equation; a possible form is $u_{tt} = u_{xx} + 2\alpha u_x u_{xx} + \frac{1}{12} u_{xxxx}$. From Boussinesq it is possible to deduce, in a convenient limit, KdV, but the Boussinesq equation itself was soon proved, in 1973, to be integrable [23]. In [23] the connection between FPU and Boussinesq is particularly emphasized.

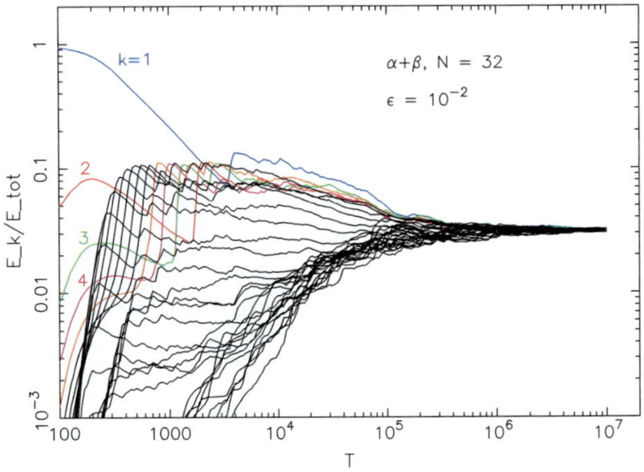

Fig. 3 Same as Fig. 1, at larger $\varepsilon = 10^{-2}$; all modes (logarithmic vertical axis, too)

In the same years, statistical physicists become interested in the Toda model [4]. As is well known, this is a Hamiltonian system with the same form as (1), V being the Toda exponential potential

$$V_T(r) = \frac{1}{\lambda^2}(e^{\lambda r} - 1 - \lambda r) .$$

In 1974 the Toda model was proved to be completely integrable, remarkably in three independent papers [5–7]. Reference [5] is particularly important for FPU, because the connection with the FPU problem is there stressed. Indeed, for $\lambda = 2\alpha$ it is

$$V_T(r) = \tfrac{1}{2}r^2 + \tfrac{1}{3}\alpha r^3 + \tfrac{1}{4}\beta_T r^4 + \tfrac{1}{5}\gamma_T r^5 + \cdots , \qquad \beta_T = \tfrac{2}{3}\alpha^2 , \ \gamma_T = \tfrac{1}{3}\alpha^3 , \ \ldots$$

so the model has a third order contact with FPU and provides an integrable approximation better than the harmonic chain. In [5] the slow stochastization of FPU is not anymore attributed to the discretization with respect to an integrable wave equation, but to the small difference, with dominating term $\frac{1}{4}(\beta - \beta_T)r^4$, between FPU and Toda. *The reference to Toda as the best integrable approximation for FPU is a considerable change of paradigm in the FPU problem.* For example, the distinction between long and short-wave initial excitation ceases to be important, although of course wave equations, definitely easier than a discrete model, remain useful in situations where only long waves are present.

The connection between FPU and Toda was proposed again, and emphasized, in 1982 [8]. In this paper, on the one hand, a striking evidence is provided that the dynamics of FPU, at low energy, is hardly distinguishable from the Toda dynamics; on the other hand, the integrability of Toda is studied very constructively, and an algorithm is proposed to compute numerically the Toda actions. We shall come back to this point in Sect. 3. The striking similarity of the FPU and Toda dynamics

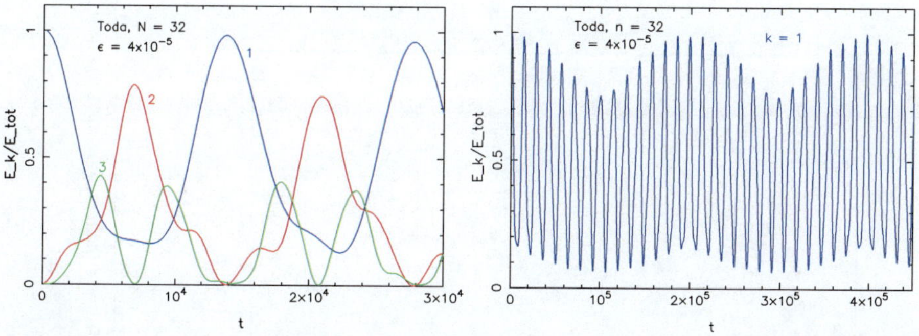

Fig. 4 Same as Fig. 2, for Toda rather than for FPU

is evident in Fig. 4, produced under the same condition as Fig. 2 but for the Toda model.

1.3 Different Phenomena at Different Time Scales

Simultaneously with the understanding of the strong connection between FPU and Toda in [8], and completely independently, a new idea entered the literature [25, 26], sometimes referred to as the "two time scales scenario." The suggestion is that at least for large N the formation of the FPU state, in which energy is shared only by a small fraction of modes as in Figs. 1 and 2, is not the end of the story, and eventually, in a possibly *much* larger time scale, statistical equilibrium is always reached: more or less, as it happens in metastable phenomena.

To illustrate such a scenario, a good way is to look at the energy profile, i.e., the distribution of energy among normal modes, at different times; raising N is also convenient. Figure 5a shows the result for $N = 1023$ and $\varepsilon = 10^{-4}$; energy was initially equidistributed among the first 10% of normal modes, with random phases. The figure shows[3] $\overline{E}_k(T)$ vs. k/N, at selected times, marked in the figure, in geometric progression. The initial profile is the black rectangle. It can be seen that already at $T \simeq 10^3$, after a transient in which the initial discontinuity is still present, a well-defined regular profile is formed, in which only some low frequency modes effectively take part to the energy sharing, the energies of the remaining ones decaying exponentially with k/N. The energy profile keeps its form nearly unchanged for a rather large time scale, $T \simeq 10^5$ or 10^6, definitely larger than the time needed to form it. Afterwards, on a much larger time scale, the dynamics slowly evolves towards energy equipartition, the high-frequency modes

[3] To be precise: time averages are computed here in a running window of width proportional to T, namely $\frac{2}{3}T \leq t \leq T$ (averages from the beginning are a little lazy), and moreover, to clean the curves, an average over 24 different choices of the phases is introduced.

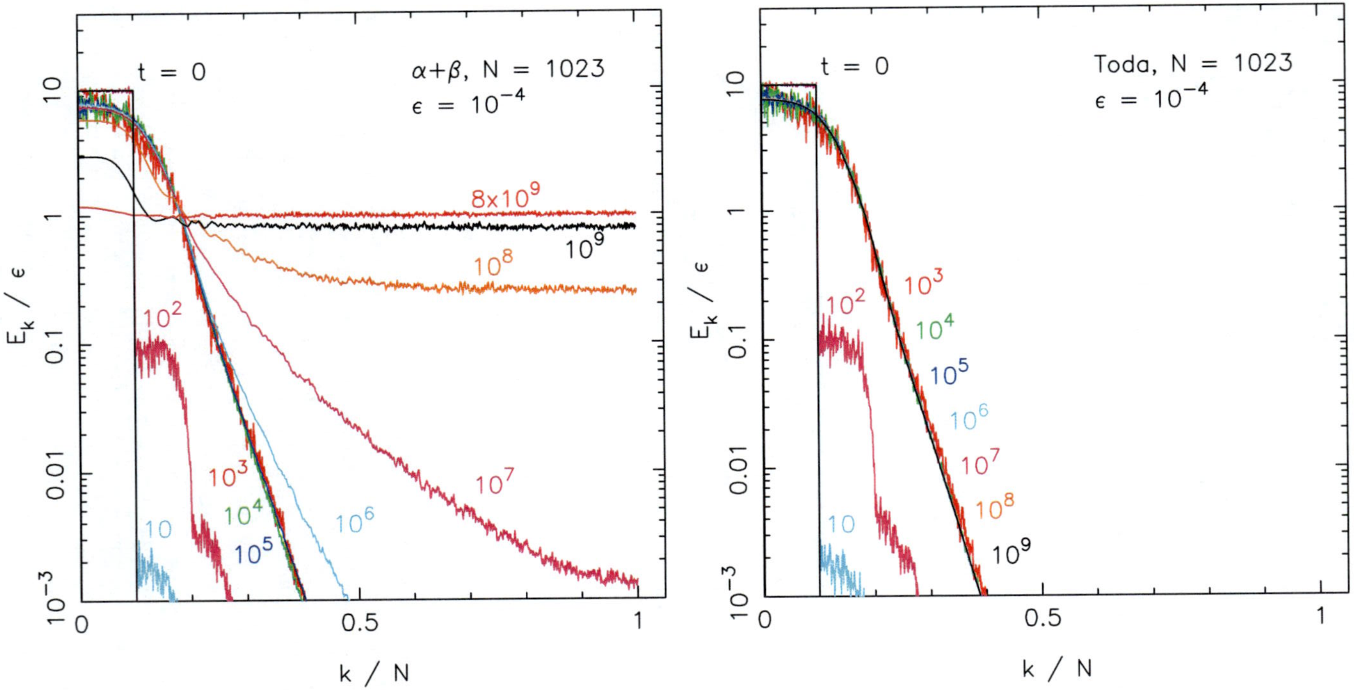

Fig. 5 The shape of the energy spectrum $\overline{E}_k(T)$ plotted vs. k/N, at selected times T (marked in the figure) in geometric progression. Left: FPU, $N = 1023$, $\alpha = 1$, $\beta = 2$, $\varepsilon = 10^{-4}$; right: the corresponding Toda model. Energy initially equidistributed among modes $0 < k/N < 0.1$, see the rectangle marked $t = 0$. Each point is the average over 24 random extractions of the initial phases

being progressively involved into the energy-sharing game; in the above conditions, equipartition requires $T \simeq 10^{10}$.

The natural conjecture, at this point (natural, but explicit in the literature only after [9], in 2011), is that the first time scale is the one in which the system behaves similarly to Toda, while on larger times the difference between the two dynamics becomes evident. To confirm such an interpretation, we can repeat the above computation for the Toda model. The result is in Fig. 5b. Quite clearly, exactly the same profile is formed, but there is no further evolution to anything different: in Toda, only the first time scale does exist and is perpetual.

We can rephrase such a view in a better language. Toda is integrable and so, for any initial datum, the motion is confined to a torus of dimension N: actions stay constant, while angles advance linearly, and generically fill the torus. Time averages on such a motion are very partial averages in the phase space, namely averages only on the angles and not the actions. For Toda, this is all. For FPU, the lack of integrability results in an additional slow drift transversal to tori, which asymptotically (according to Fig. 5) results in a diffusion throughout the phase space, and makes possible statistical equilibrium. Figure 6 shows very symbolically the situation.

Both phenomena, that is the filling of the Toda torus in the dynamics common to FPU and Toda, and the diffusion across tori possibly leading to normal statistical equilibrium, are worth to be investigated. We shall start with the latter, devoting to it the next Sect. 2, discussing instead the former in Sect. 3.

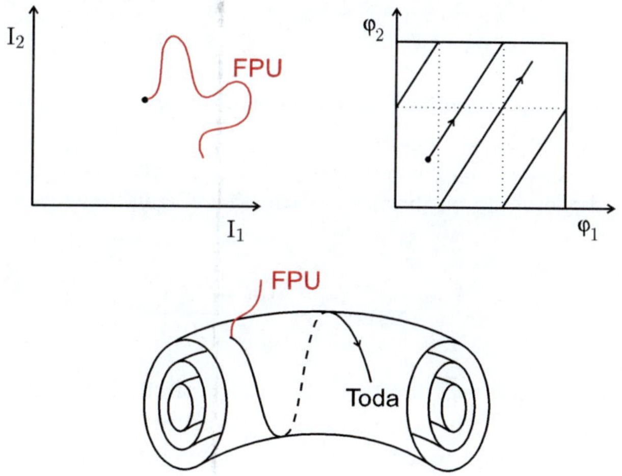

Fig. 6 Symbolically illustrating FPU as a perturbed Toda system

2 The Long-Time Motion Across Toda Tori

In the previous section we focused the attention on very special initial data, in which only a few normal modes share energy. The corresponding region of the phase space is extremely small and atypical. Measuring the rate of approach to equilibrium in such an exceptional situation can be done, see, for example, [9], but it is certainly more interesting to consider generic initial data, in which the energy is distributed randomly among normal modes with microcanonical distribution, and to study the drift of FPU trajectories across Toda tori in such a generic situation.

Such a study has been performed in [10], looking at the correlation time of the Toda constants of motion in the FPU dynamics. The Toda constants of motion, for a system with N degrees of freedom and fixed ends as we are dealing with, can be explicitly written by making reference to a larger system with $2(N+1)$ degrees of freedom and periodic boundary conditions, restricting the attention to skew-symmetric states such that

$$q_{N+1+i} = -q_{N+1-i}\ , \qquad p_{N+1+i} = -p_{N+1-i}\ , \qquad i = 0, \ldots, N\ , \tag{3}$$

which are easily seen to form an invariant submanifold. In turn, the constants of motion of the periodic chain are the eigenvalues of the Lax matrix $L(p, q)$ associated with the systems, or equivalently (much easier), the traces of the powers of L; see Sect. 3.2 for the expression of L. In the submanifold (3), precisely N constants of motion

$$F_s(p, q) = \mathrm{Tr}\, L^{2s}(p, q)\ , \qquad s = 1, \ldots, N\ ,$$

are independent and nontrivial, the odd powers of L having vanishing trace. F_1 turns out to be the total energy of the system.

For given initial data, denote shortly $F_s(t)$ for $F_s(p(t), q(t))$, and let $\langle\, .\, \rangle$ denote microcanonical averaging on the initial data. The correlation function $\mathcal{G}_s$ of F_s is defined as

$$\mathcal{G}_s(t) = \frac{\langle F_s(t)\, F_s(0)\rangle - \langle\, F_s\,\rangle^2}{\langle F_s(0)^2\rangle - \langle\, F_s(0)\,\rangle^2}\ ; \tag{4}$$

the decay time of such functions provides the desired time scale of the motion transversal to tori, in a generic situation. It is worthwhile to remark that looking at the decay of correlation functions means looking at mixing, and this is fully in the spirit of the original FPU paper.[4] Practically, the microcanonical distribution

[4] From [1]: "Instead of a gradual increase of all the higher modes, the energy is exchanged, essentially, among only a certain few. It is, therefore, very hard to observe the rate of 'thermalization' or mixing in our problem, and this was the initial purpose of the calculation." For a previous study of

in (4) is approximated by a Gaussian distribution of the normal modes coordinates P_k, Q_k, rescaled so as to fit the desired energy.

Figure 7a shows the decay of $\mathcal{G}_s(t)$, $s = 2, \ldots, 12$, for FPU with $N = 1023$, $\beta = 2$, $\varepsilon = 8 \times 10^{-4}$. Quite clearly, curves accumulate on a line $\mathcal{G}^*(t)$, which in the semi-log scale of the figure corresponds to an exponential

$$\mathcal{G}^*(t) = e^{-t/t^*} \ ;$$

the inverse slope t^*, to be thought of as depending, in principle, on N, β, and ε, provides the time scale of the motion transversal to tori we are looking at, ideally the mixing time that Fermi and coworkers aimed to observe. A similar accumulation is observed at different N and ε, $\mathcal{G}_{12}$ always appearing as a reasonable approximation of the limit curve $\mathcal{G}^*$. In the following, $\mathcal{G}^*$ will be identified with $\mathcal{G}_{12}$.

Figure 7b reports $\mathcal{G}^*(t)$ as function of t at fixed $\beta = 2$ and $\varepsilon = 2 \times 10^3$, for different N between 127 and 2047. The stability in N is quite evident. To investigate the dependence on β, computations have been repeated for fixed $N = 511$ and $\varepsilon = 2 \times 10^{-3}$, at different values of $\Delta\beta = \beta - \beta_T$ in the range $-1/3 \le \Delta\beta \le 8/3$. The result is in Fig. 8: indeed Fig. 8a reports the lines as they are, showing a rather dramatic dependence of the slope of $\mathcal{G}^*(t)$ on $\Delta\beta$ (although, remarkably, lines are identical for $\Delta\beta = \pm 1/3$); Fig. 8b shows the same curves, reported, however, as functions of the rescaled time

$$t' = \tfrac{16}{9} \Delta\beta^{-2} t$$

(no rescaling for $\beta = 2$, $\Delta\beta = 4/3$). Curves, although roughly, collapse into one, thus indicating a rough growth of t^* as $\Delta\beta^{-2}$. In a very similar way, one observes an ε-dependence of t^* which approximately follows the power law $\varepsilon^{-5/2}$, see [10] for details. So, the overall behavior of the correlation time looks

$$t^* \sim (\beta - \beta_T)^{-2} \varepsilon^{-5/2} \ .$$

Unfortunately, there is no theory at all in support of such a law. From the point of view we are exploiting in this paper, the most important point is the dependence of t^* on the difference $\beta - \beta_T$, and thus, so to speak, on the distance between FPU and Toda.[5] It is worthwhile to mention that the analysis made in [9] of the time scale of the phenomenon illustrated in Fig. 5a gives exactly the same dependence on $\Delta\beta$, although with a slightly different dependence on ε, namely with exponent $9/4$ in place of $5/2$ (a slightly faster phenomenon, for small ε).

---

the decay properties of the correlation functions of normal modes energies, done in the same spirit and actually inspiring [10], see [27].

[5] The divergence of t^* for $\Delta\beta \to 0$ should not be taken literally: for vanishing β, FPU has a higher order contact with Toda, but the difference between the two Hamiltonians does not vanish, and a crossover to a different dependence of t^* on ε, faster for small ε, is expected; see [9] for a very similar situation.

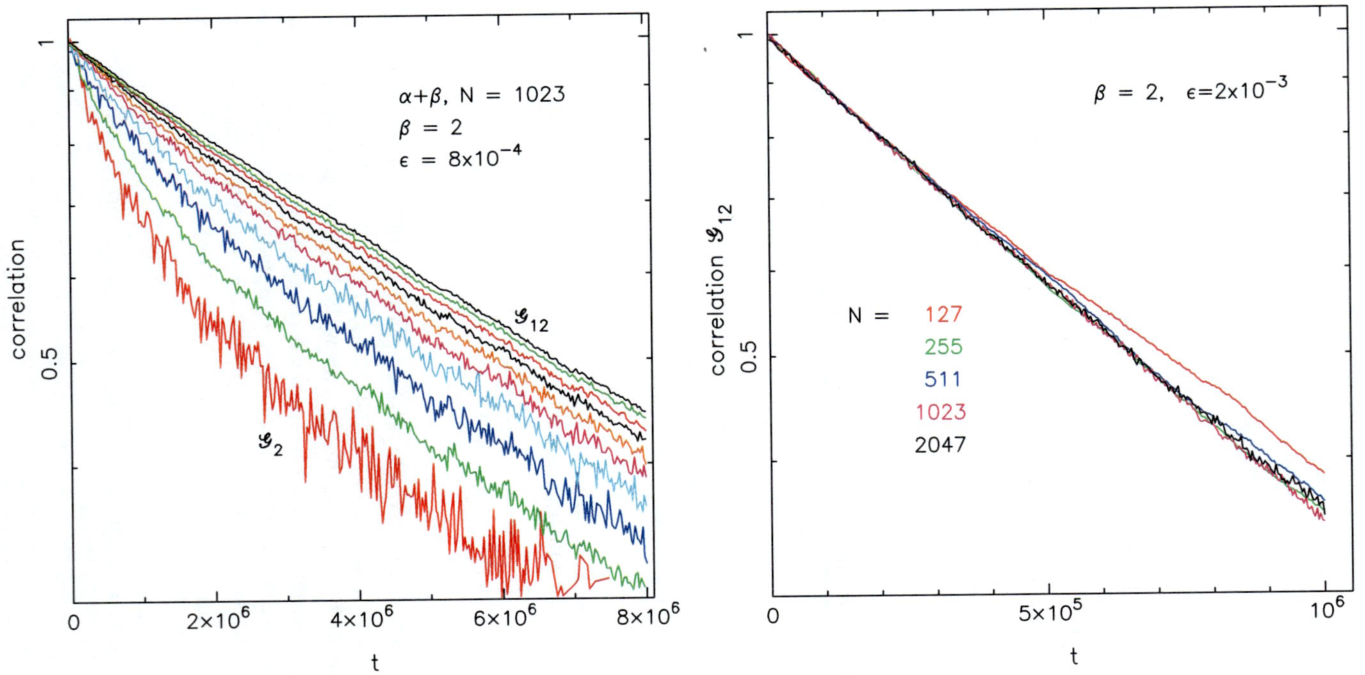

Fig. 7 Left: the time correlations $\mathcal{G}_s(t)$, $s = 2, \ldots, 12$, for FPU with $N = 1023$, $\beta = 2$, $\varepsilon = 8 \times 10^{-4}$; semi-log scale. Gaussian random extraction of 20,000 initial data. Right: the time correlation $\mathcal{G}_{12}(t)$ of F_{12}, for $\beta = 2$, $\varepsilon = 2 \times 10^{-3}$, $N = 127, \ldots, 2047$

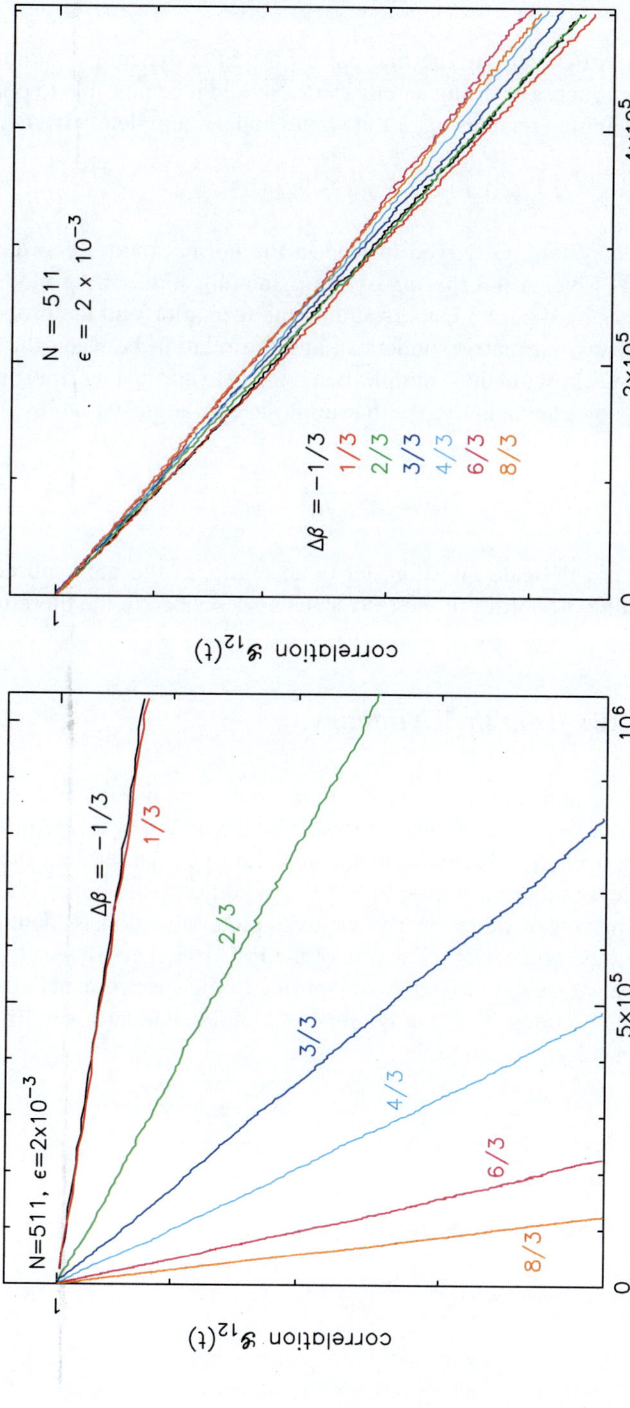

Fig. 8 The time correlations $\mathcal{G}_{12}(t)$ of F_{12}, for $N = 511$, $\varepsilon = 2 \times 10^{-3}$, $\Delta\beta = \beta - \beta_T$ as marked in the figure. Left: no rescaling. Right: time axis rescaled by a factor $\frac{9}{16}\Delta\beta^2$

3 Investigating the FPU State

The formation of the FPU state, in the first part of the FPU dynamics common to FPU and Toda, is the process of filling a Toda torus. Should it be possible to observe it in the Toda action-angle variables $(\mathcal{I}, \varphi)$, it would appear completely trivial, i.e.,

$$(\mathcal{I}^0, \varphi^0) \mapsto (\mathcal{I}^0, \varphi^0 + \omega(\mathcal{I}^0)t) \ , \tag{5}$$

as for any integrable system. Observed instead in the normal modes coordinates, it appears as a progressive partial sharing of energy among some of the modes, as illustrated in Figs. 1, 2, 3, 4, and 5 Understanding the formation and the properties of the FPU state means, ultimately, understanding the relation between the Toda action-angle coordinates, with their simple behavior (5), and the normal modes coordinates (P, Q), or equivalently, the harmonic action-angle variables (I, θ), related to (P, Q) by

$$P_k = \sqrt{2\omega_k I_k} \cos\theta_k \ , \qquad Q_k = \sqrt{2I_k/\omega_k} \sin\theta_k \ , \qquad E_k = \omega_k I_k \ .$$

Before entering such a delicate question, let us examine, in the next subsection, some important scaling properties of the FPU state, as described in the literature.

3.1 *Scalings Laws from the Dynamics*

Figure 9 refers to Toda and shows a process similar to the one in Fig. 5b. Figure 9a differs from Fig. 5b only for the higher energy $\varepsilon = 2.5 \times 10^{-4}$; the profiles are similar, but for higher energy the width of the spectrum gets larger, i.e., the FPU state includes a larger number of modes. In Fig. 9b, in addition, the initial state (the black rectangle) is narrower; quite clearly, the asymptotic situation is identical to the previous one, but the process of formation of the FPU state gets slower.

To be quantitative, we need to assign to any profile of the spectrum an "effective number" M of excited modes. This can be done in a rather standard way: if $\overline{E}_k$ is the energy spectrum at a certain time, let

$$h = -\sum_{k=1}^{N} p_k \log p_k \ , \qquad p_k = \frac{\overline{E}_k}{\sum_j \overline{E}_j} \ , \tag{6}$$

denote the so-called spectral entropy, $0 \le h \le \log N$; then

$$M = e^h \ , \qquad 1 \le M \le N \ . \tag{7}$$

In support to the definition, it is worthwhile to observe that in a situation in which exactly M modes equally share energy, while the others are at rest, the definition

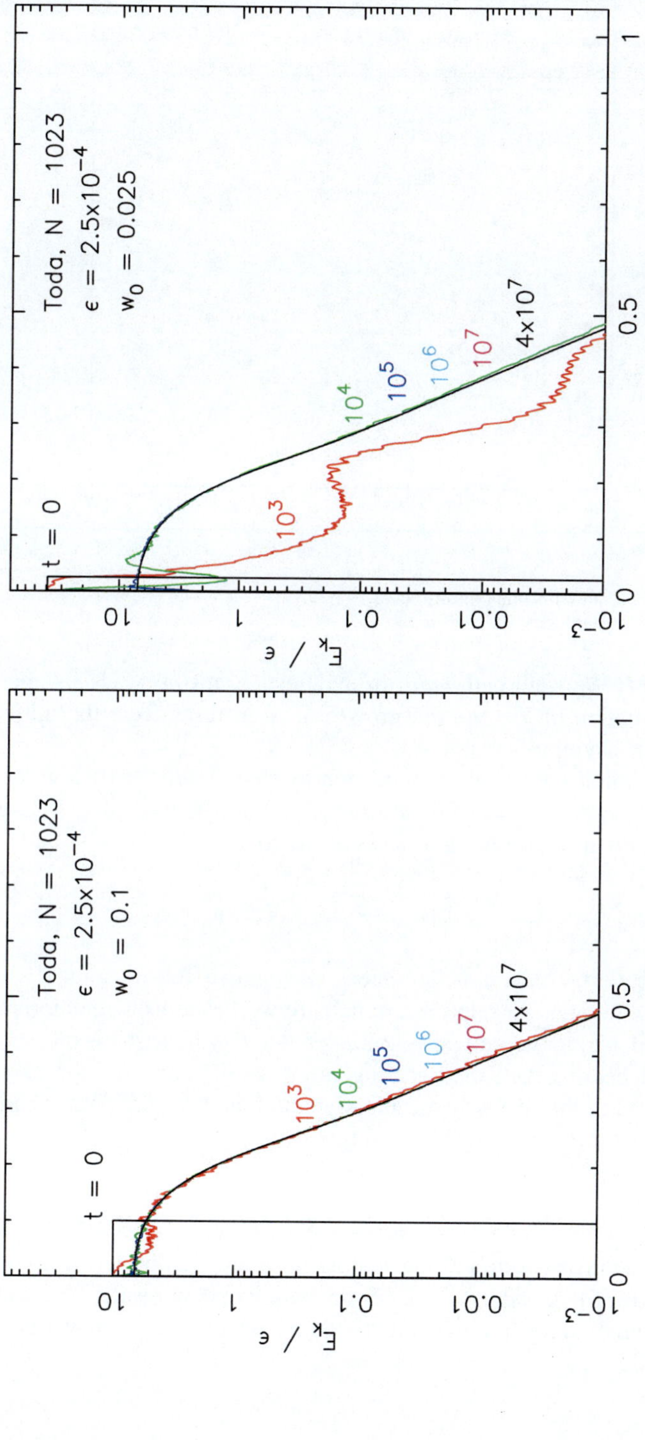

Fig. 9 Similar to Fig. 5b, higher $\varepsilon = 2.5 \times 10^{-4}$. Left: initial excitation of lower 10% of modes; right: lower 2.5%

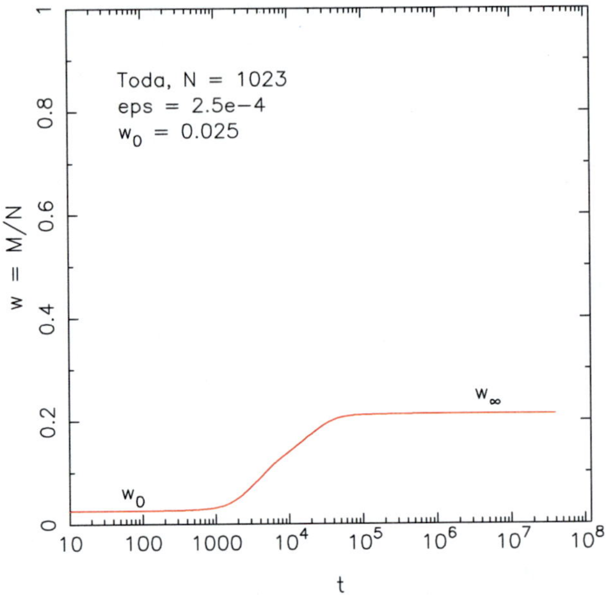

Fig. 10 The growth of $w(T)$, same conditions as Fig. 9b

gives precisely M. We shall call *width* of a state the ratio $w = M/N$. Figure 10 shows, in the situation of Fig. 9b, the growth of w in time, from the initial value $w_0 = 0.025$ to the asymptotic value $w_\infty \simeq 0.21$.

For FPU-like initial states, i.e., with the energy shared initially by low frequency modes, the width w of the spectrum is a function, in principle, of t, N, ε and the initial width w_0. Figure 10 shows a process of the form

$$w(t, N, \varepsilon, w_0) \overset{t\to\infty}{\longrightarrow} w_\infty(N, \varepsilon, w_0) .$$

In several papers, $w(t, N, \varepsilon, w_0)$ has been observed to follow some elementary scaling laws, or homogeneity relations, which are well established numerically and also partially understood theoretically, although at a very heuristic level.

The first and better established scaling law [28–30] concerns the asymptotic width w_∞, and states that if w_0 is sufficiently small, and N sufficiently large, then w_∞ is independent of both N and w_0, and it is

$$w_\infty \sim \varepsilon^{1/4} . \tag{8}$$

A more systematic investigation of the scaling laws satisfied by $w(t, N, \varepsilon, w_0)$ can be found in [31] for FPU, and in [14] for Toda, with identical results. The width w is there shown to satisfy three homogeneity relations, which reduce the variables from four to only one. The resulting scaling law depends on whether the phases θ_k^0 of the initially excited modes are chosen randomly or are coherent.

- For random initial phases, the law is

$$w(t, N, \varepsilon, w_0) = \varepsilon^{1/4} \mathcal{G}(\varepsilon^{3/8} w_0^{3/2} t) , \tag{9}$$

  $\mathcal{G}$ being a suitable function of a single variable, with a sigmoid profile as in Fig. 10.[6] This holds also for w_0 very small, including the case of a fixed small number of excited modes; in such a case the assumption of random initial phases is obviously meaningless and in fact unnecessary.
- Instead for coherent initial phases, for example, equal to each other or following some easy pattern (see [14, 31] for details), then the total energy $E = N\varepsilon$ rather than the specific energy is relevant, and (9) is replaced by

$$w(t, N, \varepsilon, w_0) = E^{1/4} w_0^{1/4} \mathcal{G}'(E^{3/8} w_0^{15/8} t) . \tag{10}$$

3.2 Scaling Laws from Toda Actions

In principle, should one know the transformation from the Toda action angle variables $(\mathcal{I}, \varphi)$ to the harmonic variables (I, θ), and conversely, one could understand everything according to the scheme

$$(I^0, \theta^0) \mapsto (\mathcal{I}^0, \varphi^0) \mapsto (\mathcal{I}^0, \varphi(t)) \mapsto (I(t), \theta(t)) . \tag{11}$$

Practically, in spite of the quite considerable theoretical progress [32–36], the relation between the Toda and the harmonic variables is not really understood, other than in the regime, very far from statistical mechanics,

$$\varepsilon \ll N^{-4} .$$

Numerically the situation is hard as well, but something can be done. Indeed, as already mentioned, there exists an algorithm to compute the Toda actions $\mathcal{I}_k$ in any configuration of the chain [8]. This is a little part of (11), but sufficient to support an "elementary" conjecture concerning the FPU state, namely:

In the FPU state, the number $\mathcal{M}$ of Toda actions which are substantially different from zero, that is the effective dimensionality of the Toda torus, scales as $N\varepsilon^{1/4}$.

---

[6] The time scale for the formation of the state has been first studied in [29], in the particular (but important) case $w_0 \sim \varepsilon^{1/4}$. The result $t \sim \varepsilon^{-3/4}$ there reported is coherent with (9).

$\mathcal{M}$ can be defined similarly to M, namely via (6) and (7), using, however, the Toda equivalent energy $\mathcal{E}_k = \omega_k \mathcal{I}_k$ in place of $\overline{E}_k$. The $\mathcal{E}_k$'s and $\mathcal{M}$ are constant in time, and can be computed at any time, including the initial state. This means *The scaling laws characterizing the FPU state are contained in the initial state, and stay in the nontrivial correspondence between harmonic actions and Toda actions.* This is a somehow innovative perspective, in which dynamics (integration of Hamilton equations) does not play a role.

According to (9) and (10), the conjecture is expected to hold for states in which one or a few harmonic actions are different from zero, or also a number proportional to N, with, however, random phases; it is instead expected to fail for coherent phases. This is precisely what we shall check in the next paragraphs A and B, devoted, respectively, to states including a single travelling wave and to states including a number of waves proportional to N, with either random or coherent phases. Concerning the algorithm to compute the actions, a quick account is provided in paragraph C.

A. States with a Single Travelling Wave Here we restrict the attention to FPU-like initial conditions, precisely to states including a single travelling wave with $k = 1$ (a preliminary account of such results, limited to smaller N, can be found in [14]). Figure 11a shows $\mathcal{M}$ vs. ε, in log-log scale, for different N ranging from 32 to 32, 768. The computed slopes, reported in the figure, indicate with *great* evidence that, for large N, $\mathcal{M}$ is indeed proportional to $\varepsilon^{1/4}$. The proportionality to N could be similarly checked, but the best way to check the conjecture is to directly plot $\mathcal{M}$ vs. $N\varepsilon^{1/4}$: if we do, see Fig. 11b, curves for different N exactly superimpose, and for large N the computed slope, see the data in the figure, is virtually 1. (One might observe that $N - 1$, rather than N, enters the abscissa of Fig. 11b. Indeed, the barycenter being at rest, the number of degrees of freedom is $N - 1$; the difference is very minor, but is visible at small N, and slightly improves the figure.)

B. States with Many Travelling Waves We studied FPU-like states with energy equidistributed among a number of waves proportional to N, namely travelling waves with $0 < k/N \leq w_0$, small w_0. The phases of the waves are chosen randomly; more precisely, before computing $\mathcal{M}$, the Toda spectrum $\mathcal{E}_k$ is averaged on several different random extractions of the phases, actually 128 of them.

Results here are encouraging, although not as satisfying as in the above case of single wave. Figure 12 shows $\mathcal{M}/(N-1)$ vs. ε, for $w_0 = 0.025$. The expected slope was $1/4$, the computed slope, for large N, is 0.24. Preliminary computations show that by decreasing w_0 the slope gets closer to $1/4$, but a systematic study (which would require larger N) has not yet been done.

Quite remarkably, however, if we pass to coherent phases θ_k, results drastically change, reflecting the difference between (10) and (9). Indeed, the independence of $\mathcal{M}/N$ on N, strongly evident in the superposition of curves in Fig. 12, gets lost; see Fig. 13a, where $\mathcal{M}/N$ is plotted vs. ε for different N (also observe the crazy behavior at large N). The choice of phases is here $\theta_k = k\pi/2$. The stability in N is, however, roughly recovered if, according to (10), $\mathcal{M}/N$ is plotted vs. the total

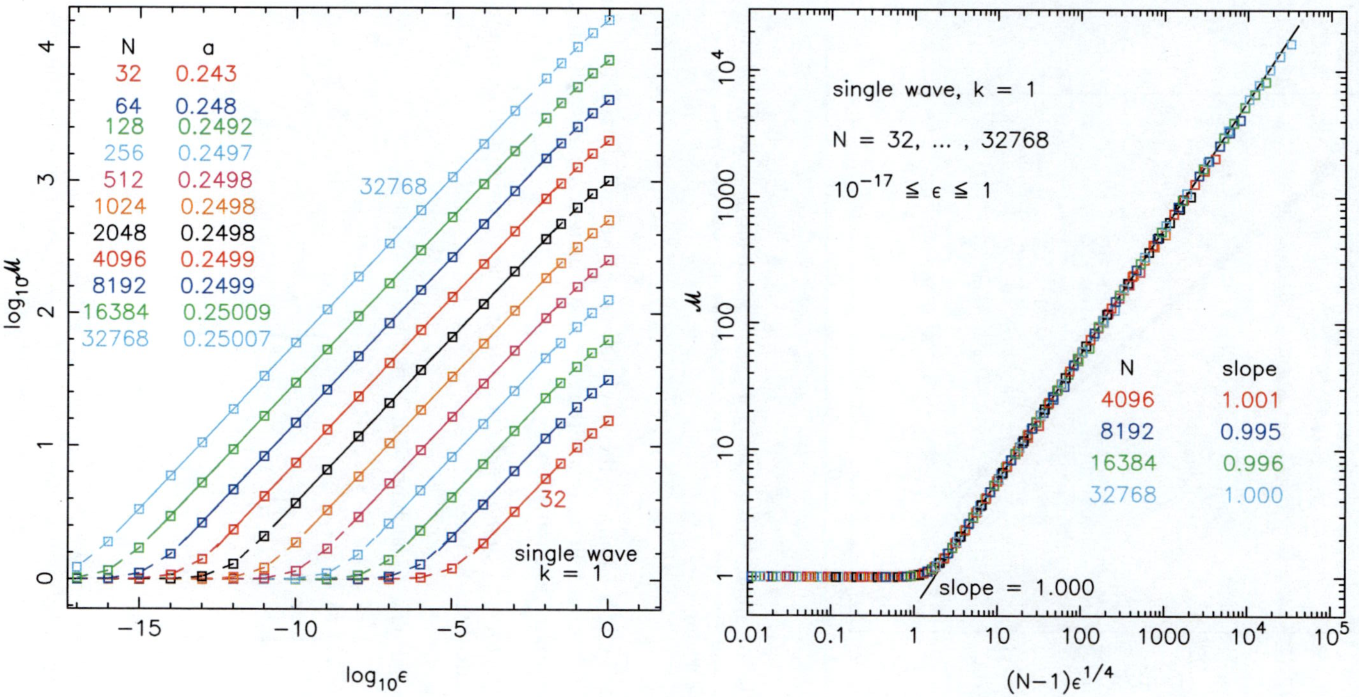

Fig. 11 Left: the effective dimensionality $\mathcal{M}$ of the Toda torus vs. ε, log-log scale, for different N. Right: the same quantity, reported as a function of $(N-1)\varepsilon^{1/4}$. Single travelling wave, $k = 1$

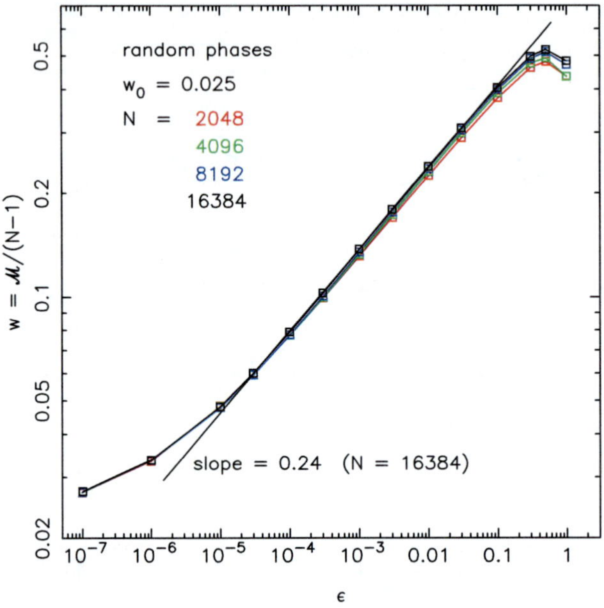

Fig. 12 The effective width $\mathcal{M}/N$ of the Toda torus, when energy is equidistributed among travelling waves with $0 < k/N \leq w_0 = 0.025$ and phases are random. Average on 128 random extractions of the phases

energy E; see Fig. 13b. Changing the way coherent phases are chosen, for example, equal to each other, or following a different "easy" pattern, changes the details of the curves, but not the phenomenon.

C. On the Algorithm to Compute Toda Actions The algorithm to compute Toda actions is essentially as follows:

– Consider a periodic Toda chain with N particles, and let

$$L = \begin{pmatrix} b_1 & a_1 & & & & a_N \\ a_1 & b_2 & a_2 & & & \\ & a_2 & b_3 & a_3 & & \\ & & & \ddots & & \\ & & & & b_{N-1} & a_{N-1} \\ a_N & & & & a_{N-1} & b_N \end{pmatrix}, \qquad \begin{aligned} a_i &= e^{\lambda(q_{i+1}-q_i)/2} \\ b_i &= -\lambda p_i \end{aligned}$$

be the associated Lax matrix (tridiagonal periodic). Let $P(x) = \det(xI - L)$ be its characteristic polynomial, and $\Delta(x) = P(x)+2$ be the so-called discriminant. When the system is at rest, Δ oscillates between -2 and 2 [8], as in Fig. 14a; for positive energies instead the shape looks as in Fig. 14b, blue curve, with "gaps,"

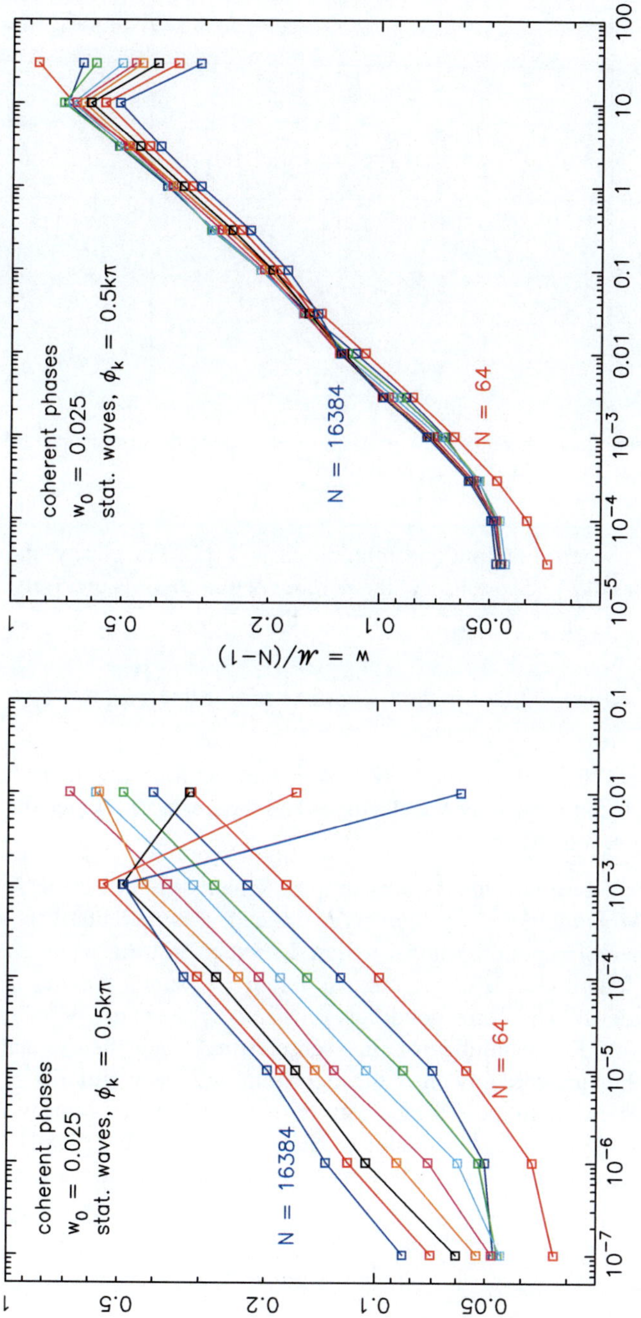

Fig. 13 Left: the fraction $\mathcal{M}/N$ vs. the specific energy ε, for $N = 64, \ldots, 16384$; stationary waves, phases $\theta_k = k\pi/2$; $w_0 = 0.05$. Right: same quantity vs. the total energy E

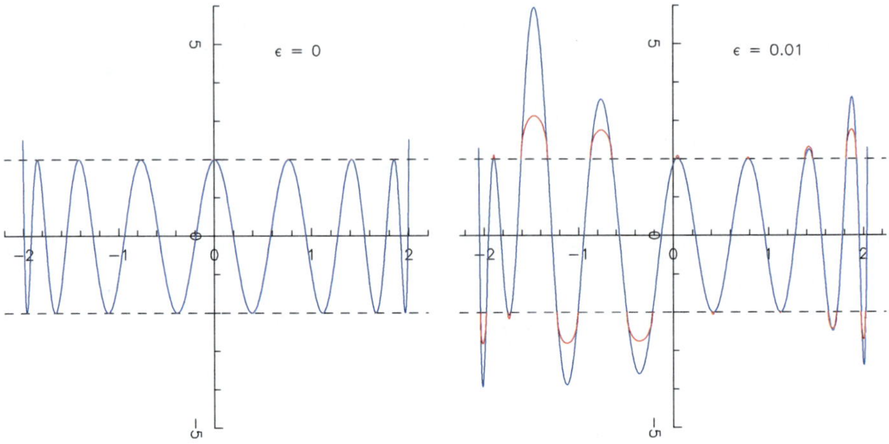

Fig. 14 Left: the discriminant $\Delta(x)$ for $N = 16$, chain at rest. Right: the discriminant (blue) and the arcs $\pm(2 + \rho)$ (red), for $\varepsilon = 10^{-2}$, in a typical situation

that is intervals where maxima and minima exceed ± 2. For clarity, the figure refers to quite small $N = 16$ (observe the number of extremals is precisely $N-1$).

– Let g_k be the k-th gap. The recipe to compute the actions is:

$$\mathcal{I}_k = \frac{1}{\pi} \int_{g_k} \rho(x)\, \mathrm{d}x \,, \qquad \rho(x) = \mathrm{acosh} \frac{|\Delta(x)|}{2} \,.$$

Red arcs in Fig. 14b are the curves $\pm(2 + \rho(x))$; the actions are precisely, up to a trivial factor, the areas between such curves and the lines ± 2. Notice that if $|\Delta|$ is large, then $\rho \simeq \log |\Delta|$.

Practically, applying the algorithm is not as simple. A main difficulty is that *for large N, unbelievably large numbers enter the game.* Indeed, assume tentatively that $\mathcal{I}_k$ is not far from the corresponding linear action I_k, which in turn, when energy is shared only by a few modes, is of the order E/ω_k. If k is small, then $\omega_k \sim N^{-1}$, so that $\mathcal{I}_k$ is of order N, and correspondingly *the peaks of $\Delta(x)$ are of the order of the exponential of N*. Large numbers are in a sense virtual, since then ρ_k and I_k are not as large, but they unavoidably enter the algorithm. Such a singularity for large N also reflects the irreducible difference between Toda actions and linear actions, unless, as in [34–36], $\varepsilon \ll N^{-4}$ is assumed. How to proceed numerically in such conditions is a little technical, and we haven't had the possibility to further discuss the question here.

3.3 A Conclusion?

The only conclusion we feel confident to draw is that considering FPU as a perturbed Toda model, rather than a perturbed linear model as is more commonly done, is very fruitful and allows to achieve a unitary view of the FPU behavior. This includes both the first time scale, with the formation of a state common to FPU and Toda, trajectories staying (almost) confined to a torus, and the second time scale, where diffusion across tori becomes important. The main point, highly non obvious, is that *FPU appears to stay close to Toda uniformly in* N; on the contrary, *both FPU and Toda look distant from the harmonic chain, no matter how small is* ε, *if* $N\varepsilon^{1/4}$ *is not small.*

Investigating the relation between normal modes, or harmonic actions, and Toda nonlinear actions, seems to us particularly important: indeed, on the one hand, normal modes are the elementary bricks of statistical mechanics, which play a key role in the equipartition theorem, and for ergodic-like systems are known to have a simple statistical behavior; on the other hand, Toda actions are very essential elements of the dynamics. As is not easy, the two points of view should be kept together. We are making some effort to continue this investigation that we consider promising.

References

1. Fermi, E., Pasta, J., Ulam, S.: Studies of Non Linear Problems, Los-Alamos Internal Report, Document LA-1940 (1955). In: Enrico Fermi Collected Papers, vol. II, The University of Chicago Press, Chicago, and Accademia Nazionale dei Lincei, Roma, 1965, pp. 977–988. Later reproduced in Lect. Appl. Math. **15**, 143–156 (1974) and in ref. [3] below
2. Chaos focus issue: The "Fermi-Pasta-Ulam" problem–the first 50 years. Chaos **15** (2005)
3. Gallavotti, G. (ed.): The Fermi-Pasta-Ulam Problem: A Status Report. Lecture Notes in Physics, vol. 728. Springer, Berlin (2008)
4. Toda, M.: Vibration of a chain with nonlinear interaction. J. Phys. Soc. Jpn. **22**, 431–436 (1967); Wave propagation in anharmonic lattices. J. Phys. Soc. Jpn. **23**, 501–506 (1967); Mechanics and statistical mechanics of nonlinear chains. J. Phys. Soc. Jpn. Suppl. **26**, 109–111 (1969); Waves in nonlinear lattice. Progr. Theor. Phys. Suppl. **45**, 174–200 (1970)
5. Manakov, S.V.: Complete integrability and stochastization of discrete dynamical systems. Sov. Phys. JEPT **40**, 269–274 (1974)
6. Hénon, M.: Integrals of the Toda lattice. Phys. Rev. B **9**, 1921–1923 (1974)
7. Flaschka, H.: The Toda lattice. II. Existence of integrals. Phys. Rev. B **9**, 1924–1925 (1974)
8. Ferguson, W.E., Flaschka, H., McLaughlin, D.W.: Nonlinear Toda modes for the Toda chain. J. Comput. Phys. **45**, 157–209 (1982)
9. Benettin, G., Ponno, A.: Time-scales to equipartition in the Fermi-Pasta-Ulam problem: finite-size effects and thermodynamic limit. J. Stat. Phys. **144**, 793–812 (2011)
10. Benettin, G., Christodoulidi, H., Ponno, A.: The Fermi-Pasta-Ulam problem and its underlying integrable dynamics. J. Stat. Phys. **152**, 195–212 (2013)
11. Benettin, G., Pasqualie, S., Ponno, A.: The Fermi-Pasta-Ulam problem and its underlying integrable dynamics: an approach through Lyapunov Exponents. J. Stat. Phys. **171**, 521–542 (2018)

12. Christodoulidi, H., Efthymiopoulos, C.: Stages of dynamics in the Fermi-Pasta-Ulam system as probed by the first Toda integral. Math. Eng. **1**, 359–377 (2019)
13. Goldfriend, T., Kurchan, J.: Equilibration of quasi-integrable systems. Phys. Rev. E **99**, 022146-1-9 (2019)
14. Benettin, G., Ponno, A.: Understanding the FPU state in FPU-like models. Math. Eng. **3**, 1–22 (2021)
15. Cercignani, C., Galgani, L., Scotti, A.: Zero-point energy in classical non-linear mechanics. Phys. Lett. A **38**, 403–404 (1972)
16. Cercignani, C.: Solitons. Theory and application. Riv. Nuovo Cim. **7**, 429–469 (1977)
17. Tuck, J.L., Menzell, M.T.: The superperiod of the nonlinear weighted string (FPU) problem. Adv. Math. **9**, 399–407 (1972). In spite of the late publication date, results go back to 1961 (see Ulam's foreward to the FPU paper in the Enrico Fermi Collected Papers [1])
18. Zabusky, N.J., Kruskal, M.D.: Interaction of solitons in a collisionless plasma and the recurrence of initial states. Phys. Rev. Lett. **15**, 240–245 (1965)
19. Gardner, C.S., Green, J.M., Kruskal, M.D., Miura, R.M.: Method for solving the Korteweg-de Vries equation. Phys. Rev. Lett. **19**, 1095–1097 (1967)
20. Lax, P.D.: Integrals of nonlinear equations of evolution and solitary waves. Comm. Pure Appl. Math. **21**, 467–490 (1968)
21. Miura, R.M., Gardner, C.S., Kruskal, M.D.: Korteweg-de Vries equation and generalization, II. Existence of conservation laws and constants of motion. J. Math. Phys. **9**, 1204–1209 (1968)
22. Zakharov, V.E., Feddeev, L.D.: Korteweg-de Vries Equation: a completely integrable Hamiltonian system. Funct. Anal. Appl. **5**, 280–286 (1971)
23. Zakharov, V.E.: On stochastization of one dimensional chains of nonlinear oscillators. Sov. Phys. JETP **38**, 108–110 (1973)
24. Izrailev, F.M., Chirikov, B.V., Statistical properties of a nonlinear string. Sov. Phys. Dokl. **11**, 30–34 (1966)
25. Fucito, E., Marchesoni, F., Marinari, E., Parisi, G., Peliti, L., Ruffo, S., Vulpiani, A.: Approach to equilibrium in a chain of nonlinear oscillators. J. Phys. **43**, 707–713 (1982)
26. Livi, R., Pettini, M., Ruffo, S., Sparpaglione, M., Vulpiani, A.: Relaxation to different stationary states in the Fermi-Pasta-Ulam model. Phys. Rev. A **28**, 3544–3552 (1983)
27. Carati, A., Galgani, L., Giorgilli, A., Paleari, S.: FPU phenomenon for generic initial data. Phys. Rev. E **76**, 022104/1-4 (2007)
28. Shepelyansky, D.L.: Low-energy chaos in the Fermi-Pasta-Ulam Problem. Nonlinearity **10**, 1331–1338 (1997)
29. Biello, J.A., Kramer, P.R., L'vov, Y.V.: Stages of energy transfer in the FPU model. In: Proceedings of the Fourth International Conference on Dynamical Systems and Differential Equations (May 24–27, 2002, Wilmington). AIMS Conference Publications 2003 (special), pp. 113–122 (2003)
30. Berchialla, L., Galgani, L., Giorgilli, A.: Localization of energy in FPU chains. Discr. Contin. Dyn. Syst. **11**, 855–866 (2004)
31. Benettin, G., Livi, R., Ponno, A.: The Fermi-Pasta-Ulam problem: scaling laws vs. initial conditions. J. Stat. Phys. **135**, 873–893 (2009)
32. Nishida, T.: A note on an existence of conditionally periodic oscillation in a one-dimensional anharmonic lattice. Mem. Fac. Engrg. Kyoto Univ. **33**, 27–44 (1971)
33. Rink, B.: Proof of Nishida's conjecture on anharmonic lattices. Comm. Math. Phys. **261**, 613–627 (2006)
34. Henrici, A., Kappeler, T.: Global action-angle variables for the periodic Toda lattice. Int. Math. Res. Not. Article ID rnn031, 1–52 (2008)
35. Henrici, A., Kappeler, T.: Global Birkhoff coordinates for the periodic Toda lattice. Nonlinearity **21**, 2731–2758 (2008)
36. Bambusi, D., Maspero, A.: Birkhoff coordinates for the Toda Lattice in the limit of infinitely many particles with an application to FPU. J. Funct. Anal. **270**, 1818–1887 (2016)

Half-Space Problems for the Boltzmann Equation of Multicomponent Mixtures

Niclas Bernhoff

Abstract Half-space problems in the kinetic theory of gases are of great importance in the study of the asymptotic behavior of solutions of boundary value problems for the Boltzmann equation for small Knudsen numbers. They provide the boundary conditions for the fluid-dynamic-type equations and Knudsen-layer corrections to the solution of the fluid-dynamic-type equations in a neighborhood of the boundary. These problems are well studied for single species, including some important contributions by Carlo Cercignani, and it is well-known that the number of additional conditions needed to be imposed depends on different regimes for the Mach number (corresponding to subsonic/supersonic evaporation/condensation). However, the case of mixtures is not as well studied in the literature. We will address some extensions of the results for half-space problems for single species to the case of multicomponent mixtures.

1 Kinetic Half-Space Problem for Mixtures

Half-space problems in the kinetic theory of gases are of great importance in the study of the asymptotic behavior of solutions of boundary value problems for the Boltzmann equation for small Knudsen numbers [26, 27]. These problems are well studied for monatomic single species [4], including seminal contributions by Carlo Cercignani [20]. However, the case of mixtures is not as well studied in the literature; among the studies for binary mixtures we mention Ref. [1] for the linearized problem, Ref. [3] for the weakly nonlinear case (with equal masses), and Ref. [7] for discrete velocity models. This chapter will address some extensions of results for half-space problems for single species to the case of multicomponent mixtures. Linearized half-space problems will be reviewed for discrete velocity models in Sect. 2 and for the full Boltzmann equation, based on a recent study of an

N. Bernhoff (✉)
Department of Mathematics and Computer Science, Karlstad University, Karlstad, Sweden
e-mail: niclas.bernhoff@kau.se

P. Barbante et al. (eds.), *From Kinetic Theory to Turbulence Modeling*, Springer INdAM Series 51, https://doi.org/10.1007/978-981-19-6462-6_4

abstractly formulated linear half-space problem [11], in Sect. 3, after an introduction in this section.

Consider a mixture of s monatomic species $\alpha_1, ..., \alpha_s$, with masses $m_{\alpha_1}, ..., m_{\alpha_s}$, respectively ($s = 1$ corresponds to the case of a monatomic single species). The distribution functions are of the form $F = \left(F_{\alpha_1}, ..., F_{\alpha_s}\right)$, where the distribution of the molecules of species α_i is given by $F_{\alpha_i} = F_{\alpha_i}(t, \mathbf{x}, \mathbf{v})$, with time $t \in \mathbb{R}_+$, position $\mathbf{x} = (x, y, z) \in \mathbb{R}^3$, and molecular velocity $\mathbf{v} = \left(v_x, v_y, v_z\right) \in \mathbb{R}^3$. The components of the equilibrium, or Maxwellian, distributions $M = \left(M_{\alpha_1}, ..., M_{\alpha_s}\right)$ are Gaussians

$$M_{\alpha_i} = n_{\alpha_i} \left(\frac{m_{\alpha_i}}{2\pi T}\right)^{3/2} e^{-m_{\alpha_i}|\mathbf{v}-\mathbf{u}|^2/(2T)},$$

where $\left\{n_{\alpha_1}, ..., n_{\alpha_s}\right\} \subset \mathbb{R}_+$, $\mathbf{u} = \left(u_x, u_y, u_z\right) \in \mathbb{R}^3$, and $T \in \mathbb{R}_+$ relate to the number densities of the species $\alpha_1, ..., \alpha_s$, the bulk velocity, and the temperature, respectively.

The steady half-space problem for the (vector) Boltzmann equation for a mixture in a slab symmetry, cf., e.g., [4, 20, 26, 27] for single species, reads

$$\begin{cases} v_x \dfrac{\partial F}{\partial x} = Q(F, F), \ F = F(x, \mathbf{v}), \ x \in \mathbb{R}_+, \\ F(0, \mathbf{v}) = M_B(\mathbf{v}) \text{ for } v_x > 0, \\ F \to M_\infty \text{ as } x \to \infty. \end{cases} \tag{1}$$

The system (1) describes the evolution of a distribution function $F = F(x, \mathbf{v})$ for $x > 0$ approaching equilibrium—in form of a Maxwellian distribution $M_\infty = M_\infty(\overline{n}, \mathbf{u}, T) = M_\infty\left(\left(n_{\alpha_1}, ..., n_{\alpha_s}\right), (u, 0, 0), T\right)$—at the far end, knowing the outgoing distribution $M_B = M_B(\mathbf{v})$ at an interface, $x = 0$. Typically, the distribution $M_B = M_B(\mathbf{v})$ is a half-Maxwellian. More general boundary conditions, where the outgoing distribution depends on the incoming distribution, can also be considered at the interface $x = 0$ [21, 22, 26, 27].

The components of the (vector) collision operator $Q = \left(Q_{\alpha_1}, ..., Q_{\alpha_s}\right)$ read

$$Q_{\alpha_i} = Q_{\alpha_i}(F, F) = \sum_{j=1}^{s} Q_{\alpha_i}^{\alpha_j}(F_{\alpha_i}, F_{\alpha_j}),$$

where (omitting the x-dependence of F) [21, 22]

$$Q_{\alpha_i}^{\alpha_j}(F_{\alpha_i}, F_{\alpha_j}) = \iint_{\mathbb{R}^3 \times S^2} \sigma_{ij} |\mathbf{v} - \mathbf{v}_*| \left(F_{\alpha_i}(\mathbf{v}') F_{\alpha_j}(\mathbf{v}'_*) - F_{\alpha_i}(\mathbf{v}) F_{\alpha_j}(\mathbf{v}_*)\right) d\mathbf{v}_* d\omega.$$

Here $\{\mathbf{v}, \mathbf{v}_*\}$ and $\{\mathbf{v}', \mathbf{v}'_*\}$ denote pre- and post-collisional velocities (or, vice versa), respectively. For hard spheres the cross sections $\sigma_{ij} = \sigma_{ji}$ are constant, and

$$\begin{cases} \mathbf{v}' = \dfrac{m_{\alpha_i}\mathbf{v} + m_{\alpha_j}\mathbf{v}_*}{m_{\alpha_i} + m_{\alpha_j}} + \dfrac{m_{\alpha_i}}{m_{\alpha_i} + m_{\alpha_j}} |\mathbf{v} - \mathbf{v}_*| \, \omega \\ \mathbf{v}'_* = \dfrac{m_{\alpha_i}\mathbf{v} + m_{\alpha_j}\mathbf{v}_*}{m_{\alpha_i} + m_{\alpha_j}} - \dfrac{m_{\alpha_i}}{m_{\alpha_i} + m_{\alpha_j}} |\mathbf{v} - \mathbf{v}_*| \, \omega \end{cases}, \omega \in S^2.$$

The set of collision invariants is spanned by $\{e_1, ..., e_s, mv_x, mv_y, mv_z, m\,|\mathbf{v}|^2\}$, where $m = (m_{\alpha_1}, ..., m_{\alpha_s})$ and $\{e_1, ..., e_s\}$ is the standard basis of $\mathbb{R}^s$, due to mass (of each species), momentum, and kinetic energy conservation.

Due to Galilean invariance, a shift, $\mathbf{v} \mapsto \mathbf{v} + \mathbf{u}$, in the velocity space ends up in

$$\begin{cases} (v_x + u) \dfrac{\partial \widetilde{F}}{\partial x} = Q(\widetilde{F}, \widetilde{F}), \\ \widetilde{F}(0, \mathbf{v}) = M_B(\mathbf{v} + \mathbf{u}) \text{ for } v_x + u > 0, \\ \widetilde{F} \to M \text{ as } x \to \infty, \end{cases} \tag{2}$$

where $\widetilde{F}(x, \mathbf{v}) = F(x, \mathbf{v} + \mathbf{u})$ and $M = M_\infty(\overline{n}, \mathbf{0}, T)$ is a non-drifting (absolute) Maxwellian. Linearizing around the Maxwellian $M = M_\infty(\overline{n}, \mathbf{0}, T)$ by denoting $\widetilde{F} = M + \sqrt{M} f$ results, neglecting quadratic terms, in

$$\begin{cases} (v_x + u) \dfrac{\partial f}{\partial x} + \mathcal{L} f = 0, \; f = f(x, \mathbf{v}), \; x \in \mathbb{R}_+, \\ f(0, \mathbf{v}) = \widetilde{M}_B(\mathbf{v}) \text{ for } v_x + u > 0, \\ f \to 0 \text{ as } x \to \infty, \end{cases} \tag{3}$$

where $\widetilde{M}_{B\alpha_i}(\mathbf{v}) = M_{\alpha_i}^{-1/2}(\mathbf{v}) \left(M_{B\alpha_i}(\mathbf{v} + \mathbf{u}) - M_{\alpha_i}(\mathbf{v}) \right)$ and

$$(\mathcal{L} f)_{\alpha_i} = \iint_{\mathbb{R}^3 \times S^2} \sigma_{ij} |\mathbf{v} - \mathbf{v}_*| M_{\alpha_j}^{1/2}(\mathbf{v}_*) \Big(M_{\alpha_i}^{1/2}(\mathbf{v}) f_{\alpha_j}(x, \mathbf{v}_*) + M_{\alpha_j}^{1/2}(\mathbf{v}_*) f_{\alpha_i}(x, \mathbf{v})$$

$$- M_{\alpha_i}^{1/2}(\mathbf{v}') f_{\alpha_j}(x, \mathbf{v}'_*) - M_{\alpha_j}^{1/2}(\mathbf{v}'_*) f_{\alpha_i}(x, \mathbf{v}') \Big) \, d\mathbf{v}_* d\omega.$$

Consider the real Hilbert space $\mathfrak{h} := \left(L^2(d\mathbf{v}) \right)^s$, with inner product

$$(f | g) = \sum_{i=1}^{s} \int_{\mathbb{R}^3} f_{\alpha_i} g_{\alpha_i} \, d\mathbf{v}, \; f, g \in \left(L^2(d\mathbf{v}) \right)^s.$$

The linearized collision operator $\mathcal{L}$ (for hard spheres) is a nonnegative, self-adjoint Fredholm operator [12], see also [1, 18, 19, 23], with the domain

$\mathrm{D}(\mathcal{L}) = \left(L^2\left((1+|\mathbf{v}|)\,d\mathbf{v}\right)\right)^s$ and the kernel

$$\ker \mathcal{L} = \operatorname{span}\left\{\sqrt{M_{\alpha_1}}e_1, ..., \sqrt{M_{\alpha_s}}e_s, \sqrt{\overline{M}}v_x, \sqrt{\overline{M}}v_y, \sqrt{\overline{M}}v_z, \sqrt{\overline{M}}\,|\mathbf{v}|^2\right\},$$

where $\overline{M} = \left(m_{\alpha_1}^2 M_{\alpha_1}, ..., m_{\alpha_s}^2 M_{\alpha_s}\right)$ and $\{e_1, ..., e_s\}$ is the standard basis of $\mathbb{R}^s$. Furthermore, the linearized collision operator $\mathcal{L}$ can be split into a positive multiplication operator $\nu = \nu(|\mathbf{v}|) = \operatorname{diag}\left(\nu_{\alpha_1}(|\mathbf{v}|), ..., \nu_{\alpha_s}(|\mathbf{v}|)\right)$ minus a compact operator K on $\mathfrak{h}$ [12] (cf. [18]; while noting that the condition on the collision kernel stated in there, rules out the hard sphere models):

$$(\mathcal{L}f)(\mathbf{v}) = \nu(|\mathbf{v}|)f(\mathbf{v}) - K(f)(\mathbf{v}),\ f \in \mathrm{D}(\mathcal{L}), \text{ with}$$
$$\nu_-\,(1+|\mathbf{v}|) \le \nu_{\alpha_i}(|\mathbf{v}|) \le \nu_+\,(1+|\mathbf{v}|) \text{ for all } i \in \{1, ..., s\} \text{ and } \mathbf{v} \in \mathbb{R}^3,$$

for some constants $0 < \nu_- < \nu_+$ and for some real number $0 < \lambda < 1$

$$(f|\mathcal{L}f) \ge \lambda\,(f|\nu(|\mathbf{v}|)f) \ge \lambda\nu_-\,(f|(1+|\mathbf{v}|)\,f)\,.$$

2 Discrete Velocity Models for Mixtures

Consider a mixture of s different species $\alpha_1, ..., \alpha_s$ with masses $m_{\alpha_1}, ..., m_{\alpha_s}$. Fix for each species α_i a set of velocities $V^{\alpha_i} = \{\mathbf{v}_1^{\alpha_i}, ..., \mathbf{v}_{N_{\alpha_i}}^{\alpha_i}\} \subset \mathbb{R}^3$ and assign the label α_i to each element in V^{α_i}. The construction results in a set of $N = N_{\alpha_1} + ... + N_{\alpha_s}$ unique pairs (velocities may be repeated for different species) [9, 14]

$$\left\{\left(\mathbf{v}_1^{\alpha_1}, \alpha_1\right), \ldots, \left(\mathbf{v}_{N_{\alpha_1}}^{\alpha_1}, \alpha_1\right), \ldots, \left(\mathbf{v}_1^{\alpha_s}, \alpha_s\right), \ldots, \left(\mathbf{v}_{N_{\alpha_s}}^{\alpha_s}, \alpha_s\right)\right\}$$
$$=: \left\{\left(\mathbf{v}_1, \alpha(1)\right), \ldots, \left(\mathbf{v}_N, \alpha(N)\right)\right\}.$$

The planar stationary system, corresponding to system (2), for the discrete Boltzmann equation reads

$$(v_{ix} + u)\,\frac{dF_i}{dx} = Q_i\,(F, F)\,, x \in \mathbb{R}_+,\ i \in \{1, ..., N\}\,,$$

with $\mathbf{v}_i = \left(v_{ix}, v_{iy}, v_{iz}\right)$, $i \in \{1, ..., N\}$, and $F = F(x) = (F_1, ..., F_N)$; or, by introducing $Q\,(F, F) = (Q_1\,(F, F)\,, ..., Q_N\,(F, F))$ and $B = \operatorname{diag}(v_{1x}, ..., v_{Nx})$,

$$B_u\frac{dF}{dx} = Q\,(F, F)\,, x \in \mathbb{R}_+, \text{ where } B_u = B + uI.$$

Discrete velocity models for mixtures can approximate the Boltzmann equation up to any order [16] (for any rational mass ratios [14]). For technical reasons, we assume that $v_{ix} + u \neq 0$ for all $i \in \{1, ..., N\}$. After some reordering we may assume that

$$v_{ix}+u > 0 \text{ for } i \in \left\{1, ..., n^{+}\right\}, \text{ while } v_{ix}+u < 0 \text{ for } i \in \left\{n^{+}+1, ..., N\right\}, n^{-} := N-n^{+}.$$

For later use, denote

$$B_u^{+} := \operatorname{diag}(v_{1x} + u, ..., v_{n^+x} + u) \text{ and } B_u^{-} := -\operatorname{diag}(v_{n^++1x} + u, ..., v_{Nx} + u),$$

and define the projections

$$R_{+}h := (h_1, ..., h_{n^+}) \text{ and } R_{-}h := \left(h_{n^++1}, ..., h_N\right) \text{ for } h = (h_1, ..., h_N) \in \mathbb{R}^N. \tag{4}$$

The collision operators are given by

$$Q_i(F, F) = \sum_{j,k,l=1}^{N} \Gamma_{ij}^{kl}(F_k F_l - F_i F_j), i \in \{1, ..., N\},$$

where the nonnegative collision coefficients $\Gamma_{ij}^{kl} = \Gamma_{ji}^{kl} = \Gamma_{kl}^{ij} \geq 0$ are zero, unless we have conservation of mass for each species, momentum, and kinetic energy;

$$\begin{aligned} \{\alpha(i), \alpha(j)\} &= \{\alpha(k), \alpha(l)\}, \\ m_{\alpha(i)}\mathbf{v}_i + m_{\alpha(j)}\mathbf{v}_j &= m_{\alpha(k)}\mathbf{v}_k + m_{\alpha(l)}\mathbf{v}_l, \\ m_{\alpha(i)} |\mathbf{v}_i|^2 + m_{\alpha(j)} \left|\mathbf{v}_j\right|^2 &= m_{\alpha(j)} |\mathbf{v}_k|^2 + m_{\alpha(l)} |\mathbf{v}_l|^2. \end{aligned} \tag{5}$$

Unlike the continuous Boltzmann equation, discrete velocity models (DVMs) may, depending on the set of velocities and permissible collisions, have additional quantities –the so-called spurious collision invariants– that are conserved under any collision; represented by additional (linearly independent to the conserved vectors induced by (5)) vectors $\varphi = (\varphi_1, ..., \varphi_N)$, such that $\varphi_i + \varphi_j = \varphi_k + \varphi_l$ for any indices such that Γ_{ij}^{kl} is nonzero. DVMs without spurious collision invariants, where the physical collision invariants (5) are linearly independent, are called normal, see [14] and references therein. Carlo Cercignani made fundamental contributions to the study of normal DVMs for mixtures, see, e.g., [17]. Normal DVMs have exactly $s + 4$ linearly independent collision invariants. That is, all collision invariants are given by

$$\phi = (\phi_1, ..., \phi_N), \text{ with } \phi_i = a_{\alpha(i)} + m_{\alpha(i)}\mathbf{b} \cdot \mathbf{v}_i + c m_{\alpha(i)} |\mathbf{v}_i|^2 \tag{6}$$

for some constant $\left\{a_{\alpha_1}, ..., a_{\alpha_s}, c\right\} \subset \mathbb{R}$, and $\mathbf{b} \in \mathbb{R}^3$.

The stationary points, or equilibrium distributions, are Maxwellians

$$M = e^{\phi}, \text{ or } M = (M_1, \dots, M_N), \text{ with } M_i = e^{\phi_i}, \tag{7}$$

where (for normal DVMs) ϕ is a collision invariant (6).

A linearization, around a non-drifting Maxwellian M (7) with collision invariants (6) having $\mathbf{b} = \mathbf{0}$, by $F = M + M^{1/2} f$, results, neglecting quadratic terms, in the linearized system

$$B_u \frac{df}{dx} + Lf = 0, \tag{8}$$

where the $N \times N$ matrix L is the linearized collision operator [5, 9, 14].

The linearized operator L is symmetric, nonnegative, and the null space $N(L)$ of L is given by

$$N(L) = \left\{ M^{1/2}\phi \mid \phi \text{ is a collision invariant} \right\}.$$

The formal solution of system (8) is $f(x) = e^{-xB_u L} f(0)$, while the bounded [resp. slowly increasing] solutions are of the form

$$f(x) = \sum_{r=1}^{m^+} \beta_r \varphi_r e^{-\lambda_r x} + \sum_{i=1}^{s+4-l} \mu_i y_i + \sum_{j=1}^{l} \eta_j z_j \; [+ \alpha_j \left(w_j - x z_j\right)], \tag{9}$$

where (here and below, $\langle \cdot, \cdot \rangle$ denote the Euclidean scalar product in $\mathbb{R}^n$ for some appropriate choice of n; here $n = N$)

$$\begin{aligned}
L\varphi_r &= \lambda_r B_u \varphi_r, \left\langle \varphi_q, B_u \varphi_r \right\rangle = \lambda_r \delta_{qr}, \text{ with } \lambda_1, \dots, \lambda_{m^+} > 0, \\
N(L) &= \operatorname{span}(y_1, \dots, y_{s+4-l}, z_1, \dots, z_l), \; L w_j = B_u z_j \in N(L)^{\perp}, \\
\langle y_i, B_u y_k \rangle &= \gamma_i \delta_{ik}, \text{ with } \gamma_1, \dots, \gamma_{k^+} > 0 \text{ and } \gamma_{k^+ + 1}, \dots, \gamma_{s+4-l} < 0, \\
\left\langle u_r, B_u w_j \right\rangle &= \left\langle y_i, B_u w_j \right\rangle = \left\langle w_j, B_u w_p \right\rangle = 0, \text{ and } \left\langle w_j, B_u z_p \right\rangle = \delta_{jp}.
\end{aligned} \tag{10}$$

Here $m^+ = n^+ - k^+ - l$ is the number of positive eigenvalues (counted with multiplicity) of the matrix $B_u^{-1}L$, while $\left(k^+, s+4-k^+-l, l\right)$ is the signature of the restriction of the quadratic form $\langle \cdot, B_u \cdot \rangle$ to the kernel of L. The values of k^+ and l are given by the table

	$u < u_-$	$u = u_-$	$u_- < u < 0$	$u = 0$	$0 < u < u_+$	$u = u_+$	$u_+ < u$
k^+	0	0	1	1	$s+3$	$s+3$	$s+4$
l	0	1	0	$s+2$	0	1	0

, (11)

assuming that the velocity set is of the form $\mathcal{V} = \mathrm{V}_{\alpha_1} \times \ldots \times \mathrm{V}_{\alpha_s}$—where each set V_{α_i} can be decomposed as, cf. [2],

$$\mathrm{V}_{\alpha_i} = \mathrm{V}_{0\alpha_i}^3 \text{ for some set } \mathrm{V}_{0\alpha_i} \subset \mathbb{R}, \text{ such that } \mathrm{V}_{0\alpha_i} = -\mathrm{V}_{0\alpha_i},$$

and represents a normal DVM for a single species—and constitutes a normal DVM. Then any non-drifting Maxwellian is of the form $M = (M_{\alpha_1}, \ldots, M_{\alpha_s})$, where $M_{\alpha_i} = \frac{n_{\alpha_i}}{S_{\alpha_i}} e^{-m_{\alpha_i}|\mathbf{v}|^2/(2\theta)}$, with $S_{\alpha_i} = \sum_{j=1}^{N_{\alpha_i}} e^{-m_{\alpha_i}|\mathbf{v}_j|^2/(2\theta)}$, and one has

$$u_{\pm} = \pm \sqrt{\frac{\sum_{i=1}^{s} n_{\alpha_i} \left\langle \frac{M_{\alpha_i}}{n_{\alpha_i}}, m_{\alpha_i} v_x^2 \right\rangle^2}{\sum_{i=1}^{s} m_{\alpha_i} n_{\alpha_i} \left\langle \frac{M_{\alpha_i}}{n_{\alpha_i}}, m_{\alpha_i} v_x^2 \right\rangle}} \sqrt{\frac{\kappa+2}{3}}, \text{ with } \kappa = \frac{\sum_{i=1}^{s} n_{\alpha_i} \left\langle \frac{M_{\alpha_i}}{n_{\alpha_i}}, m_{\alpha_i}^2 v_x^4 \right\rangle}{\sum_{i=1}^{s} n_{\alpha_i} \left\langle \frac{M_{\alpha_i}}{n_{\alpha_i}}, m_{\alpha_i} v_x^2 \right\rangle^2}.$$

At the interface $x = 0$ we consider the general boundary condition

$$R_+ f(0) = C R_- f(0) + h_0, \tag{12}$$

where $R_\pm$ are the projections (4), C is a given $n^+ \times n^-$ real matrix, and $h_0 \in \mathbb{R}^{n^+}$, and obtain the following results, cf. [5, 14]:

Theorem 1

(i) Assume that $C^T B_u^+ C \leq B_u^-$. Then by imposing $k^+ + l$ conditions on h_0, the problem (8),(12) has a unique solution with exponential decay in x.

(ii) Assume that $C^T B_u^+ C < B_u^-$, or, if $l = 0$, that $C^T B_u^+ C \leq B_u^-$. Then the problem (8),(12) has a unique bounded [resp. slowly increasing] solution with asymptotic flow $f_A = \sum_{i=1}^{s+4-l} \mu_i y_i + \sum_{j=1}^{l} \eta_j z_j \, [+ \alpha_j (w_j - x z_j)]$, if the $s+4-k^+-l$ [resp. $s+4-k^+$] parameters $\mu_{k^++1}, \ldots, \mu_k$ [and $\alpha_1, \ldots, \alpha_l$] are prescribed.

Here, see expression (9), slowly increasing, means that the solution is either bounded or linearly increasing in x as $x \to \infty$.

The exponential speed of convergence in Theorem 1 is given by a leading term with the factor $e^{-\sigma x}$, where $\sigma = \sigma_u = \min_{1 \leq i \leq m^+} \lambda_i > 0$ for fixed u, cf. relations (10). However, as u tends to a value u_0, for which $l \geq 1$, from the left, that is, $u \to u_0^-$, l positive eigenvalues λ_i [or, equivalently, l negative γ_i (10)] will approach zero, that is, $\lambda_i \to 0^+$ [$\gamma_i \to 0^-$], as $u \to u_0^-$, and then become negative [positive] as u passes u_0. Indeed, the exponential speed of convergence is not uniform in u as u tends to u_0 from the left, $u \to u_0^-$. However, by imposing l extra conditions on the indata h_0 for $u < u_0$ one can obtain uniform exponential speed of convergence in a neighborhood of $u = u_0$.

The results in Theorem 1 can be extended: (i) in a natural way (cf. [5, 6]), to yield also for singular matrices B_u, if $N(L) \cap N(B_u) = \{0\}$; (ii) to an inhomogeneous case with some restrictions on the inhomogeneity $S = S(x)$ (on the right hand side of system (8)) [5, 9]; (iii) to the weakly (for "small" indata h_0) nonlinear case for a function vanishing at infinity [9]. Extensions to polyatomic molecules are addressed in [8, 9], see also [10].

3 Half-Space Problem for the Full Boltzmann Equation

Consider the continuous velocity case for a mixture of s species, and an inhomogeneous version of system (3), with more general boundary conditions at the interface:

$$\begin{cases} (v_x + u)\dfrac{\partial f}{\partial x} + \mathcal{L}f = S \\ f(0,\mathbf{v}) = RPf(0,\mathbf{v}) + f_b(\mathbf{v}) \text{ for } v_x + u > 0 \end{cases} \tag{13}$$

for given indata $f_b \in \mathfrak{h}_+ \cap \mathrm{D}(\mathcal{L})$ with $\mathfrak{h}_+ := \mathfrak{h}|_{v_x+u>0}$ at the interface, where $e^{\sigma x}S(x,\mathbf{v}) \in L^2(\mathbb{R}_+;\mathfrak{h})$ and $e^{\sigma x}f = e^{\sigma x}f(x,\mathbf{v}) \in L^2(\mathbb{R}_+;\mathfrak{h})$ for some positive number $\sigma > 0$, and $S = S(x,\mathbf{v}) \in (\ker\mathcal{L})^\perp$ for all $x \in \mathbb{R}_+$. Here $Pf(x,\mathbf{v}) = f(x,\mathbf{v}_-)$, where $\mathbf{v}_- = \mathbf{v} - (2(v_x+u),0,0)$, while R is a general linear (boundary) operator on $\mathfrak{h}$ fulfilling, cf. [21, p. 164], the conditions:

$$(Rh|\,(v_x+u)\,g)_+ = ((v_x+u)\,h|\,Rg)_+$$

$$(Rg|\,(v_x+u)\,Rg)_+ \le (g|\,(v_x+u)\,g)_+\,,$$

where $(\cdot|\cdot)_+ = (\cdot|\cdot)|_{v_x+u>0}$. For single species, especially the case of complete absorption, $R = 0$, at the interface, is well studied in the literature, see, e.g., the review [4] and the references therein. Substituting $f = e^{-\sigma x}g$ results in

$$\begin{cases} (v_x + u)\dfrac{\partial g}{\partial x} + \mathcal{L}g - \sigma\,(v_x+u)\,g = e^{\sigma x}S \\ g(0,\mathbf{v}) = RPg(0,\mathbf{v}) + f_b,\ v_x + u > 0, \end{cases} \tag{14}$$

for given $f_b \in \mathfrak{h}_+ \cap \mathrm{D}(\mathcal{L})$, where $g = g(x,\mathbf{v}) \in L^2(\mathbb{R}_+;\mathfrak{h})$, $S = S(x,\mathbf{v}) \in \mathrm{Im}\mathcal{L}$ for all $x \in \mathbb{R}_+$, and $e^{\sigma x}S(x,\mathbf{v}) \in L^2(\mathbb{R}_+;\mathfrak{h})$ for some positive number $\sigma > 0$.

Define $\Phi := \{\phi_1, ..., \phi_{s+4}\} \subset \mathfrak{h}$, with

$$\phi_{1,\alpha_i} = \frac{m_{\alpha_i}}{\sqrt{2\rho T}}\sqrt{M_{\alpha_i}}\left(\sqrt{\frac{\rho}{15nT}}\,|\mathbf{v}|^2 + v_x\right);\ \phi_{2,\alpha_i} = \sqrt{\frac{m_{\alpha_i}}{nT}}\sqrt{M_{\alpha_i}}\,v_y;$$

$$\phi_{3,\alpha_i} = \sqrt{\frac{m_{\alpha_i}}{nT}}\sqrt{M_{\alpha_i}}\,v_z;\ \phi_{4+j} = \frac{\widetilde{\phi}_{4+j}}{\|\widetilde{\phi}_{4+j}\|},\ j \in \{0, ..., s-1\};$$

$$\phi_{4+s,\alpha_i} = \frac{m_{\alpha_i}}{\sqrt{2\rho T}}\sqrt{M_{\alpha_i}}\left(\sqrt{\frac{\rho}{15nT}}\,|\mathbf{v}|^2 - v_x\right); \rho = \sum_{k=1}^{s} m_{\alpha_k} n_{\alpha_k},\ n = \sum_{k=1}^{s} n_{\alpha_k};$$

$$\widetilde{\phi}_{4+j} = \widehat{\phi}_{4+j} - \sum_{k=1}^{j}\left(\widehat{\phi}_{4+j}\middle|\,\phi_{3+k}\right)\phi_{3+k};\ \widehat{\phi}_{4+j,\alpha_i} = \sqrt{M_{\alpha_i}}\left(\frac{m_{\alpha_i} n_{\alpha_j}}{5nT}\,|\mathbf{v}|^2 - \delta_{ij}\right),$$

and denote

$$\Psi_+^u := \{\phi_i \,|(\phi_i|\,(v_x+u)\,\phi_i) > 0\} = \{\phi_1, ..., \phi_{k^+}\},$$
$$\Psi_0^u := \{\phi_i \,|(\phi_i|\,(v_x+u)\,\phi_i) = 0\} = \{\psi_1, ..., \psi_l\}.$$

The set Φ is an orthonormal basis of $\ker\mathcal{L}$ that is orthogonal with respect to the form $(\cdot|\,(v_x+u)\,\cdot)$ and $\left(k^+, s+4-k^+-l, l\right)$ is the signature of the restriction of the quadratic form $(\phi|\,(v_x+u)\,\phi)$ to the kernel of $\mathcal{L}$. The values of k^+ and l with respect to u are given by table (11) with $u_\pm = \pm\sqrt{\frac{5nT}{3\rho}}$. More specific, we have that

$$\Psi_0^{u_-} = \{\phi_1\};\ \Psi_0^{0} = \{\phi_2, ..., \phi_{s+3}\};\ \Psi_0^{u_+} = \{\phi_{s+4}\};\ \text{with } u_\pm = \pm\sqrt{\frac{5nT}{3\rho}},$$

while $\Psi_0^u = \emptyset$ if $u \notin \left\{0, \pm\sqrt{\frac{5nT}{3\rho}}\right\}$, and

$$\Psi_+^u = \emptyset \text{ if } u \le u_-;\ \Psi_+^u = \{\phi_1\} \text{ if } u_- < u \le 0;$$
$$\Psi_+^u = \{\phi_1, ..., \phi_{s+3}\} \text{ if } 0 < u \le u_+;\ \Psi_+^u = \{\phi_1, ..., \phi_{s+4}\} \text{ if } u > u_+.$$

Theorem 2 ([11]) *Imposing $k^+ + l$ conditions on f_b, there exists a unique solution $f = f(x, \mathbf{v})$ to the problem (13), such that $e^{\sigma x} f(x, \mathbf{v}) \in L^2(\mathbb{R}_+; \mathfrak{h})$ for some $\sigma > 0$.*

Corollary 1 *There exists a unique solution f to the problem (13), such that $e^{\sigma x}(f(x, \mathbf{v}) - f_\infty) \in L^2(\mathbb{R}_+; \mathfrak{h})$, with $f_\infty = \lim_{x\to\infty} f(x, \mathbf{v}) \in \ker\mathcal{L}$, for some $\sigma > 0$, if the $s+4-\left(k^+ + l\right)$ parameters $\left(f_\infty|\,\phi_{k^++l+1}\right), \ldots, (f_\infty|\,\phi_{s+4})$ are prescribed.*

The exponential speed of convergence $e^{-\sigma x}$ is determined by $\sigma = \sigma_u$ for fixed u. However, the speed of convergence is not uniform in u as u tends to some degenerate value $u_0 \in \left\{0, \pm\sqrt{\frac{5nT}{3\rho}}\right\}$, for which $l \ge 1$, from the left; it will appear l slowly varying mode(s) as $u \to u_0^-$ (cf. [13] and the references therein). Indeed, for $u < u_0 \in \left\{0, \pm\sqrt{\frac{5nT}{3\rho}}\right\}$:

$$((v_x+u)\,\phi_i|\,\phi_i) = ((v_x+u_0)\,\phi_i|\,\phi_i) + u - u_0 = u - u_0 \to 0^- \text{ as } u \to u_0^-$$

exactly for $i \in \left\{k^+ + 1, ..., k^+ + l\right\}$, inducing l slowly varying mode(s) as $u \to u_0^-$. However, by imposing l extra conditions on the indata for u less than u_0 the slowly varying modes present as $u \to u_0^-$ can be removed [11], i.e., uniform exponential speed of convergence can be obtained in a neighborhood of u_0.

The corresponding problem for monatomic single species in a neighborhood of $u = 0$, in the nonlinear context, is considered in more detail in [13], see also [25].

The results may be extended to the weakly nonlinear case applying methods similar to the ones in [13, 24].

3.1 Brief Outline of Main Steps of the Proof of Theorem 2

If $u \in \left\{0, \pm\sqrt{\frac{5nT}{3\rho}}\right\}$, then for each ψ_r, $r \in \{1, ..., l\}$, there exists $\varphi_r \in \mathrm{D}(\mathcal{L})$, such that (without loss of generality, cf. [15])

$$\mathcal{L}\varphi_r = (v_x + u)\,\psi_r \in (\ker \mathcal{L})^{\perp} = \mathrm{Im}\mathcal{L},$$
$$\big((v_x + u)\,\psi_r \,|\, \varphi_j\big) = \alpha_r \delta_{rj} \geq 0 \text{ and } \big((v_x + u)\,\varphi_r \,|\, \phi_i\big) = \big((v_x + u)\,\varphi_r \,|\, \varphi_j\big) = 0$$
$$\text{for all } \{r, j\} \subseteq \{1, \dots, l\}, i \in \left\{1, \dots, k^+, k^+ + l + 1, \dots, s + 4\right\}.$$

Introduce the linear operators

$$\Pi_+ := \sum_{i=1}^{k^+} (\cdot \,|\, \phi_i)\,\phi_i;\; \Pi_0 := \sum_{r=1}^{l} \frac{(\cdot \,|\, \varphi_r)}{(\varphi_r \,|\, \mathcal{L}\varphi_r)^2}\varphi_r,$$
$$\Lambda := \mathcal{L} - \sigma\,(v_x + u) + \alpha\Pi_+\,(v_x + u) + \beta\,(v_x + u)\,\Pi_0\,(v_x + u) \tag{15}$$

and consider the **penalized problem** (cf. [13, 24, 28]):

$$\begin{cases} (v_x + u)\,\dfrac{\partial g}{\partial x} + \Lambda g = e^{\sigma x} S,\; x > 0, \\ g(0, \mathbf{v}) = RPg(0, \mathbf{v}) + f_b,\, v_x + u > 0 \end{cases}. \tag{16}$$

Then one can obtain the following result [11]:

Lemma 1 *For appropriately chosen positive constants α, β, and σ there exists a positive number $\mu = \mu\,(\alpha, \beta, \sigma) > 0$ such that*

$$\left(\Lambda^{\star} f \,\middle|\, f\right) = (\Lambda f \,|\, f) \geq \mu\,(f \,|\, f) \text{ for all } f \in \mathrm{D}(\mathcal{L}),$$

where

$$\Lambda^{\star} = \mathcal{L} - \sigma\,(v_x + u) + \alpha\,(v_x + u)\,\Pi_+ + \beta\,(v_x + u)\,\Pi_0\,(v_x + u)\,,$$

is the adjoint operator of Λ.

Define the linear operator $\mathcal{T}g := (v_x + u)\dfrac{\partial g}{\partial x} + \Lambda g$, with domain

$$\mathrm{D}(\mathcal{T}) := \left\{\{g(x,\mathbf{v}), (v_x+u)\frac{\partial g}{\partial x}(x,\mathbf{v}), \mathcal{L}g(x,\mathbf{v})\} \subset L^2(\mathbb{R}_+;\mathfrak{h}),\ \widetilde{R}g(0,\mathbf{v}) = 0\right\},$$

where $\widetilde{R} := \mathbf{1}_{v_x+u>0} - R\mathbf{1}_{v_x+u<0}$.

Then the adjoint operator of $\mathcal{T}$ is given by the operator

$$\mathcal{T}^*g = -(v_x+u)\frac{\partial g}{\partial x} + \Lambda^* g,$$

with domain

$$\mathrm{D}(\mathcal{T}^*) = \left\{\{g(x,\mathbf{v}), (v_x+u)\frac{\partial g}{\partial x}(x,\mathbf{v}), \mathcal{L}g(x,\mathbf{v})\} \subset L^2(\mathbb{R}_+;\mathfrak{h}),\ \widetilde{R}^*g(0,\mathbf{v}) = 0\right\},$$

where $\widetilde{R}^* = R\mathbf{1}_{v_x+u>0} - \mathbf{1}_{v_x+u<0}$.

Moreover, one can obtain the following result [11]:

Lemma 2 *For some $\alpha, \beta, \sigma > 0$ there exists a positive number $\mu = \mu(\alpha,\beta,\sigma) > 0$ such that*

$$\|\mathcal{T}g\|_{L^2(\mathbb{R}_+;\mathfrak{h})} \geq \mu \|g\|_{L^2(\mathbb{R}_+;\mathfrak{h})} \text{ for all } g \in \mathrm{D}(\mathcal{T}),$$
$$\|\mathcal{T}^*g\|_{L^2(\mathbb{R}_+;\mathfrak{h})} \geq \mu \|g\|_{L^2(\mathbb{R}_+;\mathfrak{h})} \text{ for all } g \in \mathrm{D}(\mathcal{T}^*).$$

In particular, $\ker \mathcal{T} = \{0\}$ *and* $\mathrm{Im}\mathcal{T} = L^2(\mathbb{R}_+;\mathfrak{h})$.

Based on this one can obtain the following result [11]:

Proposition 1 *There exists a unique solution $g(x,\mathbf{v}) \in L^2(\mathbb{R}_+;\mathfrak{h})$ to the penalized problem (16).*

Denoting by $\mathcal{I}: \mathfrak{h}_+ \to \mathfrak{h}$ the solution operator

$$\mathcal{I}(f_b) = g(0,\mathbf{v}),$$

where $g(x,\mathbf{v}) \in L^2(\mathbb{R}_+;\mathfrak{h})$ is the unique solution of the penalized problem (16) in Proposition 1, one can obtain the following result [11]:

Lemma 3 *The solution of the penalized problem (16) is a solution of the problem (14) if and only if*

$$\Pi_+((v_x+u)\mathcal{I}(f_b)) = \Pi_0((v_x+u)\mathcal{I}(f_b)) = 0, \textit{ or, equivalently,}$$

$$\Pi_+((v_x+u)\mathcal{I}(f_b)) = \widetilde{\Pi}_0((v_x+u)\mathcal{I}(f_b)) = 0;\ \widetilde{\Pi}_0 := \sum_{r=1}^{l}(\cdot|\psi_r)\psi_r.$$

Finally, one can conclude that [11]:

Theorem 3 *There exists a unique solution* $g(x, \mathbf{v}) \in L^2(\mathbb{R}_+; \mathfrak{h})$ *to the problem* (14) *imposing* $k^+ + l$ *conditions on* f_b.

Acknowledgments The author is indebted to Prof. F. Golse for valuable discussions and kind hospitality during visits to Paris and acknowledges the support by French Institute in Sweden (through the FRÖ program in 2016) and SveFUM (in 2017 and 2019) for his visits.

References

1. Aoki, K., Bardos, C., Takata, S.: Knudsen layer for gas mixtures. J. Stat. Phys. **112**, 629–655 (2003)
2. Babovsky, H.: Shocks in the light of discrete velocity models. AIP Conf. Proc. **2132**, 060002 (2019)
3. Bardos, C., Yang, X.: The classification of well-posed kinetic boundary layer for hard sphere gas mixtures. Commun. Partial Differ. Equ. **37**, 1286–1314 (2012)
4. Bardos, C., Golse, F., Sone, Y.: Half-space problems for the Boltzmann equation: a survey. J. Stat. Phys. **124**, 275–300 (2006)
5. Bernhoff, N.: On half-space problems for the linearized discrete Boltzmann equation. Riv. Mat. Univ. Parma **9**, 73–124 (2008)
6. Bernhoff, N.: On half-space problems for the weakly non-linear discrete Boltzmann equation. Kinet. Relat. Models **3**, 195–222 (2010)
7. Bernhoff, N.: Boundary layers and shock profiles for the discrete Boltzmann equation for mixtures. Kinet. Relat. Models **5**, 1–19 (2012)
8. Bernhoff, N.: Discrete velocity models for multicomponent mixtures and polyatomic molecules without nonphysical collision invariants and shock profiles. AIP Conf. Proc. **1786**, 040005 (2016)
9. Bernhoff, N.: Boundary layers for discrete kinetic models: multicomponent mixtures, polyatomic molecules, bimolecular reactions, and quantum kinetic equations. Kinet. Relat. Models **10**, 925–955 (2017)
10. Bernhoff, N.: Discrete velocity models for polyatomic molecules without nonphysical collision invariants. J. Stat. Phys. **172**, 742–761 (2018)
11. Bernhoff, N.: Linear half-space problems in kinetic theory: abstract formulation and regime transitions (2022). arXiv: 2201.03459
12. Bernhoff, N.: Linearized Boltzmann collision operator: I. Polyatomic molecules modeled by a discrete internal energy variable and multicomponent mixtures (2022). arXiv: 2201.01365
13. Bernhoff, N., Golse, F.: On the boundary layer equations with phase transition in the kinetic theory of gases. Arch. Ration. Mech. Anal. **240**, 51–98 (2021)
14. Bernhoff, N., Vinerean, M.C.: Discrete velocity models for multicomponent mixtures without nonphysical collision invariants. J. Stat. Phys. **165**, 434–453 (2016)
15. Bobylev, A.V., Bernhoff, N.: Discrete velocity models and dynamical systems. In: Lecture Notes on the Discretization of the Boltzmann Equation, pp. 203–222. World Scientific, Singapore (2003)
16. Bobylev, A.V., Cercignani, C.: Discrete velocity models for mixtures. J. Stat. Phys. **91**, 327–341 (1998)
17. Bobylev, A.V., Cercignani, C.: Discrete velocity models without non-physical invariants. J. Stat. Phys. **97**, 677–686 (1999)
18. Boudin, L., Grec, B., Pavić, M., Salvarani, F.: Diffusion asymptotics of a kinetic model for gaseous mixtures. Kinet. Relat. Models **6**, 137–157 (2013)

19. Briant, M., Daus, E.S.: The Boltzmann equation for a multi-species mixture close to global equilibrium. Arch. Ration. Mech. Anal. **222**, 1367–1443 (2016)
20. Cercignani, C.: Half-space problems in the kinetic theory of gases. In: Trends in Applications of Pure Mathematics to Mechanics, pp. 35–50. Springer, Berlin (1986)
21. Cercignani, C.: The Boltzmann Equation and its Applications. Springer, Berlin (1988)
22. Cercignani, C.: Rarefied Gas Dynamics. Cambridge University Press, Cambridge (2000)
23. Daus, E.S., Jungel, A., Mouhot, C., Zamponi, S.: Hypocoercivity for a linearized multispecies Boltzmann system. SIAM J. Math. Anal. **48**, 538–568 (2016)
24. Golse, F.: Analysis of the boundary layer equation in the kinetic theory of gases. Bull. Inst. Math. Acad. Sin. **3**, 211–242 (2008)
25. Liu, T.P., Yu, S.H.: Invariant manifolds for steady Boltzmann flows and applications. Arch. Ration. Mech. Anal. **209**, 869–997 (2013)
26. Sone, Y.: Kinetic Theory and Fluid Dynamics. Birkhauser, Basel (2002)
27. Sone, Y.: Molecular Gas Dynamics. Birkhauser, Basel (2007)
28. Ukai, S., Yang, T., Yu, S.H.: Nonlinear boundary layers of the Boltzmann equation: I. Existence. Commun. Math. Phys. **236**, 373–393 (2003)

BGK Model for a Mixture with Two Reversible Reactions

Marzia Bisi and Romina Travaglini

Abstract We construct a kinetic model of BGK type for a mixture of eight polyatomic gases undergoing two separate bimolecular and reversible chemical reactions. Maxwellian attractors depend on suitable auxiliary parameters, and we prove that they may be uniquely determined imposing preservation of Boltzmann collision invariants. The mass action laws of the two reactions and the total energy conservation turn out to be a set of transcendental equations in the auxiliary fields that has to be carefully investigated in order to prove uniqueness of solution.

1 Introduction

BGK-type descriptions for gas mixtures have been proposed both in inert and reactive frames. For inert mixtures, there are relaxation models with a unique relaxation operator for each species [1], as well as other more complicated models involving a sum of binary relaxation operators, preserving thus the structure of the original Boltzmann system [8, 12, 13, 15]. The same two kinds of models are available for mixtures subject to simple chemical reactions, see, for instance, [5, 11] for models with a single BGK operator per species and [9, 14] for models with sums of BGK terms. The simplest formulation involving a unique relaxation operator for each species [1] has been recently generalized even to mixtures of monatomic and polyatomic gases, each one with its own number of internal energy levels [3, 4]. In reactive models it is usually considered a very basic physical setting, made by only four species subject to a bimolecular and reversible chemical reaction. The following step we would like to take is to deal with a number of gas species greater than four, involved in different chemical reactions. In this case the equilibrium configuration is governed by a higher number of mass action laws, leading to a higher number of auxiliary quantities to be determined and of transcendental

M. Bisi (✉), R. Travaglini
Department of Mathematical, Physical and Computer Sciences, University of Parma, Parma, Italy
e-mail: marzia.bisi@unipr.it; romina.travaglini@unipr.it

P. Barbante et al. (eds.), *From Kinetic Theory to Turbulence Modeling*, Springer INdAM Series 51, https://doi.org/10.1007/978-981-19-6462-6_5

equations to be managed. Our intent is to show that, in a case of eight polyatomic gases undergoing two disjoint reactions, i.e., no gas species is involved in more than one reaction, it is possible to write one of the auxiliary densities in terms of the other ones and of the auxiliary temperature; successively we express everything in terms of the auxiliary temperature, and we prove that it can be detected as the unique admissible solution of a transcendental equation. In Sect. 2 we present our physical setting and the basic kinetic approach, while Sect. 3 is devoted to the detailed construction of a consistent BGK model. Finally, Sect. 4 contains some conclusions and perspectives.

2 Physical Setting and Kinetic Approach

We consider a mixture of eight gas species, G^i, $i = 1, \ldots, 8$, and we suppose that they are involved in two separate reversible chemical reactions:

$$G^1 + G^2 \leftrightarrows G^3 + G^4, \tag{1}$$

$$G^5 + G^6 \leftrightarrows G^7 + G^8. \tag{2}$$

Reactions (1) and (2) involve different gases; more complicated situations with at least a species involved in both reactions will be dealt with in future research. Each gas species G^i has its own mass m^i and, due to conservation laws, particle masses satisfy the relations $m^1 + m^2 = m^3 + m^4$ and $m^5 + m^6 = m^7 + m^8$. As usual in the kinetic description of polyatomic gases based on discrete energy levels [10, 16], each gas species can be seen composed by a certain number $L^i \geq 1$ of components, C^i_j, $j = 1, \ldots, L^i$, each one corresponding to a discrete internal energy level E^i_j, with (without loss of generality) $E^i_j < E^i_k$ for any $1 \leq j, k \leq L^i$, with $j < k$. The particular case $L^i = 1$ obviously corresponds to a monatomic gas. Distribution functions for each energy component C^i_j are

$$f^i_j(t, \mathbf{x}, \mathbf{v}), \quad i = 1, \ldots, 8, \quad j = 1, \ldots, L^i. \tag{3}$$

Encounters among particles of the mixture are modeled as binary instantaneous collisions and can be generically written as

$$\begin{gathered} C^i_j + C^h_k \to C^l_m + C^n_p, \qquad 1 \leq i, h, l, n \leq 8, \\ 1 \leq j \leq L^i, \qquad 1 \leq k \leq L^h, \qquad 1 \leq m \leq L^l, \qquad 1 \leq p \leq L^n. \end{gathered} \tag{4}$$

The superscript i of component C^i_j refers to the gas G^i which it belongs to, while the subscript j identifies its energy level E^i_j. These encounters are elastic if $i = l$,

$h = n$, $j = m$, $k = p$, since particles do not change their species and energy level, while they are inelastic if $i = l$, $h = n$, $j \neq m$ or $k \neq p$, since particles maintain their chemical species but pass from one energy level to another (the subscripts change). Moreover, such collisions describe chemical reactions when $(i, h) \neq (l, n)$ (the superscripts identifying the species change), with $(i, h), (l, n) \in \{(1, 2), (3, 4)\}$ or $(i, h), (l, n) \in \{(5, 6), (7, 8)\}$. In the last two cases the collision is endothermic if $E^l_m + E^n_p - E^i_j - E^h_k \geq 0$ or exothermic if $E^l_m + E^n_p - E^i_j - E^h_k < 0$.

Moments of components C^i_j we are interested in are: number densities n^i_j (providing the total density of G^i as $n^i = \sum_{j=1}^{L^i} n^i_j$), drift velocities $\mathbf{u}^i_j$, and kinetic temperatures T^i_j. They are defined as

$$n^i_j = \int_{\mathbb{R}^3} f^i_j(\mathbf{v})\, d\mathbf{v}, \qquad \mathbf{u}^i_j = \frac{1}{n^i_j} \int_{\mathbb{R}^3} \mathbf{v}\, f^i_j(\mathbf{v})\, d\mathbf{v}, \qquad T^i_j = \frac{m^i}{3\, n^i_j} \int_{\mathbb{R}^3} |\mathbf{v} - \mathbf{u}^i_j|^2 f^i_j(\mathbf{v})\, d\mathbf{v}.$$

During the evolution of the mixture we have conservation of six independent combinations of gas species number densities that can be chosen as $n^1 + n^3$, $n^1 + n^4$, $n^2 + n^4$, $n^5 + n^7$, $n^5 + n^8$, $n^6 + n^8$, of global momentum and of total energy. Specifically, the sum $n^1 + n^3$ is preserved since, in each direct reaction (1), to the disappearance of a particle G^1 there corresponds the creation of a particle G^3, and vice versa in the reverse reaction; analogous properties hold for the other combinations.

As usual, at equilibrium the distribution functions are of Maxwellian type, with all components sharing a common mean velocity $\mathbf{u}$ and temperature T,

$$f^i_{jM}(\mathbf{v}) = n^i_j \left(\frac{m^i}{2\pi T}\right)^{3/2} \exp\left(-\frac{m^i}{2T} |\mathbf{v} - \mathbf{u}|^2\right), \qquad i = 1, \ldots, 8, \quad j = 1, \ldots, L^i. \tag{5}$$

Moreover, at equilibrium, number density of each component n^i_j is related to the total density n^i of the corresponding gas as $n^i_j = n^i \exp\left(-\frac{E^i_j - E^i_1}{T}\right) / \mathcal{Z}^i(T)$ and we also have the two mass action laws of chemistry regulating the ratio between reacting species densities:

$$\frac{n^i n^j}{n^h n^k} = \left(\frac{m^i m^j}{m^h m^k}\right)^{\frac{3}{2}} \frac{\mathcal{Z}^i(T)\mathcal{Z}^j(T)}{\mathcal{Z}^h(T)\mathcal{Z}^k(T)} \exp\left(\frac{\Delta E^{hk}_{ij}}{T}\right), \tag{6}$$

with $(i, j, h, k) = (1, 2, 3, 4)$ or $(5, 6, 7, 8)$, $\mathcal{Z}^i(T) = \sum_{k=1}^{L^i} \exp\left(-\frac{E^i_k - E^i_1}{T}\right)$, $\Delta E^{34}_{12} := \Delta E^* = E^3_1 + E^4_1 - E^2_1 - E^1_1$ and $\Delta E^{78}_{56} := \Delta E^{**} = E^7_1 + E^8_1 - E^5_1 - E^6_1$. In next section we shall prove that our BGK model is well defined for any choice for the signs of the two chemical energy gaps ΔE^* and ΔE^{**}.

3 BGK Model

The structure of the BGK model we are showing here is the same as in [2–4], thus we write the set of $L^1 + \ldots + L^8$ kinetic BGK-type equations as

$$\frac{\partial f_j^i}{\partial t} + \mathbf{v} \cdot \nabla_{\mathbf{x}} f_j^i = \nu_j^i (\mathcal{M}_j^i - f_j^i), \qquad i = 1, \ldots, 8, \qquad j = 1, \ldots, L^i, \tag{7}$$

in which the collision operator is a relaxation term, with ν_j^i macroscopic collision frequencies and $\mathcal{M}_j^i$ Maxwellian attractors as in [5],

$$\mathcal{M}_j^i(\mathbf{v}) := \tilde{n}_j^i \left(\frac{m^i}{2\pi \tilde{T}} \right)^{3/2} \exp\left[-\frac{m^i}{2\tilde{T}} |\mathbf{v} - \tilde{\mathbf{u}}|^2 \right], \qquad \begin{array}{l} i = 1, \ldots, 8, \\ j = 1, \ldots, L^i, \end{array} \tag{8}$$

depending on auxiliary parameters $\tilde{n}_j^i$ ($i = 1, \ldots, 8$, $j = 1, \ldots, L^i$), $\tilde{\mathbf{u}}$, $\tilde{T}$. In addition, we require that fictitious parameters fulfill all other equilibrium conditions [16], namely that auxiliary densities satisfy relations valid for the final configuration of components corresponding to each energy level

$$\tilde{n}_j^i = \frac{\tilde{n}^i}{\mathcal{Z}^i(\tilde{T})} \exp\left(-\frac{E_j^i - E_1^i}{\tilde{T}} \right), \qquad \begin{array}{l} i = 1, \ldots, 8, \\ j = 1, \ldots, L^i, \end{array} \tag{9}$$

and that fictitious total densities $\tilde{n}^i$ satisfy the mass action laws

$$\frac{\tilde{n}^1 \tilde{n}^2}{\tilde{n}^3 \tilde{n}^4} = \left(\frac{m^1 m^2}{m^3 m^4} \right)^{3/2} \frac{\mathcal{Z}^1(\tilde{T}) \mathcal{Z}^2(\tilde{T})}{\mathcal{Z}^3(\tilde{T}) \mathcal{Z}^4(\tilde{T})} \exp\left(\frac{\Delta E^*}{\tilde{T}} \right), \tag{10}$$

$$\frac{\tilde{n}^5 \tilde{n}^6}{\tilde{n}^7 \tilde{n}^8} = \left(\frac{m^5 m^6}{m^7 m^8} \right)^{3/2} \frac{\mathcal{Z}^5(\tilde{T}) \mathcal{Z}^6(\tilde{T})}{\mathcal{Z}^7(\tilde{T}) \mathcal{Z}^8(\tilde{T})} \exp\left(\frac{\Delta E^{**}}{\tilde{T}} \right). \tag{11}$$

Our purpose is to show that auxiliary parameters $\tilde{n}_j^i$, $\tilde{\mathbf{u}}$, $\tilde{T}$ can be expressed in terms of the actual quantities n_j^i, $\mathbf{u}_j^i$, T_j^i of the mixture. This is possible imposing that the BGK model preserves the conservation of the same collision invariants of the Boltzmann equations. More specifically, we impose conservation of the six combinations of number densities expressed before that give

$$\sum_{j=1}^{L^h} \nu_j^h \int_{\mathbb{R}^3} (\mathcal{M}_j^h - f_j^h) d\mathbf{v} + \sum_{j=1}^{L^k} \nu_j^k \int_{\mathbb{R}^3} (\mathcal{M}_j^k - f_j^k) d\mathbf{v} = 0, \tag{12}$$

with (h, k) varying in the set $\{(1, 3), (1, 4), (2, 4), (5, 7), (5, 8), (6, 8)\}$. Then we have to require conservation of global momentum and of total energy, provided by

$$\sum_{i=1}^{8} \sum_{j=1}^{L^i} \nu_j^i m^i \int_{\mathbb{R}^3} \mathbf{v}\,(\mathcal{M}_j^i - f_j^i) d\mathbf{v} = \mathbf{0}, \tag{13}$$

$$\sum_{i=1}^{8} \sum_{j=1}^{L^i} \nu_j^i \int_{\mathbb{R}^3} \left(\frac{1}{2} m^i |\mathbf{v}|^2 + E_j^i\right) (\mathcal{M}_j^i - f_j^i) d\mathbf{v} = 0. \tag{14}$$

Relations (12) can be expressed as

$$\sum_{j=1}^{L^i} \nu_j^i \tilde{n}_j^i = \sum_{j=1}^{L^i} \nu_j^i n_j^i + \lambda^i \sum_{j=1}^{L^1} \nu_j^1 (\tilde{n}_j^1 - n_j^1), \qquad i = 1, \ldots, 4, \tag{15}$$

$$\sum_{j=1}^{L^i} \nu_j^i \tilde{n}_j^i = \sum_{j=1}^{L^i} \nu_j^i n_j^i + \lambda^i \sum_{j=1}^{L^5} \nu_j^5 (\tilde{n}_j^5 - n_j^5), \qquad i = 5, \ldots, 8, \tag{16}$$

with $\lambda^1 = \lambda^2 = -\lambda^3 = -\lambda^4 = \lambda^5 = \lambda^6 = -\lambda^7 = -\lambda^8 = 1$. Bearing in mind (9), from (15) and (16) we can express total fictitious densities $\tilde{n}^2, \tilde{n}^3, \tilde{n}^4$ and $\tilde{n}^6, \tilde{n}^7, \tilde{n}^8$ in terms of $\tilde{n}^1$ and $\tilde{n}^5$, respectively:

$$\frac{\tilde{n}^i}{\mathcal{Z}^i(\tilde{T})} = \left[\sum_{j=1}^{L^i} \nu_j^i \exp\left(-\frac{E_j^i - E_1^i}{\tilde{T}}\right)\right]^{-1} \left\{\sum_{j=1}^{L^i} \nu_j^i n_j^i - \lambda^i \sum_{j=1}^{L^1} \nu_j^1 n_j^1 \right.$$
$$\left. + \lambda^i \left[\sum_{j=1}^{L^1} \nu_j^1 \exp\left(-\frac{E_j^1 - E_1^1}{\tilde{T}}\right)\right] \frac{\tilde{n}^1}{\mathcal{Z}^1(\tilde{T})}\right\}, \qquad i = 1, \ldots, 4, \tag{17}$$

$$\frac{\tilde{n}^i}{\mathcal{Z}^i(\tilde{T})} = \left[\sum_{j=1}^{L^i} \nu_j^i \exp\left(-\frac{E_j^i - E_1^i}{\tilde{T}}\right)\right]^{-1} \left\{\sum_{j=1}^{L^i} \nu_j^i n_j^i - \lambda^i \sum_{j=1}^{L^5} \nu_j^5 n_j^5 \right.$$
$$\left. + \lambda^i \left[\sum_{j=1}^{L^5} \nu_j^5 \exp\left(-\frac{E_j^5 - E_1^5}{\tilde{T}}\right)\right] \frac{\tilde{n}^5}{\mathcal{Z}^5(\tilde{T})}\right\}, \qquad i = 5, \ldots, 8, \tag{18}$$

observing that for $i = 1$ and $i = 5$ we get trivial identities.

Analogously to [2, 3], conservation of total momentum (13) explicitly provides auxiliary mean velocity as follows

$$\tilde{\mathbf{u}} = \left(\sum_{i=1}^{8} \sum_{j=1}^{L^i} \nu^i_j \, m^i \, n^i_j \, \mathbf{u}^i_j \right) \Big/ \left(\sum_{i=1}^{8} \sum_{j=1}^{L^i} \nu^i_j \, m^i \, n^i_j \right). \tag{19}$$

Conservation of total energy (14), instead, leads to

$$\frac{3}{2} \sum_{i=1}^{8} \sum_{j=1}^{L^i} \nu^i_j \, n^i_j \tilde{T} + \sum_{i=1}^{8} Y^i(\tilde{n}^i, \tilde{T}) \mathcal{E}^i(\tilde{T}) = \Lambda, \tag{20}$$

where we have set

$$Y^i(\tilde{n}^i, \tilde{T}) = \frac{\tilde{n}^i}{\mathcal{Z}^i(\tilde{T})} \sum_{j=1}^{L^i} \nu^i_j \exp\left(-\frac{E^i_j - E^i_1}{\tilde{T}} \right), \qquad i = 1, \ldots, 8, \tag{21}$$

$$\mathcal{E}^i(\tilde{T}) = \sum_{j=1}^{L^i} \nu^i_j \, E^i_j \exp\left(-\frac{E^i_j - E^i_1}{\tilde{T}} \right) \Big/ \left[\sum_{k=1}^{L^i} \nu^i_k \exp\left(-\frac{E^i_k - E^i_1}{\tilde{T}} \right) \right], \tag{22}$$

and the right hand side Λ (not reported here for brevity) contains only actual densities n^i_j, velocities $\mathbf{u}^i_j$, temperatures T^i_j and energies E^i_j. We note that even if the two chemical reactions are disjoint, all the eight species interact energetically, thus all the components are involved in the total energy conservation. This adds a further constraint in determining auxiliary parameters.

Relations (17)-(18) may thus be written as

$$Y^i = \lambda^i Y^1 + \alpha^i, \qquad i = 1, \ldots, 4, \tag{23}$$

$$Y^i = \lambda^i Y^5 + \beta^i, \qquad i = 5, \ldots, 8, \tag{24}$$

with quantities

$$\alpha^i = \sum_{j=1}^{L^i} \nu^i_j n^i_j - \lambda^i \sum_{j=1}^{L^1} \nu^1_j n^1_j, \qquad \beta^i = \sum_{j=1}^{L^i} \nu^i_j n^i_j - \lambda^i \sum_{j=1}^{L^5} \nu^5_j n^5_j. \tag{25}$$

Consequently, we have that (20) turns into the following expression

$$\mathcal{D}(\tilde{T}) \left(Y^1 - \sum_{j=1}^{L^1} \nu^1_j n^1_j \right) + \mathcal{B}(\tilde{T}) \left(Y^5 - \sum_{j=1}^{L^5} \nu^5_j n^5_j \right) = \mathcal{N}(\tilde{T}), \tag{26}$$

with functions

$$\mathcal{D}(\tilde{T}) = \sum_{i=1}^{4} \lambda^i \, \mathcal{E}^i(\tilde{T}), \qquad \mathcal{B}(\tilde{T}) = \sum_{i=5}^{8} \lambda^i \, \mathcal{E}^i(\tilde{T}), \tag{27}$$

$$\mathcal{N}(\tilde{T}) = \Lambda - \sum_{i=1}^{8} \left(\sum_{m=1}^{L^i} \nu_m^i \, n_m^i \right) \left[\frac{3}{2} \tilde{T} + \mathcal{E}^i(\tilde{T}) \right]. \tag{28}$$

The rest of the paper will be devoted to prove the following result.

Theorem 1 *The system of equations constituted by (26) and by the fictitious mass action laws (10)–(11), depending on three variables $\tilde{n}^1$, $\tilde{n}^5$, $\tilde{T}$, has a unique solution guaranteeing positivity of all auxiliary fields.*

The proof will be divided into several technical lemmas. First, we note that Eq. (26) gives a relation between the auxiliary densities $\tilde{n}^1$, $\tilde{n}^5$ and $\tilde{T}$, and, in particular, it allows to express $\tilde{n}^5$ in terms of $\tilde{n}^1$ and $\tilde{T}$. In order to show that these quantities are uniquely determined and that are physically admissible, we can exploit the two fictitious mass action laws (10)–(11) (a system of two coupled transcendental equations). By inserting the expressions (23) into (10) we get

$$\mathcal{G}^*(\tilde{T}, Y^1) = \left(\frac{m^1 m^2}{m^3 m^4} \right)^{3/2}, \tag{29}$$

with $\mathcal{G}^*(\tilde{T}, Y^1) = \mathcal{G}_1(Y^1) \cdot \mathcal{G}_2(\tilde{T}) \cdot \mathcal{G}_3(\tilde{T})$, being

$$\mathcal{G}_1(Y^1) = \frac{Y^1 \left[Y^1 + \alpha^2 \right]}{\left[-Y^1 + \alpha^3 \right] \left[-Y^1 + \alpha^4 \right]}, \tag{30}$$

$$\mathcal{G}_2(\tilde{T}) = \frac{\sum_{k=1}^{L^3} \nu_k^3 \exp\left(-\frac{E_k^3 - E_1^3}{\tilde{T}} \right) \sum_{k=1}^{L^4} \nu_k^4 \exp\left(-\frac{E_k^4 - E_1^4}{\tilde{T}} \right)}{\sum_{k=1}^{L^1} \nu_k^1 \exp\left(-\frac{E_k^1 - E_1^1}{\tilde{T}} \right) \sum_{k=1}^{L^2} \nu_k^2 \exp\left(-\frac{E_k^2 - E_1^2}{\tilde{T}} \right)}, \tag{31}$$

$$\mathcal{G}_3(\tilde{T}) = \exp\left(-\frac{\Delta E^*}{\tilde{T}} \right). \tag{32}$$

The set of admissible values for Y^1, for which auxiliary densities $\tilde{n}^i$, $i = 1, \ldots, 4$, are positive, is

$$A_1 = \left\{ Y^1 > 0 \, : \, \max\left(0, -\alpha^2\right) < Y^1 < \min\left(\alpha^3, \alpha^4\right) \right\}, \tag{33}$$

which is a connected interval of $\mathbb{R}$. Equation (29) may be rewritten as

$$L^*(\tilde{T}, Y^1) = 0, \tag{34}$$

with

$$L^*(\tilde{T}, Y^1) = \mathcal{G}_1(Y^1) - M(\tilde{T}), \quad M(\tilde{T}) = \left[\mathcal{G}_2(\tilde{T}) \cdot \mathcal{G}_3(\tilde{T})\right]^{-1} \cdot \left(\frac{m^1 m^2}{m^3 m^4}\right)^{3/2}, \tag{35}$$

and we can state the following result.

Lemma 1 *For any value $\bar{T} > 0$ there exists a unique positive value $\bar{Y}^1$ in the admissible set A_1 such that (34) is satisfied. Moreover, there exists an open interval I with $\bar{T} \in I$ and a unique function $Y^1(\tilde{T}) : I \to (0, +\infty)$ such that $Y^1(\bar{T}) = \bar{Y}^1$ and $L^*(\tilde{T}, Y^1(\tilde{T})) = 0$ for any $\tilde{T} \in I$. Moreover $Y^1(\tilde{T})$ is differentiable on I and the sign of $\dfrac{dY^1(\tilde{T})}{d\tilde{T}}$ is the same of $\mathcal{D}(\tilde{T})$, with function $\mathcal{D}$ defined in (27).*

Proof We compute

$$\frac{d\mathcal{G}_1}{dY^1} = \mathcal{G}_1(Y^1)\left[\frac{1}{Y^1} + \frac{1}{Y^1 + \alpha^2} + \frac{1}{-Y^1 + \alpha^3} + \frac{1}{-Y^1 + \alpha^4}\right], \tag{36}$$

and we observe that $\dfrac{d\mathcal{G}_1}{dY^1}$ is strictly positive in the admissible set A_1. It follows that the function $\mathcal{G}_1$ is strictly monotonically increasing in A_1 and it ranges between

$$\lim_{Y^1 \to \max(0, -\alpha^2)} \mathcal{G}_1(Y^1) = 0 \quad \text{and} \quad \lim_{Y^1 \to \min(\alpha^3, \alpha^4)} \mathcal{G}_1(Y^1) = +\infty. \tag{37}$$

Thus, for any positive value $\bar{T}$ there is a unique admissible value $\bar{Y}^1$ for which Eq. (34) is satisfied and we are able to express it explicitly in terms of $\bar{T}$:

$$\bar{Y}^1 = \frac{-\alpha^2 - M(\bar{T})(\alpha^3 + \alpha^4) + \sqrt{\left[\alpha^2 + M(\bar{T})(\alpha^3 + \alpha^4)\right]^2 + 4\alpha^3\alpha^4 M(\bar{T})(1 - M(\bar{T}))}}{2\left(1 - M(\bar{T})\right)}. \tag{38}$$

Being $\dfrac{\partial L^*}{\partial Y^1} = \dfrac{d\mathcal{G}_1}{dY^1}$ strictly positive and being $L^*(\tilde{T}, Y^1)$ differentiable with respect to $\tilde{T}$ for any $\tilde{T}$, we can apply the implicit function theorem that allows us to write the derivative

$$(Y^1)'(\tilde{T}) = -\left(\frac{\partial L^*(\tilde{T}, Y^1)}{\partial \tilde{T}}\right)\left(\frac{\partial L^*(\tilde{T}, Y^1)}{\partial Y^1}\right)^{-1}. \tag{39}$$

We compute the derivative of function $L^*(\tilde{T}, Y^1)$ with respect to $\tilde{T}$, getting

$$\frac{\partial L^*(\tilde{T}, Y^1)}{\partial \tilde{T}} = -M'(\tilde{T}) = -\left[\mathcal{G}_2(\tilde{T}) \cdot \mathcal{G}_3(\tilde{T})\right]^{-1} \cdot \left(\frac{m^1 m^2}{m^3 m^4}\right)^{3/2} \cdot \frac{\mathcal{D}(\tilde{T})}{\tilde{T}^2}, \tag{40}$$

and we see that the sign of (39) is the same of $\mathcal{D}(\tilde{T})$. □

Now, combining (24) and (26), and taking into account the results of previous lemma, we may express quantities $Y^5, \ldots, Y^8$ in terms of $\tilde{T}$ only, getting

$$Y^i(\tilde{T}) = \lambda^i \mathcal{S}(\tilde{T}) + \sum_{j=1}^{L^i} \nu^i_j n^i_j, \qquad i = 5, \ldots, 8, \tag{41}$$

with

$$\mathcal{S}(\tilde{T}) = \frac{\hat{\mathcal{N}}(\tilde{T})}{\mathcal{B}(\tilde{T})}, \qquad \hat{\mathcal{N}}(\tilde{T}) = \mathcal{N}(\tilde{T}) + \mathcal{D}(\tilde{T})\left(\sum_{j=1}^{L^1} \nu^1_j n^1_j - Y^1(\tilde{T})\right). \tag{42}$$

Consequently, the mass action law (11) turns out to be a transcendental equation in $\tilde{T}$ that may be cast as

$$\mathcal{G}^{**}(\tilde{T}) = \left(\frac{m^5 m^6}{m^7 m^8}\right)^{3/2}, \tag{43}$$

with $\mathcal{G}^{**}(\tilde{T}) = \mathcal{G}_4(\tilde{T}) \cdot \mathcal{G}_5(\tilde{T}) \cdot \mathcal{G}_6(\tilde{T})$, being

$$\begin{aligned} \mathcal{G}_4(\tilde{T}) = {} & \left(\textstyle\sum_{j=1}^{L^5} \nu^5_j n^5_j + \mathcal{S}(\tilde{T})\right) \left(\textstyle\sum_{j=1}^{L^6} \nu^6_j n^6_j + \mathcal{S}(\tilde{T})\right) \\ & \times \left(\textstyle\sum_{j=1}^{L^7} \nu^7_j n^7_j - \mathcal{S}(\tilde{T})\right)^{-1} \left(\textstyle\sum_{j=1}^{L^8} \nu^8_j n^8_j - \mathcal{S}(\tilde{T})\right)^{-1}, \end{aligned} \tag{44}$$

$$\mathcal{G}_5(\tilde{T}) = \frac{\sum_{k=1}^{L^7} \nu^7_k \exp\left(-\frac{E^7_k - E^7_1}{\tilde{T}}\right) \sum_{k=1}^{L^8} \nu^8_k \exp\left(-\frac{E^8_k - E^8_1}{\tilde{T}}\right)}{\sum_{k=1}^{L^5} \nu^5_k \exp\left(-\frac{E^5_k - E^5_1}{\tilde{T}}\right) \sum_{k=1}^{L^6} \nu^6_k \exp\left(-\frac{E^6_k - E^6_1}{\tilde{T}}\right)}, \tag{45}$$

$$\mathcal{G}_6(\tilde{T}) = \exp\left(-\frac{\Delta E^{**}}{\tilde{T}}\right). \tag{46}$$

We show that is possible to find a unique solution of (43) in terms of the actual parameters of the mixture, in the admissible set

$$A_2 = \left\{ \tilde{T} > 0 \,:\, \max\left(-\sum_{j=1}^{L^5} \nu_j^5 n_j^5, -\sum_{j=1}^{L^6} \nu_j^6 n_j^6 \right) < \mathcal{S}(\tilde{T}) < \min\left(\sum_{j=1}^{L^7} \nu_j^7 n_j^7, \sum_{j=1}^{L^8} \nu_j^8 n_j^8 \right) \right\}, \tag{47}$$

guaranteeing positivity of auxiliary densities $\tilde{n}^5$, $\tilde{n}^6$, $\tilde{n}^7$, $\tilde{n}^8$. To do so, we shall exploit the fact that the behavior of function $\mathcal{G}^{**}$ is strictly related to the properties of functions $\mathcal{S}$ and $\mathcal{B}$, which will be explored in turn through Lemmas 2, 3 and 4.

Lemma 2 *Let $I = (\tilde{T}_1, \tilde{T}_2) \subseteq A_2$ be any interval in which the function $\mathcal{B}(\tilde{T})$ given in (27) is strictly negative (positive), then the function $\mathcal{S}(\tilde{T})$ is strictly monotonically increasing (decreasing) in I with respect to $\tilde{T}$.*

Proof We may compute the derivatives of quantities appearing in (42), and we get

$$\mathcal{D}'(\tilde{T}) = \sum_{i=1}^{4} \lambda^i \mathcal{F}^i(\tilde{T}), \qquad \mathcal{B}'(\tilde{T}) = \sum_{i=5}^{8} \lambda^i \mathcal{F}^i(\tilde{T}), \tag{48}$$

$$\begin{aligned}\hat{\mathcal{N}}'(\tilde{T}) = &-\tfrac{3}{2} \textstyle\sum_{i=1}^{8} \left(\sum_{j=1}^{L^i} \nu_j^i n_j^i \right) - \sum_{i=5}^{8} \left(\sum_{j=1}^{L^i} \nu_j^i n_j^i \right) \mathcal{F}^i(\tilde{T}) \\ &- \textstyle\sum_{i=1}^{4} \left[\sum_{j=1}^{L^i} \nu_j^i n_j^i + \lambda^i \left(Y^1(\tilde{T}) - \sum_{j=1}^{L^1} \nu_j^1 n_j^1 \right) \right] \mathcal{F}^i(\tilde{T}) - \mathcal{D}(\tilde{T}) (Y^1)'(\tilde{T}),\end{aligned} \tag{49}$$

where functions $\mathcal{F}^i(\tilde{T})$ stand for

$$\mathcal{F}^i(\tilde{T}) = \frac{\sum_{j=1}^{L^i} \sum_{k=1}^{L^i} \frac{\nu_j^i \nu_k^i}{2\tilde{T}^2} \left[E_j^i - E_k^i \right]^2 \exp\left(-\frac{E_j^i + E_k^i - 2E_1^i}{\tilde{T}} \right)}{\left[\sum_{k=1}^{L^i} \nu_k^i \exp\left(-\frac{E_k^i - E_1^i}{\tilde{T}} \right) \right]^2} \geq 0, \quad i = 1, \ldots, 8. \tag{50}$$

Notice that, since the quantity $Y^1(\tilde{T})$ belongs to A_1, the derivative $\hat{\mathcal{N}}'(\tilde{T}) < 0$. Combining these results, it may be easily checked that the signs of $\mathcal{S}'(\tilde{T})$ and $\mathcal{B}(\tilde{T})$ are opposite. □

We prove another result concerning the behavior of $\mathcal{S}(\tilde{T})$.

Lemma 3 *The function $\mathcal{S}(\tilde{T})$ admits a unique positive value T^* such that $\mathcal{S}(T^*) = 0$.*

Proof Since $\mathcal{S}(\tilde{T}) = \hat{\mathcal{N}}(\tilde{T}) / \mathcal{B}(\tilde{T})$, we show that the function $\hat{\mathcal{N}}(\tilde{T})$ defined in (42) vanishes for a unique value of $\tilde{T}$. We discuss the result depending on the sign of the quantity ΔE^*. If $\Delta E^* < 0$, recalling the definition of function $M(\tilde{T})$ given in (35), we have $\lim_{\tilde{T}\to 0^+} M(\tilde{T}) = 0$, and from expression (38) we also have $\lim_{\tilde{T}\to 0^+} Y^1(\tilde{T}) = 0$. If, instead, $\Delta E^* > 0$, the limit for $T \to 0^+$ of $M(\tilde{T})$ is $+\infty$, so the computation of the limit of $Y^1(\tilde{T})$ is not straightforward; skipping details, rationalizing (38) we get

$$\lim_{\tilde{T}\to 0^+} Y^1(\tilde{T}) = \frac{2\alpha^3\alpha^4}{\alpha^3 + \alpha^4 + |\alpha^3 - \alpha^4|} = \alpha^K, \tag{51}$$

with $K = 4$ if $\alpha^3 > \alpha^4$ or $K = 3$ if $\alpha^4 > \alpha^3$. Therefore, keeping in mind the expressions for quantities α^i given in (25),

$$\lim_{\tilde{T}\to 0^+} \mathcal{D}(\tilde{T}) \left[\sum_{j=1}^{L^1} \nu_j^1 n_j^1 - Y^1(\tilde{T}) \right] = \begin{cases} -\Delta E^* \sum_{j=1}^{L^1} \nu_j^1 n_j^1 & \text{if } \Delta E^* < 0 \\ \Delta E^* \sum_{j=1}^{L^K} \nu_j^K n_j^K & \text{if } \Delta E^* > 0, \end{cases} \tag{52}$$

that is, in both cases, a positive quantity. Since also $\lim_{\tilde{T}\to 0^+} \mathcal{N}(\tilde{T}) > 0$, we can state that the function $\hat{\mathcal{N}}(\tilde{T})$ tends to a positive value as $\tilde{T} \to 0^+$; on the other hand, $\lim_{\tilde{T}\to +\infty} \hat{\mathcal{N}}(\tilde{T}) = -\infty$ thus, recalling that $\hat{\mathcal{N}}'(\tilde{T}) < 0$ and being the function $\hat{\mathcal{N}}(\tilde{T})$ continuous, it is null in only one $\tilde{T} = T^*$. □

Moreover, we can prove the following result.

Lemma 4 *On every interval* $\left(\tilde{T}_1, \tilde{T}_2\right) \subseteq A_2$ *the sign of* $\mathcal{B}(\tilde{T})$ *does not change.*

Proof Being the limits

$$\lim_{\tilde{T}\to 0^+} \mathcal{B}(\tilde{T}) = -\Delta E^{**}, \qquad \lim_{\tilde{T}\to +\infty} \mathcal{B}(\tilde{T}) = \sum_{i=5}^{8} \lambda^i \left(\sum_{j=1}^{L^i} \nu_j^i E_j^i \Big/ \sum_{j=1}^{L^i} \nu_j^i \right), \tag{53}$$

and since the derivative of $\mathcal{B}(\tilde{T})$ expressed in (48) may change sign in relation to internal energy levels and collision frequencies, it could exist a positive value $T^\#$ root of $\mathcal{B}(\tilde{T})$. We shall suppose that the choice of initial data, internal energies and collision frequencies is such that $T^\# \neq T^*$, since in this case we would have only algebraic equations. We note thus that $\lim_{\tilde{T}\to \tilde{T}^\#} \mathcal{S}(\tilde{T}) = \pm\infty$, getting a neighborhood of $\tilde{T}^\#$ not contained in A_2. Thus $\mathcal{B}(\tilde{T})$ does not vanish on any interval $\left(\tilde{T}_1, \tilde{T}_2\right)$ of A_2. □

Now, let $\left(\tilde{T}_a, \tilde{T}_b\right)$ be a connected component of A_2. The previous result has as a consequence the fact that on $\left(\tilde{T}_a, \tilde{T}_b\right)$ the sign of $\mathcal{S}'(\tilde{T})$ does not change. This means that $\mathcal{S}(\tilde{T})$ varies monotonically from the minimum to the maximum value that it can assume in A_2, which have different sign. Consequently it has a root in $\left(\tilde{T}_1, \tilde{T}_2\right)$. Also in the case in which $\tilde{T}_a = 0$, since $\lim_{\tilde{T}\to 0} \mathcal{B}(\tilde{T}) < 0$, $\mathcal{S}(\tilde{T})$ is monotonically increasing from $\lim_{\tilde{T}\to 0} \mathcal{S}(\tilde{T}) < 0$ to its upper bound that is positive, having a root as well. Since in Lemma 3 we proved that $\mathcal{S}(\tilde{T})$ has a unique positive root, it follows that $\left(\tilde{T}_a, \tilde{T}_b\right)$ is the only connected component of A_2, then the set A_2 is connected. This fact allows us to prove the following statement.

Lemma 5 *There exists a unique value* $\bar{T} \in A_2$ *such that*

$$L^{**}(\bar{T}) = \mathcal{G}^{**}(\bar{T}, \bar{Y}^1(\bar{T})) - \left(\frac{m^5 m^6}{m^7 m^8}\right)^{3/2} = 0. \tag{54}$$

Proof The derivative with respect to $\tilde{T}$ of the function $L^{**}(\tilde{T})$ turns out to be

$$\frac{dL^{**}(\tilde{T})}{d\tilde{T}} = L^{**}(\tilde{T}) \left\{ \mathcal{S}'(\tilde{T}) \left[\sum_{i=5}^{8} \left(\sum_{j=1}^{L^i} \nu^i_j n^i_j + \lambda^i \mathcal{S}(\tilde{T}) \right)^{-1} \right] - \frac{1}{\tilde{T}^2} \mathcal{B}(\tilde{T}) \right\}. \tag{55}$$

The term in square brackets is positive in A_2 and, as proved in Lemma 2, $\mathcal{S}'(\tilde{T})$ and $-\mathcal{B}(\tilde{T})$ have the same sign. It follows that $L^{**}(\tilde{T})$ is strictly monotonically increasing or decreasing in A_2, that is an interval of type $\left(\tilde{T}_{min}, \tilde{T}_{max}\right)$. In particular, when $\tilde{T} \to \tilde{T}_{min}$, $\mathcal{G}^{**}(\tilde{T}) \to 0$ and when $\tilde{T} \to \tilde{T}_{max}$, $\mathcal{G}^{**}(\tilde{T}) \to +\infty$ or vice versa. This holds also in the case $\tilde{T}_{min} = 0$ if we have $\Delta E^{**} > 0$; if, instead, $\Delta E^{**} < 0$, it is needed to take the inverse in both sides of Eq. (43) and repeat all the calculations defining $L^{**}(\bar{T})$ as $\left[\mathcal{G}^{**}(\bar{T}, \bar{Y}^1(\bar{T}))\right]^{-1} - \left[(m^7 m^8)/(m^5 m^6)\right]^{3/2}$. Thus, we can find a unique value $\bar{T}$ for which $L^{**}(\bar{T}) = 0$. □

This result provides the existence and uniqueness of an admissible auxiliary temperature $\tilde{T}$ that can be expressed in terms of actual parameters of the mixture. By inserting it in (38) we get the admissible quantity Y^1, then from (23) and (41) we obtain the remaining Y^i, $i = 2, \ldots, 8$. This allows us to construct the Maxwellian attractors $\mathcal{M}^i_j$ given in (8). The equilibrium configuration of our BGK model is given by distributions coinciding with Maxwellian attractors, thus equilibrium mean velocities, temperatures, and number densities of each component will be exactly the auxiliary parameters individuated previously: $\mathbf{u}^i_j = \tilde{\mathbf{u}} = \mathbf{u}$, $T^i_j = \tilde{T} = T$, $n^i_j = \tilde{n}^i_j$, $(i = 1, \ldots, 8,\ j = 1, \ldots, L^i)$, with global density of each gas n^i related to the number densities of its components through the partition function, and densities and temperature bound together by the two mass action laws of chemistry (6). By the same arguments used in [3, 4], we are able to prove that our model also fulfills the classical H-theorem, guaranteeing thus the stability of collision equilibria.

4 Conclusions

In this paper we have built up a BGK model for a mixture of eight species undergoing two bimolecular and reversible chemical reactions. The eight constituents may be monatomic (with a fixed internal energy) or polyatomic (with a proper number of internal energy levels). This work generalizes the BGK models proposed in [2] for a mixture of four reacting gases with the same number of levels, and in [3] for an inert mixture of monatomic and polyatomic particles. The additional difficulties arisen with respect to the inert frame or to the case of four gases subject to one reversible reaction [4] have been commented on. We remark that the two considered reactions involve different species; the construction of relaxation-type models in case of two non-independent reactions (with at least a gas involved in both of them) would give rise to more complicated relations for conservation of sums of number densities, and therefore to more cumbersome transcendental equations. This problem is worth to be investigated in a near future.

Another way of modeling the non-translational degrees of freedom of polyatomic particles assumes the internal energy to be a continuous positive variable I, and even in this case a BGK model is available [6]. In such formulation, the different polyatomic nature of each constituent may be modeled by means of suitable weight functions $\varphi^i(I)$ in the Boltzmann collision kernels. The classical options $\varphi^i(I) = I^{\alpha_i}$ correspond to polytropic gases, with a caloric equation of state linearly depending on temperature. In our discrete energy levels formulation, the specific heat at constant volume is in general a nonlinear function of the temperature, corresponding thus to non-polytropic gases, and the polytropic case may be recovered introducing a suitable linear approximation in the internal energy function [7]. The same procedure could be applied also in the hydrodynamic equations corresponding to the present BGK model and will be matter of future investigation.

Acknowledgments This work has been performed in the frame of activities supported by University of Parma, by INdAM-GNFM, and by the Italian PRIN Project *Multiscale phenomena in Continuum Mechanics: singular limits, off-equilibrium, and transitions* (2017YBKNCE).

References

1. Andries, P., Aoki, K., Perthame, B.: A consistent BGK–type model for gas mixtures. J. Stat. Phys. **106**, 993–1018 (2002)
2. Bisi, M., Caceres, M.J.: A BGK relaxation model for polyatomic gas mixtures. Commun. Math. Sci. **14**, 297–325 (2016)
3. Bisi, M., Travaglini, R.: A BGK model for mixtures of monoatomic and polyatomic gases with discrete internal energy. Phys. A **547**, 124441 (2020)
4. Bisi, M., Travaglini, R.: A kinetic BGK relaxation model for a reacting mixture of polyatomic gases. In: Salvarani, F. (ed.) Recent Advances in Kinetic Equations and Applications. INdAM Series **48**, 69–92, Springer, Berlin (2021).

5. Bisi, M., Groppi, M., Spiga, G.: Kinetic Bhatnagar–Gross–Krook model for fast reactive mixtures and its hydrodynamic limit. Phys. Rev. E **81**, 036327, 1–9 (2010)
6. Bisi, M., Monaco, R., Soares, A.J.: A BGK model for reactive mixtures of polyatomic gases with continuous internal energy. J. Phys. A Math. Theor. **51**, 125501, 1–29 (2018)
7. Bisi, M., Ruggeri, T., Spiga, G.: Dynamical pressure in a polyatomic gas: interplay between kinetic theory and Extended Thermodynamics. Kinet. Relat. Models **11**, 71–95 (2018)
8. Bobylev, A.V., Bisi, M., Groppi, M., Spiga, G., Potapenko, I.F.: A general consistent BGK model for gas mixtures. Kinet. Relat. Models **11**, 1377–1393 (2018)
9. Brull, S., Schneider, J.: Derivation of a BGK model for reacting gas mixtures. Commun. Math. Sci. **12**, 1199–1223 (2014)
10. Giovangigli, V.: Multicomponent flow modeling. In: Series on Modeling and Simulation in Science, Engineering and Technology. Birkhuser, Boston (1999)
11. Groppi, M., Spiga, G.: A Bhatnagar–Gross–Krook–type approach for chemically reacting gas mixtures. Phys. Fluids **16**, 4273–4284 (2004)
12. Haack, J.R., Hauck, C.D., Murillo, M.S.: A conservative, entropic multispecies BGK model. J. Stat. Phys. **168**, 826–856 (2017)
13. Klingenberg, C., Pirner, M., Puppo, G.: A consistent kinetic model for a two-component mixture with an application to plasma. Kinet. Relat. Models **10**, 445–465 (2017)
14. Kremer, G.M., Pandolfi Bianchi, M., Soares, A.J.: A relaxation kinetic model for transport phenomena in a reactive flow. Phys. Fluids **18**, 037104, 1–15 (2006)
15. McCormack, F.J.: Construction of linearized kinetic models for gaseous mixtures and molecular gases. Phys. Fluids **16**, 2095–2105 (1973)
16. Rossani, A., Spiga, G.: A note on the kinetic theory of chemically reacting gases. Physica A **272**, 563–573 (1999)

On a Class of Self-Similar Solutions of the Boltzmann Equation

A.V. Bobylev

Abstract We consider a class of distribution functions having the form $f(v,t) = e^{-dat}F(ve^{-at})$, $a =$ const., where $v \in \mathbb{R}^d, d \geq 2$ and $t \in \mathbb{R}_+$ denote the particle velocity and the time. This class of self-similar solutions to the spatially homogeneous Boltzmann equation (BE) for Maxwell molecules was studied by Bobylev and Cercignani in early 2000s. The solutions are positive but have an infinite second moment (energy). However, the same class of distribution functions with finite energy appears to be closely connected with quite different class of group-invariant solutions of the spatially inhomogeneous BE. This is a motivation for considering the so-called modified spatially homogeneous BE, which contains an extra force term proportional to a given matrix A. The modified BE was recently studied under the assumption of "sufficient smallness of norm $\|A\|$" without explicit estimates of the smallness. We fill this gap and prove that all important facts related to self-similar solutions remain valid also for $\|A\| = O[10^{-1}]$ in appropriate dimensionless units.

1 Introduction

Self-similar solutions of nonlinear equations of mathematical physics are always interesting for both physicists and mathematicians. The simplest example of such solution for the classical Boltzmann equation can be described as follows. We consider the spatially homogeneous case of this equation for Maxwell molecules (with or without cut-off). The solution $f(v,t)$ of this equation depends on the velocity $v \in \mathbb{R}^3$ and the time $t \in \mathbb{R}_+$. It is easy to show that the Boltzmann equation admits (at least formally) a particular class of solutions having the form

$$f(v,t) = e^{-3ct}F(ve^{-ct}), \quad c = const. \tag{1}$$

A.V. Bobylev (✉)
Keldysh Institute of Applied Mathematics RAS, Moscow, Russia
e-mail: alexander.bobylev@kau.se

P. Barbante et al. (eds.), *From Kinetic Theory to Turbulence Modeling*, Springer INdAM Series 51, https://doi.org/10.1007/978-981-19-6462-6_6

Then we obtain the time-independent equation for the self-similar profile $F(v)$ and some conditions on the constant c. This is what we did jointly with Carlo Cercignani approximately 20 years ago [5–7]. By that time I already had some experience with related classes of group—invariant solutions to this equation [1, 2]. On the other hand, Carlo Cercignani had an idea to use the Pomeau [24] approach to the structure of infinitely strong shock wave. The connection with this approach is briefly explained in [6]. Unfortunately we did not manage to produce something useful for shock waves. However, the results of [5–7] for the spatially homogeneous case were interesting. In fact we studied only isotropic (depending on $|v|$) solutions having the form (1). We have proved that these solutions are positive and describe a large-time asymptotics for certain classes of initial data. Integral representations of two particular solutions were constructed in explicit form. They represent two non-trivial examples of eternal solutions of the Boltzmann equations, which exist for all real values of time t (for more information see the book [4]).

The self-similar solutions of the form (1) and related questions were studied more deeply in some interesting mathematical papers, see, e.g., [11, 23]. Of course, a serious drawback of solutions (1) is that they obviously contradict to the energy (the second moment of F) conservation. The contradiction can be avoided only if the second moment of the positive function $F(v)$ is infinite. It is really so for above discussed solutions, they all have an infinite second moment. Thus, it is a very unusual class of solutions for rarefied gas dynamics. However, there is a quite different class of spatially inhomogeneous solutions to the Boltzmann equation, which can also lead to distribution functions of the form (1.1). These are the so-called homoenergetic affine flows introduced in 1950s independently by Galkin [17] and Truesdell [25]. There are many related references, we just mention two books [18, 26], and contributions of Cercignani to these areas [12–14]. Recently these flows were considered in detail in a series of papers by James, Nota and Velazquez [19–21], see also [10, 15]. Roughly speaking, these are spatially inhomogeneous flows of gas having the linear (with respect to spatial variable $x \in \mathbb{R}^3$) profile of the bulk velocity. This assumption allows to reduce the problem to the modified spatially homogeneous Boltzmann equation. The general form of this equation introduced in [10] is also considered in the present paper. Our aim is to study solutions of this equation having the form (1). The interest to this problem is partly caused by the first proof of existence of self-similar solutions of that kind in [19] and [10]. The modified Boltzmann equation depends on a small parameter. This small parameter is assumed in these papers to be "as small as we want". One of the goals of the present paper is to weaken this assumption and to prove that all main results of our previous work [10] remain valid for moderately small values of the parameter (roughly 10% of contribution of collisions) and to show why the proof cannot be extended to larger values. Unfortunately the restricted volume of the paper does not allow to include details of the proofs, they will be published elsewhere. Below we confine ourselves to statement of the problem in Sect. 2, eigenvalue problem for matrices in Sect. 3 and formulation of main results in Sect. 4.

2 Statement of the Problem

We consider a modified Boltzmann equation [10] for the distribution function $f(v, t)$, where $v \in \mathbb{R}^d$, $d \geq 2$, and $t \in \mathbb{R}_+$ denotes, respectively, the particle velocity and time. The equation reads

$$\partial_t f - div_v (Avf) = Q(f, f), \tag{2}$$

where $A \in M_{d\times d}(\mathbb{R})$ is a constant real matrix and $Q(f, f)$ is the collision integral for Maxwell molecules

$$Q(f, f)(v) = \int\limits_{\mathbb{R}^d \times S^{d-1}} dwdng\left(\hat{u} \cdot n\right)\left[f(v')f(w') - f(v)f(w)\right], \; n \in S^{d-1} \tag{3}$$

$$u = v - w, \quad \hat{u} = u/|u|, \quad v' = \frac{1}{2}(v + w + |u|n), \; w' = \frac{1}{2}(v + w - |u|n).$$

We assume that

$$f(v, 0) = f_0(v) \geq 0, \quad \int_{\mathbb{R}^d} dv\; f_0(v)v = 0, \quad \int_{\mathbb{R}^d} dv\; f_0(v) = 1. \tag{4}$$

The kernel $g(\eta)$ in (3), with $\eta \in [-1, 1]$, is non-negative and normalized by unity

$$\int_{S^{d-1}} dng(\omega \cdot n) = 1, \quad \omega \in S^{d-1}. \tag{5}$$

The motivation for considering the modified Boltzmann equation (2) is discussed in detail in [10]. In particular, we note that the well-known shear flow for the Boltzmann equation is described by Eq. (2) with nilpotent matrix A such that

$$A = \{a_{ij}; \quad i, j = 1, ..., d\}, \tag{6}$$

where $a_{12} = a = \text{const.}$, $a \neq 0$ and all other elements of A are zeros. Many related details and references can be found in [8, 12–14, 18, 19, 26].

Formally the second term in (2) describes the action of the external force

$$F = -Av, \quad A \in M_{d\times d}(\mathbb{R}), \tag{7}$$

which looks like the anisotropic *friction* force proportional to components of the particle velocity. Let us consider, e.g., the simplest case of Eq. (2) with

$$A = aI, \quad a \in \mathbb{R}, \tag{8}$$

where I is the unit matrix, a is a constant with any sign. If $a > 0$ this is just a regular friction force $F = -av, a > 0$. The solution $f(v, t)$ of (2)–(5) under assumption (8) leads to the following behaviour of the second moment (energy):

$$d\mathcal{E}(t)/dt = -2a\mathcal{E}(t) \Rightarrow \mathcal{E}(t) = \mathcal{E}(0)e^{-2at},$$

where

$$\mathcal{E}(t) = \int_{\mathbb{R}^d} dv f(v, t)|v|^2.$$

This equality gives an idea to consider the Eq. (2) in self-similar variables by substitution

$$f(v, t) = e^{dat} \tilde{f}(\tilde{v}, t), \quad \tilde{v} = ve^{at}. \tag{9}$$

Then after simple calculations, we obtain the familiar spatially homogeneous Boltzmann equation for $\tilde{f}(\tilde{v}, t)$

$$\tilde{f}_t = Q(\tilde{f}, \tilde{f}), \quad \tilde{f}|_{t=0} = f_0(\tilde{v}). \tag{10}$$

If, in addition,

$$\mathcal{E}(0) = \int_{\mathbb{R}^d} dv f_0(v)|v|^2 = d,$$

then we know (H-theorem for (10)) that

$$\tilde{f}(\tilde{v}, t) \rightarrow \tilde{f}_M(\tilde{v}) = (2\pi)^{-d/2} e^{-|\tilde{v}|^2/2}, \quad \text{as } t \rightarrow \infty.$$

Hence, coming back to initial variables, we (1) obtain the simplest self-similar solution of (2), (8), namely

$$f_{s-s}(v, t) = \left(2\pi e^{-2at}\right)^{-d/2} \exp\left(-|ve^{at}|^2/2\right), \tag{11}$$

and (2) show that this particular solution is an attractor for various classes of initial data. It is obvious that this simple example is valid also for $a < 0$ in (8) (accelerating forces) and for arbitrary kernel (not necessary the Maxwellian one) in the collision integral.

Roughly speaking, our task is to prove that the situation is, to some extent, similar in the case of arbitrary matrix A in (2) provided that its norm is not too large. In fact, all proofs were already done in our previous paper [10] with standard formulations of results like as "There exists such $\varepsilon_0 > 0$ that the following property holds under assumption that $\|A\| \leq \varepsilon_0$... ". This approach allows to avoid some technical work, but it does not show true limits (in terms of $\|A\|$) of the results. The main aim of this paper is to partly clarify this question. Here and below we use the so-called operator norm for matrices [22]. Its properties are discussed in the next section.

Following [10] we pass to the Fourier-representation [3] of the Eq. (2) and introduce the characteristic function $\varphi(k,t)$ (cf. [16])

$$\varphi(k,t) = \int_{\mathbb{R}^d} dv\, f(v,t)e^{-ik\cdot v}, \quad k \in \mathbb{R}^d. \tag{12}$$

Then we obtain

$$\partial_t \varphi + (Ak)\cdot\partial_k\varphi = \mathcal{I}^+(\varphi,\varphi) - \varphi_{|k=0}\varphi\,, \tag{13}$$

where

$$\mathcal{I}^+(\varphi,\varphi)(k) = \int\limits_{S^{d-1}} dng\left(\hat{k}\cdot n\right)\varphi(k_+)\varphi(k_-), \quad k_\pm = \frac{1}{2}\left(k \pm |k|n\right), \; \hat{k} = \frac{k}{|k|}. \tag{14}$$

The initial condition becomes

$$\varphi(k,0) = \varphi_0(k) = \int_{\mathbb{R}^d} dv\, f_0(v)e^{-ik\cdot v}, \quad \varphi_0(0) = 1.$$

Note that (2) implies the mass conservation. Therefore

$$\varphi(0,t) = \varphi_0(0) = 1, \tag{15}$$

and we obtain from (13)

$$\partial_t\varphi + \varphi + (Ak)\cdot\partial_k\varphi = \mathcal{I}^+(\varphi,\varphi) = \Gamma(\varphi). \tag{16}$$

For brevity we consider below the self-similar solution only. Following [10], we look for such solution in the form

$$\varphi_{s-s}(k,t) = \Psi(ke^{\beta t}), \; \beta \in \mathbb{R}. \tag{17}$$

Note that it corresponds to the distribution function (11), where $a = -\beta$. The parameter β will be defined below.

Then we pass to self-similar variables in (16) by substitution

$$\varphi(k,t) = \tilde{\varphi}(\tilde{k},t), \quad \tilde{k} = ke^{\beta t}, \tag{18}$$

and obtain omitting tildes

$$\partial_t\varphi + \varphi + \left(A_\beta k\right)\cdot\partial_k\varphi = \Gamma\left[\varphi\right], \quad A_\beta = A + \beta I. \tag{19}$$

It is clear that the self-similar solution (17) of Eq. (16) becomes a stationary solution for Eq. (19). The differential form of the stationary solution is obvious from (19). Its integral form can be obtained at the formal level from the operator identity

$$\int_0^\infty dt e^{-t\left(1+\hat{D}\right)} = \left(1+\hat{D}\right)^{-1}, \tag{20}$$

where $\hat{D}$ is an abstract operator. We refer to [10] for conditions of equivalence of these integral and differential forms of equation for $\Psi(k)$. The integral equation reads [10]

$$\Psi(k) = \int_0^\infty dt E_\beta(t)\Gamma[\Psi(k)], \tag{21}$$

where $\Gamma[\Psi(k)] = \mathcal{I}^+(\Psi, \Psi)$ is given in (14),

$$E_\beta(t) = \exp\left[-t\left(1 + A_\beta k \cdot \partial_k\right)\right]. \tag{22}$$

It is easy to see that the action of the operator $E_\beta(t)$ on any function $\varphi(k)$ is given by formula

$$E_\beta(t)\varphi(k) = e^{-t}\varphi\left[e^{-\beta t}\left(e^{-tA}k\right)\right]. \tag{23}$$

Equation (21) will be solved below with all necessary estimates. We begin in the next section with the definition of β and some preliminary estimates.

3 Eigenvalue Problem for Matrices

We can apply the operator $\left(1 + A_\beta\, k \cdot \partial_k\right)$ to the Eq. (21) and obtain the equation for $\Psi(k)$ in differential form (see also (19))

$$\left(1 + \beta k \cdot \partial_k + Ak \cdot \partial_k\right)\Psi(k) = \Gamma[\Psi](k). \tag{24}$$

It is always assumed below that $\Psi(k)$ is a characteristic function (the Fourier transform of a probability measure in $\mathbb{R}^d$) and has the following asymptotic behaviour for small $|k|$:

$$\Psi(k) = 1 - \frac{1}{2}B : k \otimes k + O\left(|k|^p\right) \tag{25}$$

for some $2 < p \leq 4$. The notation $B = \{b_{ij};\ i, j = 1, \cdots, d\}$ is used for symmetric positively defined matrix. We also denote for brevity

$$B : k \otimes k = \sum_{i,j=1}^{d} b_{ij} k_i k_j .$$

The formula (25) means that the corresponding distribution function, i.e., the inverse Fourier transform of $\Psi(k)$, has finite moments of the order $2+\varepsilon$, $\varepsilon > 0$ (see [10] for details). It can be shown that the matrix B and the parameter β satisfy the following equation (see Eq. (8) in [10]):

$$\beta B + \theta \left(B - \frac{\mathrm{Tr} B}{d} I \right) + \langle BA \rangle = 0, \tag{26}$$

where

$$\theta = \frac{qd}{4(d-1)}, \quad q = \int_{S^{d-1}} dn g(\omega \cdot n)[1 - (\omega \cdot n)^2], \quad \omega \in S^{d-1};$$
$$\mathrm{Tr} B = \sum_{i=1}^{d} b_{ii}, \quad \langle BA \rangle = \frac{1}{2}[BA + (BA)^T], \tag{27}$$

here the upper index T denotes the transposed matrix. This equation can be easily obtained by substitution of (25) into Eq. (24). We are interested in solution (β, B) of the eigenvalue problem (26) such that the eigenvalue β has the largest (as compared to other eigenvalues) real part. In addition, the real symmetric matrix B must have only positive eigenvalues. The existence of such solution (β, B) was proved in Lemma 7.3 in [10] under assumption that $\|A\| \leq \varepsilon_0$ for sufficiently small $\varepsilon_0 > 0$, where

$$\|A\| = \sup_{|k|=1} |Ak|, \quad k \in \mathbb{R}^d. \tag{28}$$

No estimate of ε_0 was given in [10]. Our aim in this paper is to fill this gap and to show that main results of that paper remain valid for moderately small values of ε_0.

4 Main Results

We begin with the eigenvalue problem from Sect. 3.

Lemma 1 *For any real $(d \times d)$ matrix A satisfying the condition*

$$\|A\| < \frac{\theta}{6}, \quad \theta = \frac{qd}{4(d-1)}, \tag{29}$$

in the notation of Eq. (27)*, the eigenvalue problem* (26) *has a unique solution* (β, B) *such that*

1. β *has the largest real part among all eigenvalues of the problem* (26) *and*
2. *the symmetric* $(d \times d)$*-matrix* B *is normalized by condition* $\mathrm{Tr}B = d$.

This solution can be represented by power series

$$\beta = \theta \sum_{n=1}^{\infty} \beta_n \varepsilon^n, \quad B = \sum_{n=1}^{\infty} B_n \varepsilon^n, \quad \varepsilon = \frac{\|A\|}{\theta}, \quad B_0 = I,$$

convergent for $|\varepsilon| \le 1/6$. *The eigenvalue* β *is real and simple. The matrix* B *is positive-definite. The following estimates are valid under condition* (29)

$$|\beta| < 2\|A\|, \quad \beta - \Re\beta' \ge \theta - 5\|A\|, \quad \|B - I\| < 1,$$

where $\beta' \neq \beta$ *is any other eigenvalue of the problem* (26).

This lemma defines the parameter β in the Eq. (21) for the self-similar profile $\Psi(k)$. It also defines the matrix B related to behaviour of $\Psi(k)$ for small $|k|$ (see Eq. (25)). The function $\Psi(k)$ can be constructed by iterations of the integral operator in (21). A convenient initial approximation is given by function

$$\Psi_0(k) = \exp\left(-\frac{1}{2} B : k \otimes k\right), \quad k \in \mathbb{R}^d.$$

We just formulate the final result.

Theorem 1 *Consider the integral equation* (21) *and assume that*

$$\|A\| < \frac{q}{24}, \quad q = \int_{S^{d-1}} dn g(\omega \cdot n)[1 - (\omega \cdot n)^2], \quad \omega \in S^{d-1}. \tag{30}$$

It is also assumed that the solution $\Psi(k)$ *of that equation has asymptotic behaviour for small* $|k|$ *in accordance with Eq.* (25) *for some* $p \in (2, 4]$.

(i) Then the symmetric matrix B *in* (25) *normalized by condition* $\mathrm{Tr}B = d$ *and the parameter* β *in* (21) *coincide with the solution* (β, B) *of eigenvalue problem* (26) *constructed in Lemma 1.*

(ii) For β *and* B *from the item (i) there is a unique characteristic function* $\Psi(k)$ *that solves Eqs.* (21)–(23) *and satisfies the asymptotic formula* (25)*. The correct value of* p *in* (25) *is* $p = 4$ *provided that condition* (30) *holds.*

We also consider the initial value problem for characteristic function $\varphi(k, t)$ in self-similar coordinates (16). The problem reads (see Eq. (19))

$$\varphi_t + A_\beta k \cdot \varphi_k + \varphi = \Gamma(\varphi), \quad \varphi|_{t=0} = \varphi_0(k), \quad k \in \mathbb{R}^d, \tag{31}$$

where $A_\beta = A + \beta I$, $\beta \in \mathbb{R}$. It is assumed that

$$\left|\varphi_0(k) - \left(1 - \frac{1}{2} G_0 : k \otimes k\right)\right| \le C_0 |k|^4, \quad C_0 = \text{const.}, \quad k \in \mathbb{R}^d, \tag{32}$$

in the notation analogous to Eq. (25). Then it is known from [10] that there exists a unique characteristic function $\varphi(k, t)$ that solves the problem (31) and satisfies the condition

$$\left|\varphi(k, t) - \left(1 - \frac{1}{2} G(t) : k \otimes k\right)\right| \le C_1 |k|^4, \quad C_1 = \text{const.}, \quad k \in \mathbb{R}^d,$$

where $G(t)$ is a time-dependent symmetric $(d \times d)$ matrix that solves the problem

$$\frac{1}{2} G_t + \beta G + \theta \left(G - \frac{\mathrm{Tr} G}{d} I\right) + \langle GA \rangle = 0, \quad G|_{t=0} = G_0, \tag{33}$$

in the notation of Eqs. (27) with $B = G$.

Hence, the matrix $G(t)$ satisfies the linear ODE with constant coefficients. We assume that the parameter β in (31) is the eigenvalue from Lemma 1. Then the matrix B from Lemma 1 is a stationary solution of Eq. (33). We can prove that

$$G(t) = c^2 B + O[\exp(-qt/12)], \quad t \ge 0,$$

for some constant $c > 0$ provided the condition (30) is satisfied. The constant c depends on G_0. The following statement shows the asymptotic role of the self-similar profile $\Psi(k)$.

Theorem 2 *Let $\varphi(k, t)$ be a solution of the problem (31), where $\|A\| < q/24$ and $\varphi_0(k)$ is a characteristic function satisfying (32). Let the parameter β in (31) and the function $\Psi(k)$ be the same as in Theorem 1. Then there exist two constants $c > 0$ and $C > 0$ such that*

$$|\varphi(k, t) - \Psi(ck)| \le C(|k|^2 + |k|^4) e^{-\mu t}, \quad \mu = \frac{q - 24\|A\|}{16}, \quad k \in \mathbb{R}^d, \ t \ge 0.$$

The above results can be expressed in terms of distribution functions $f(v, t)$ in the same way as in [10].

5 Conclusions

We have considered the modified spatially homogeneous Maxwell–Boltzmann equation (2). The equation contains an additional force term $\mathrm{div} Avf$, where $v \in \mathbb{R}^d$, A is an arbitrary constant $(d \times d)$-matrix. Applications of this equation are connected with well-known homoenergetic solutions to the spatially inhomogeneous Boltzmann equation studied by many authors since 1950s. The self-similar solutions and related questions for Eq. (2) were recently considered in detail in [10] by using the Fourier transform and some properties of the Boltzmann collision operator in the Fourier representation [9]. Main results of [10] were obtained under assumption of "sufficiently small norm of A" in (2) without explicit estimates of this "smallness". Our aim in this paper was to fill this gap and to prove that most of the results related to self-similar solutions remain valid for moderately small matrices A with norm $\|A\| = O(10^{-1})$ in dimensionless units. This is important for applications because it shows boundaries for the approach based on the perturbation theory. The main results of the paper are formulated in Theorems 1 and 2 from Sect. 4. These theorems extend the corresponding results of [10] to moderate values of $\| A\|$. The main idea of proofs of new estimates is based on detailed study of the eigenvalue problem (26), see Lemma 1 from Sect. 4. A by-product result is the proof of existence of the bounded fourth moment of the self-similar profile for moderate values of $\| A\|$. The question of existence of *all* moments for the self-similar profile $F(v)$ remains open even in the class of arbitrarily small norm of A.

Acknowledgments The work is dedicated to memory of Carlo Cercignani. The research was supported by Russian Science Foundation Grant No. 18-11-00238-П. The author thanks Alessia Nota and Juan Velazquez for valuable discussions.

References

1. Bobylev, A.V.: On exact solutions of the Boltzmann equation. Dokl. Acad. Nauk SSSR **225**, 1296–1299 (1975). English translation in Sov. Phys. Dokl. **20**, 822–824 (1976)
2. Bobylev, A.V.: A class of invariant solutions of the Boltzmann equation. Dokl. Acad. Nauk SSSR **231**, 571–574 (1976). English translation in Sov. Phys. Dokl. **21**, 632–635 (1976)
3. Bobylev, A.V.: Fourier transform method in the theory of the Boltzmann equation for Maxwell molecules. Dokl. Acad. Nauk SSSR **225**, 1041–1044 (1975). English translation in Sov. Phys. Dokl. **20**, 820–822 (1976)
4. Bobylev, A.V.: Kinetic Equations, Volume 1: Boltzmann Equation, Maxwell Models and Hydrodynamics Beyond Navier–Stokes. De Gruyter, Berlin/Boston (2020)
5. Bobylev, A.V., Cercignani, C.: Self-similar solutions of the Boltzmann equation and their applications. J. Stat. Phys. **106**(5-6), 1039–1071 (2002)
6. Bobylev, A.V., Cercignani, C.: Exact eternal solutions of the Boltzmann equation. J. Stat. Phys. **106**(5-6), 1019–1038 (2002)
7. Bobylev, A.V., Cercignani, C.: Self-similar asymptotics for the Boltzmann equation with elastic and inelastic interactions. J. Stat. Phys. **110**, 333–375 (2003)
8. Bobylev, A.V., Caraffini, G.L., Spiga, G.: On group invariant solutions of the Boltzmann equation. J. Math. Phys. **37**, 2787–2795 (1996)
9. Bobylev, A.V., Cercignani, C., Gamba I.M.: On the self-similar asymptotics for generalized non-linear kinetic Maxwell models. Commun. Math. Phys. **291**, 599–644 (2009)
10. Bobylev, A.V., Nota, A., Velazquez, J.L.: Self-similar asymptotics for a modified Maxwell–Boltzmann equation in systems subject to deformations. Commun. Math. Phys. **380**, 409–448 (2020)

11. Cannone, M., Carch, G.: Infinite energy solutions to the homogeneous Boltzmann equation. Commun. Pure Appl. Math. **63**(6), 747–778 (2010)
12. Cercignani, C.: Existence of homogeneous affine flows for the Boltzmann equation. Arch. Ration. Mech. Anal. **105**(4), 377–387 (1989)
13. Cercignani, C.: Shear flow of a granular material. J. Stat. Phys. **102**(5), 1407–1415 (2001)
14. Cercignani, C.: The Boltzmann equation approach to the shear flow of a granular material. Phil. Trans. R. Soc. **360**, 437–451 (2002)
15. Duan, R., Liu, Sh.: The Boltzmann equation for uniform shear flow. arXiv: 2008. 0255 IvI [math. AP] 6 Aug. 2020
16. Feller, W.: An Introduction to Probability Theory and Applications. Wiley, New York (1971)
17. Galkin, V.S.: On a class of solutions of Grade's moment equation. PMM **20**, 445–446 (1956)
18. Garzo, V., Santos, A.: Kinetic Theory of Gases in Shear Flow, Nonlinear Transport. Kluwer Academic Publishers, Dordrech (2003)
19. James, R.D., Nota, A., Velazquez, J.J.L.: Self-similar profiles for homo-energetic solutions of the Boltzmann equation: particle velocity distribution and entropy. Arch. Ration. Mech. Anal. **231**(2), 787–843 (2019)
20. James, R.D., Nota A., Velazquez, J.J.L.: Long-time asymptotics for homoenergetic solutions of the Boltzmann equation: collision-dominated case. J. Nonlin. Sci. **29**(5), 1943–1973 (2019)
21. James, R.D., Nota, A., Velazquez, J.J.L.: Long-time asymptotics for homoenergetic solutions of the Boltzmann equation: hyperbolic - dominated case. Nonlinearity **33**(8), 3781–3815 (2020)
22. Kato, T.: Perturbation Theory for Linear Operators, Classics in Mathematics. Springer (1976)
23. Morimoto, Y., Yang, T., Zhao, H.: Convergence to the self-similar solutions to the spatially homogeneous Boltzmann equation. J. Eur. Math. Soc. **19**, 2041–2067 (2017)
24. Pomeau, Y.: Shock at very large Mach number in simple gases: a physicist approach. Transp. Theory Stat. Phys. **16**, 727–734 (1987)
25. Truesdell, C.: On the pressures and flux of energy in a gas according to Maxwell's kinetic theory I. J. Ration. Mech. Anal. **5**, 55–128 (1956)
26. Truesdell, C., Muncaster, R.G.: Fundamentals of Maxwell's Kinetic Theory for a Simple Monatomic Gas. Academic Press, London (1980)

The Einstein Classical Program, the Wheeler-Feynman Reabsorption and Kirchhoff's Law

A. Carati and L. Galgani

Abstract The Einstein "Classical Program" consists in trying to recover Quantum Mechanics (undoubtedly the "good" theory) within a "realistic" theory. Here we address the extreme form of the program in which the realistic theory is just classical electrodynamics of charges, with their Newtonian trajectories. First of all we remove the objection that in a classical frame "electrons fall on nuclei and ions come to rest", because of radiation emission by the accelerated charges. Indeed this is not proved for charges in bulk, which is the case of interest for atomic physics. On the other hand we use a generic cancellation property of the single-particle emissions for matter in bulk, which was proposed by Wheeler and Feynman in 1945, and is actually proven. We also point out that such cancellation explains two fundamental laws of physics, whose microscopic explanation is apparently lacking. Namely, Kirchhoff's law for the energy radiated by a hot body (emission proportional to surface rather than to volume), and reabsorption in plasmas. Finally, some examples of implementation of the extreme Einstein program are mentioned, one of which, the existence of polaritons in ionic crystals, is a phenomenon not yet explained in a quantum frame.

1 Introduction

The Einstein Classical Program, informally presented in the last part (Reply to Authors) of the book "Albert Einstein Philosopher Scientist" [1], amounts to try to prove that Quantum Mechanics (admittedly the correct theory) may be recovered as some kind of theorem within a comprehensive theory having some "realistic" character. An extreme form of the program consists in considering as realistic theory classical electrodynamics of point charges with their trajectories, or rather, as we shall see, classical electrodynamics of point charges *in bulk*. Such program was

A. Carati (✉) · L. Galgani
Department of Mathematics, Università degli Studi di Milano, Milano, Italy
e-mail: andrea.carati@unimi.it; luigi.galgani@unimi.it

P. Barbante et al. (eds.), *From Kinetic Theory to Turbulence Modeling*, Springer INdAM Series 51, https://doi.org/10.1007/978-981-19-6462-6_7

also conceived by Carlo Cercignani, to whose memory the present Conference is dedicated: see papers [2, 3] and [4], as well as the contribution of Fabrizio Gangemi to this conference, in connection with zero-point energy [5].

Now, the common opinion, expressed in all available handbooks, is that such extreme form of the program is a priori impossible, because of a presumed "unsurmountable" obstacle. Namely that, due to radiation emission by accelerated charges, persistent motions could not exist in a classical framework for atomic physics: "electrons would fall on nuclei and ions would come to rest". We dare to challenge such objection, and show that it is wrong. The proof of this statement is the main content of the present paper. It will be given both along the lines of the 1945 Wheeler and Feynman paper [6] (i.e., "in general", with suitable hypotheses), and also by direct verification in particular cases.

In order to go to the heart of the problem, it may be useful to turn the question around, and ask where a proof of this objection can be found. The embarrassing answer is that such proof does not exist. The point is that here we are facing with a misunderstanding. Indeed, it is true that, according to the familiar Larmor formula, an accelerated charge emits power proportional to its acceleration squared, but this is only proven for a single charge, whereas, considering a typical counterexample, a macroscopic current in a metallic ring (a current constituted by an essentially infinite number of charges) is proven not to emit energy at all.[1] In other terms, the fields created by the charges of a system add up, but in general their emissions do not, and what occurs for a many-body system of partially correlated charges is not at all obvious. Rather, in their 1945 paper, addressing such problem in a classical framework, Wheeler and Feynman [6] advanced the opposite conjecture, namely that, in general (we will come back later to this point), in the case of charges in bulk, a full compensation (or cancellation, or reabsorption) is met, so that no emission proportional to volume occurs, i.e., the bulk acts, they say, as a *perfect or complete absorber*.

It is a curious fact that such qualitatively relevant result seems to have been essentially ignored by the scientific community. Perhaps, for what concerns its possible application to the foundations of quantum mechanics, one reason may be found in the impact of the examples in which quantization was originally performed, since they were all concerned with single atoms or molecules (think of the cases of H or H_2). This was emphasized by Fokker in the first lines of article [7], so often quoted by Wheeler and Feynman. In any case, if their result were correct, then the main objection to the extreme Einstein program would be removed, and one would be authorized to pursue it. Furthermore, one would have available a theoretical explanation of two relevant macroscopic phenomena apparently considered still unexplained, both in the classical and in the quantum framework. Namely, Kirchhoff law (the radiant energy emitted by a hot body is proportional to surface rather than to volume), and the reabsorption phenomenon of plasmas (persistence of the gyration

[1] This is proposed as an exercise in Jackson's handbook.

motions of the electrons in a strong magnetic field).[2] For a microscopic approach to plasmas, see the contribution [10] to this conference.

Two are the contributions we aim to bring to the fore with the present paper, on the basis of works performed in our group since the year 2003. The first one is of a didactic character, consisting in an attempt at producing a concise, hopefully clear, formulation of the Wheeler-Feynman "theory", which is presented by them in an extremely verbose, rather uneasy, style. The second contribution concerns their main result, i.e., the complete reabsorption property of the bulk. The way they formulate such property may appear at first sight a little ambiguous, to the point that one might wonder whether it is a proof or a conjecture (this possibly being the reason for the essential neglect of their paper by the scientific community). In this connection, working at a general level as they do, we first of all present a sufficient condition for the reabsorption property, expressed in terms of the "material current" (i.e., the electric current defined by the motions of the charges), which may appear even more perspicuous than their own hypothesis (validity of the Wheeler-Feynman identity). Moreover, we report results for a concrete realistic model of ionic crystals, in which the only hypothesis introduced is the very system of equations defining the Wheeler-Feynman model (the "standard" model, as they call it). Then no explicit hypothesis is anymore required, and the Wheeler-Feynman identity, which formally expresses the complete reabsorption property, shows up as a theorem.

In Sect. 2 we introduce the Wheeler-Feynman dynamical system, i.e., the equations of motion (involving damping forces) postulated by them, along the lines of Dirac, for the dynamics of a system of N charges. Additionally. we mention their main idea of the reabsorption property, namely, that the damping force acting on each charge may be canceled by the electric forces due to all other charges in the bulk, thus making undamped motions possible. In Sect. 3 we show how they formulate the *reabsorption property* in mathematical terms, and also add a significant sufficient condition for it. Moreover, we introduce a complementary sufficient condition, that might be even more perspicuous.

In Sect. 4 we illustrate the results obtained for concrete examples, particularly a realistic model of ionic crystals, for which the reabsorption property is checked directly, with no need of further hypotheses. We also show how this leads, for the ionic crystal model, to the proof of relevant physical properties. First of all the existence of polaritons, a phenomenon, due to retardation, not yet proven in a quantum framework, and furthermore a very good reproduction of the infrared spectra, expressed in terms of the time-correlations of the Newtonian trajectories of the ions.

[2] Concerning plasma reabsorption, the lack of an explanation at the microscopic level of particle trajectories is generally pointed out, in particular by Artsimovich [8], the "father of tokamak". However, at the macroscopic level of dynamics of fluids an explanation is apparently given in a long paper by Bornatici et al. [9].

2 The Wheeler-Feynman Approach to the Electrodynamics of Charges in Bulk: The Idea of a Reabsorption or Cancellation Making Undamped Motions Possible

2.1 *The Dirac-Wheeler-Feynman Equations of Motion*

The Wheeler-Feynman electrodynamics is first of all, as they repeatedly point out, nothing but *standard electrodynamics of point charges*, as formulated by Dirac in his celebrated paper [11] "*Classical theory of radiating electrons*", written in 1938, i.e., ten years after his formulation of Quantum Electrodynamics.

In the non relativistic approximation we consider in this paper, this amounts to Newton equations for a system of N point particles, with mutual retarded electric forces, and with the radiation reaction (or damping) force acting on each charge, described by the familiar term proportional to the time derivative of acceleration, introduced by Planck and discussed by Abraham and Lorentz. So one has the system of equations[3]

$$m_j \ddot{\mathbf{x}}_j = e_j \sum_{l \neq j} \mathbf{E}_l^{ret} + \frac{2e_j^2}{3c^3} \dddot{\mathbf{x}}_j \ , \qquad (1)$$

where m_j, e_j and $\mathbf{x}_j$ are mass, charge and position vector of the j-th charge, $\mathbf{E}_l^{ret}$ the retarded field due to the l-th charge evaluated at the position of the j-th charge, while c the speed of light in a vacuum.[4]

2.1.1 Conceptual Problems for the Dirac Equation: The Idea of Wheeler and Feynman

The Dirac equation, even in the case of a single charge in an external field, is often regarded with some suspicion, since its solutions are known to present peculiar unexpected features. Some strange features were pointed out by Dirac himself, in

[3] We don't discuss here the role of an *incident field*, as Dirac calls it, that could be added at the right hand side. Indeed, it was shown for example in [12] how susceptibility, the relevant quantity for the computation of the spectra (which in principle involves external fields), can actually be obtained as a property of the equilibrium fluctuations of the isolated system.

[4] Notice that, from the analytic point of view, because of retardation these equations present formidable problems. For example, for an infinitely extended system the computation of accelerations requires knowing the motions of all particles from minus infinity up to the initial time. Thus essentially nothing is known in general even about existence of solutions. An analytic approach is however available, which leads to particular significant solutions (traveling waves), for a linearized version of the model describing ionic crystals. As we shall see later, this approach has made possible the description of polaritons, a phenomenon which is not yet amenable to a quantum description.

a very positive way, up to the point of writing the following comment: (page 157, bottom): "*This will lead us to the most beautiful feature of the theory*".

Notice that the presence of a damping force might suggest that in a classical framework persistent motions (for example periodic ones), that are the relevant ones for atomic physics, would be altogether impossible. Here, however, the main contribution of Wheeler and Feynman comes in, someway implementing the premonition of Dirac. Their contribution can be described in the simplest way by making reference to the symmetry properties of the equations with respect to time reversal. The point is that Eq. (1) do not present such invariance, since they contain the third-order damping effective force, which takes into account the radiation any single charge would emit if it were isolated, in addition to containing retarded forces. Rather, what Wheeler and Feynman conceived is that the damping forces acting on each charge are effectively necessary for persistent motions to occur. Indeed, they understood that in general the forces that act on a charge due to the remaining charges of the bulk, in fact build up (in the limit of an essentially infinite number of charges) a term with a modulus equal to that of the damping force, but with opposite sign. Thus the damping force is exactly canceled, whereas an increase of energy would take place if the damping force were missing in the equations of motion. So in the end one remains with a time-reversal invariant system, thus allowing for the occurrence of persistent motions.

We met with such incredible cancellation when studying a toy model [13] for the dynamics of a black body, in which the cancellation showed up unexpectedly in the analytic expression of the solutions. Only later could we understand that this is indeed the content of the Wheeler-Feynman "theory", of which we were somehow aware without having really understood it.[5]

3 The Reabsorption Property, or Cancellation, as the Main Content of the Wheeler-Feynman Electrodynamics for Matter in Bulk

3.1 First Step: The Oseen Identity

The simplest way of illustrating the procedure used by Wheeler and Feynman is the following. One introduces the obvious identity[6]

$$\mathbf{E}_l^{ret} = \frac{1}{2}\Big(\mathbf{E}_l^{ret} + \mathbf{E}_l^{adv}\Big) + \frac{1}{2}\Big(\mathbf{E}_l^{ret} - \mathbf{E}_l^{adv}\Big). \tag{2}$$

[5] It may be worth recalling that such work was performed in connection with the foundations of quantum mechanics. Indeed it was motivated by a criticism to the original papers of Planck on the black-body. See work [14], presented in occasion of the sixtieth birthday of Francesco Guerra.

[6] Notice that an explicit introduction of advanced fields is not necessary. When we first met with the cancellation in our toy model, the advanced fields entered only later, in a reinterpretation of the result. See the contribution [10] at this conference.

so that equation (1) takes the form

$$m_j\ddot{\mathbf{x}}_j - \frac{1}{2}e_j\sum_{l\neq j}\left(\mathbf{E}_l^{ret} + \mathbf{E}_l^{adv}\right) = \frac{2e_j^2}{3c^3}\dddot{\mathbf{x}}_j + \frac{1}{2}e_j\sum_{l\neq j}\left(\mathbf{E}_l^{ret} - \mathbf{E}_l^{adv}\right) \tag{3}$$

in which the distribution of the terms between the two sides was performed so that they are symmetric and respectively anti symmetric with respect to time reversal. Now, Wheeler and Feynman say that the bulk is *a perfect or complete absorber* if the right hand side vanishes identically. Additionally they give arguments supporting the thesis that, *in general*, for large numbers N of charges the bulk should actually be a perfect absorber. In any case, for a perfect absorber the right hand side of Eq. (3) vanishes, i.e., one has the identity

$$\frac{2e_j^2}{3c^3}\dddot{\mathbf{x}}_j + \frac{1}{2}e_j\sum_{l\neq j}\left(\mathbf{E}_l^{ret} - \mathbf{E}_l^{adv}\right)\Big|_{\mathbf{x}_j} = 0\,, \tag{4}$$

which plays the role of some kind of kinematic property. Thus, eventually, the role of equations of motion is played by the remaining equations

$$m_j\ddot{\mathbf{x}}_j = \frac{1}{2}e_j\sum_{l\neq j}\left(\mathbf{E}_l^{ret} + \mathbf{E}_l^{adv}\right)\Big|_{\mathbf{x}_j}\,, \tag{5}$$

which enjoy time-reversal symmetry. This is the way in which the authors recover time-reversal invariance by virtue of suitable properties of the bulk (an apparently strange suggestion, indeed). In their words (page 170, top left): "*We find in the case of an absorbing universe a complete equivalence of the theory of Schwarzschild and Fokker*" (Eq. 5) "*on the one hand, and the usual formulation of electrodynamics*" (Eq. 1) "*on the other*".

The symmetric equations (5) remind of the occurrence of the Feynman (or causal) propagator (semi sum of advanced and retarded propagators) in quantum electrodynamics. In particular, if delay is neglected, one recovers the familiar Hamiltonian model usually considered in quantum mechanics, in which one behaves as if no damping force were conceived at all.[7]

Notice that the structure of the identity (4) shows how the name "complete absorber" introduced by Wheeler and Feynman for such bulk is appropriate. Indeed it says that the power that would be emitted by any charge j (the term proportional to $\dddot{\mathbf{x}}_j$) if it were isolated and not immersed in a bulk is *exactly* canceled by a suitable

[7] A reading of Born's book of 1933 (see [15], page 431) shows that the relevance of the classical radiation reaction force was well clear to him. However, he had to take into account that it was not easy to fit such force within quantum theory. In his very words: *"Diese ganze klassische Theorie der Strahlungdampfung ist natürlich mit der heutigen Quantumtheorie des Licht und der Materie nicht verträglich"*.

joint action (semi difference of retarded and advanced forces) of all other charges: Energy emission proportional to volume is thus excluded, and only one proportional to surface might be allowed (see [16]). This is indeed the **reabsorption property** which removes the main objection to the use of Newtonian trajectories in atomic physics, and at the same time explains Kirchhoff's law and reabsorption in plasmas. One should point out, however, that the reabsorption property only excludes the presence of radiation having a speed equal to c, the speed of light in a vacuum. It will be explained in the next section how a perfect absorber may be transparent to light propagating with a refractive index n different from 1.

One can also understand why such identity at first appeared to us to be some kind of an incredible miracle, because it looks as if *all the remaining* charges with index $l \neq j$ make a conspiracy exactly suited to cancel the emission of the considered charge j, irrespective of where it is located. Additionally, this should occur for any one of the charges.

To the cancellation property expressed by Eq. (4) we gave the name of **Oseen identity**, since it was first proposed by such an esteemed Swedish mathematical-physicist[8] in a paper published in the *Annalen der Physik* in 1916. However, with the authority of a well known encyclopedia of physics [17] such result was estimated to be *irrig*, i.e., wrong, and Oseen decided to leave electrodynamics and work in the field of hydrodynamics, where he obtained relevant acknowledged results.

3.1.1 A Global Formulation: The Wheeler-Feynman Identity Inspired to Dirac's Radiation Field

The elimination of the idea itself of a "conspiracy" is performed by Wheeler a Feynman through a formulation of global type, as follows. Using the relation, found by Dirac,

$$\frac{2e_j^2}{3c^3}\dddot{\mathbf{x}}_j = (\mathbf{E}_j^{ret} - \mathbf{E}_j^{adv})\Big|_{\mathbf{x}_j}$$

the Oseen identity (4) takes the form

$$\sum_{all}(\mathbf{E}_l^{ret} - \mathbf{E}_l^{adv})\Big|_{\mathbf{x}_j} = 0$$

i.e. (with $\mathbf{E}^{ret} = \sum_{all}\mathbf{E}_l^{ret}$, $\mathbf{E}^{adv} = \sum_{all}\mathbf{E}_l^{adv}$)

$$(\mathbf{E}^{ret} - \mathbf{E}^{adv})\Big|_{\mathbf{x}_j} = 0\ ,$$

[8] Well known for his "extinction theorem", relevant for the macroscopic theory of Optics in solids.

namely, the total advanced field coincides with the total retarded one at the positions of the particles. On the other hand, this is certainly guaranteed if one assumes, more in general, that such fields are equal at all points $\mathbf{x}$ of space, i.e., if one assumes as a hypothesis the relation

$$(\mathbf{E}^{ret} - \mathbf{E}^{adv})\big|_{\mathbf{x}} = 0 \quad \text{for all } \mathbf{x}\ , \tag{6}$$

which we call *the Wheeler-Feynman identity.*

One might now recall how Dirac had already pointed out that the vector $(\mathbf{E}_l^{ret} - \mathbf{E}_l^{adv})$, being a solution of Maxwell's equations without sources, has the physical meaning of the "*field of radiation*" produced by charge l at the space-time point $(\mathbf{x}, t)$. Thus the Wheeler-Feynman identity (6) appropriately expresses, in Dirac's terminology, the property that the field of radiation created by the whole bulk actually vanishes, i.e., in the Wheeler-Feynman terminology, that the bulk is a complete absorber.

3.1.2 A Further Little Step: A Sufficient Condition for the Reabsorption Property, as an Uncorrelation Property of the "material" Electric Current

Thus, Wheeler and Feynman proved that the identity (6) is a *sufficient condition* for the bulk to be a perfect absorber, so that the time-reversal invariant form (5) of the equations of motion actually holds. In their paper they give qualitative arguments that appear to support the idea that the perfect-absorber property of the bulk should be somehow "generic". However our impression is that it is not an easy task to understand whether one is dealing here with a theorem or with a conjecture. We have no doubt that such uncertainty is due to our inabilities, but we dare complement here their exposition by adding a sufficient condition that, in our opinion, makes the generic validity of the Wheeler-Feynman identity more plausible. The new condition for the perfect-absorber property of the bulk makes reference to the electric current density $\mathbf{j}$. This seems to be significant, since it is obvious that the currents usually employed in producing radio emissions should not occur in a perfect absorber. On the other hand these currents present peculiar correlation properties. So it is clear that suitable uncorrelation properties of the current (i.e. of the motions of the charges) are expected to constitute significant characteristic features of a perfect absorber. A sufficient uncorrelation property is given below.

Preliminarily we show that the original Wheeler-Feynman identity (6) is equivalent to a condition involving the material current, namely that the latter should propagate with a (phase) velocity different from the velocity c of light in a vacuum. This is seen (and made clear) as follows. Recall that one has

$$\mathbf{E}^{ret} = \int \mathbf{j}(x^\mu - y^\mu)\mathcal{D}^{ret}(y^\mu)\,\mathrm{d}y$$

$$\mathbf{E}^{adv} = \int \mathbf{j}(x^\mu - y^\mu)\mathcal{D}^{adv}(y^\mu)\,\mathrm{d}y$$

where $\mathcal{D}^{ret}$ and $\mathcal{D}^{adv}$ are the retarded and the advanced propagators (or Green functions) respectively,[9] the integration being extended over the whole space-time. Now, $\mathcal{D}_- = \mathcal{D}^{ret} - \mathcal{D}^{adv}$ is a solution of the homogeneous wave-equation, so that the support of its Fourier-transform $\hat{\mathcal{D}}_-(\omega, \mathbf{k})$ is totally contained in the light-cone $\frac{\omega^2}{c^2} - k^2 = 0$,[10] i.e., $\hat{\mathcal{D}}_-(\omega, \mathbf{k})$ vanishes outside the light-cone. Since the Fourier transform of a correlation is the product of the Fourier-transforms, the Fourier-transform of the difference $(\mathbf{E}^{ret} - \mathbf{E}^{adv})$ is given by

$$\hat{\mathbf{E}}^{ret} - \hat{\mathbf{E}}^{adv} = \hat{\mathbf{j}}(\omega, \mathbf{k})\hat{\mathcal{D}}_-(\omega, \mathbf{k}) ,$$

and thus can vanish everywhere (as required by the Wheeler-Feynman condition) if and only if $\hat{\mathbf{j}}(\omega, \mathbf{k})$ vanishes on the light-cone. In conclusion, an equivalent condition for the perfect-absorber property of the bulk is that the Fourier transform $\hat{\mathbf{j}}(\omega, \mathbf{k})$ of the current density vanishes on the light cone, i.e., that the current velocity be different from the velocity c of light in a vacuum.

Finally, we give now a sufficient condition involving a correlation property of the current. This was proved in the Appendix to paper [18]. Referring to such paper for details, the result can be summarized as follows. One considers the four-current density j^ν truncated in a box of volume V, and defines its auto correlation by

$$C_{j^\nu}(s, t, \mathbf{x}) = \lim_{V\to\infty} \frac{1}{V} \int_V j^\nu(s, \mathbf{y}) j^\nu(s + t, \mathbf{y} - \mathbf{x})) d\mathbf{y} .$$

Then one has

$$\lim_{V\to\infty} \frac{1}{V} \int_V \left| \mathbf{A}_{ret} - \mathbf{A}_{adv} \right|^2 d\mathbf{x} = 0 ,$$

if the causality condition

$$C_{j^\nu}(s, t, \mathbf{x}) = 0 \quad \text{for} \quad c^2 t^2 - \mathbf{x} \cdot \mathbf{x} \le 0$$

holds. Here $\mathbf{A}_{ret}$ and $\mathbf{A}_{adv}$ are the retarded and the advanced potentials, solutions of the wave equations with the given current. So the theorem says that for causal currents the retarded and the advanced potentials (and thus also the corresponding fields) are almost everywhere equal in space-time, i.e., they differ at most in a set having vanishing relative measure. Namely, they satisfy a form of the Wheeler-Feynman identity.

[9] Namely, the advanced and the retarded solutions of the wave equation having as source a four-dimensional delta.

[10] This simply follows remarking that, in terms of Fourier-transforms, the homogeneous wave equation $\partial_\mu \partial^\mu \mathcal{D}_- = 0$ reads as $(\frac{\omega^2}{c^2} - k^2)\hat{\mathcal{D}}_- = 0$.

Thus for a current satisfying the causality condition the bulk is a perfect absorber, Kirchhoff's law holds, plasma reabsorption occurs, and the extreme form of the Einstein Classical Program can be pursued.

4 The Wheeler-Feynman Reabsorption Property Checked in Particular Models: The Case of Ionic Crystals

The turning point, with a passage from general reasonings to realistic models of Wheeler-Feynman type, occurred for us through our dear late friend Giuseppe Pastori Parravicini, an authority among the solid-state-physics theoreticians in Italy. Indeed, he had come to know of our result for the toy model [13] of a black-body, in which the Wheeler-Feynman identity had been checked in a direct way, without the need of introducing any assumption (apart form the *Ansatz* of looking for normal modes).[11] Probably stimulated also by the very last sentence of our paper:("*In our opinion, the status of the microscopic foundation of Optics .... should perhaps be reconsidered*"), he suggested to us that we might be able to explain the existence of polaritons. This is a phenomenon concerning ionic crystals and due to retardation of the forces, that the solid state theoreticians were unable to explain in quantum mechanical terms. However, it is illustrated below how the phenomenon was explained in a classical framework [18], thus going even beyond the Einstein program.

4.1 *The Ionic Crystals Model*

For what concerns ionic crystals (the prototype of which is Lithium Fluoride, LiF), here we limit ourselves to recall that since the times of Born they are described as a system of point ions with mutual retarded electric interactions, plus an effective phenomenological inter ionic potential, which implicitly takes into account the contribution of the electrons, whose dynamics is neglected. Obviously, in the spirit of the Dirac-Wheeler-Feynman approach, we also introduce the damping force acting on each ion, i.e., we are assuming the equations of motion (1). For details see [18].

4.1.1 Proof of the Wheeler-Feynman Identity, and the Existence of Dispersion Relations

The Wheeler-Feynman identity is proved by checking the equivalent property previously illustrated, namely that the phase waves of the fields have a velocity

[11] The computations for a minor variant of the model suited for describing fusion machines in plasmas are illustrated in this Conference in the contribution [10].

different from c. This is seen by a computation of the dispersion relations for a linearization of the model.

If one looks for traveling wave solutions $\mathbf{Q}e^{i(\omega t-\mathbf{k}\cdot\mathbf{x}_j)}$ of equations (5), one meets with a secular equation which, due to the non instantaneous character of the forces involved, is not algebraic in ω^2, being instead transcendent. A good approximation of the equation to be satisfied by $\mathbf{Q}$ is given, at least for the infrared frequencies, by the formula

$$-\omega^2 \hat{\mathcal{M}}\mathbf{Q} = \hat{\mathcal{F}}(\mathbf{k})\mathbf{Q} + c_1 \frac{\mathbf{k}\times\mathbf{Q}}{\omega^2 - c^2k^2}$$

where $\hat{\mathcal{M}}$ is the diagonal mass matrix, $\hat{\mathcal{F}}(\mathbf{k})$ is a matrix depending both on the phenomenological inter ionic forces and on the electric field, whereas the last term is explicitly due to the finiteness of the propagation velocity of the electric field. The constant c_1 depends in general on the type of crystal. The relevant point is that the last term[12] has the factor $\omega^2 - c^2k^2$ as denominator. So traveling waves with $\omega = \pm ck$ cannot exist and thus, as just recalled, the Wheeler-Feynman identity holds, i.e., the bulk is proven to be a perfect absorber, in their sense.[13]

Notice that the fact that the crystal is a perfect absorber does not mean that light cannot propagate inside of it. Rather, the crystal even is completely transparent (apart from a very restricted interval of frequencies). The point is that, in the medium, light propagates with a velocity that is different from c, and it can be shown to coincide with the phase velocity of the "material" waves (see again paper [18]). One thus explains, by the way, the occurring of a refractive index different from 1. Instead, what does not propagate in the crystal is the "free" electromagnetic field (i.e., the solution of the homogeneous wave equations), in agreement with the Ewald–Oseen extinction theorem of optics.

4.1.2 Existence of Polaritons

We now illustrate how the existence of polaritons is proved. First of all, one notices that there is a difference between the longitudinal waves (those with $\mathbf{k}$ parallel to $\mathbf{Q}$) and the transversal ones (those with $\mathbf{k}$ orthogonal to $\mathbf{Q}$). In the first case, the secular equation is analogous to that of the purely mechanical (i.e. Coulomb) one; in the

[12] This is the dominant term of the series over $\mathbf{m}$ (regarding the reciprocal lattice) appearing in formula (15) of the quoted paper, i.e., the term with $\mathbf{m} = 0$, for which one has $\mathbf{q_m} = 0$.

[13] By the way, also the very existence of a dispersion relation is made possible by such identity. Indeed, due to the delay, the dispersion relation turns out to be complex. The imaginary part, however, vanishes identically just by virtue of the Wheeler-Feynman cancellation. Thus one remains with a unique real equation involving two "unknowns", ω and $\mathbf{k}$, so that ω is defined in terms of the "parameter" $\mathbf{k}$, i.e., the curve $\omega(\mathbf{k})$ turns out to be defined. This is indeed the way we first happened to meet with the Wheeler-Feynman identity, when studying our toy model of a black body [13].

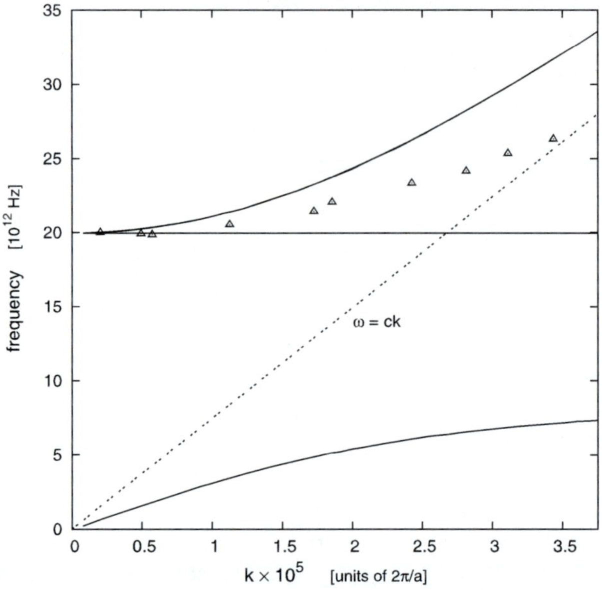

Fig. 1 Dispersion relation for Lithium Fluoride (solid line). The three acoustic branches are not shown. The horizontal line is the longitudinal branch, while the two transverse ones are doubly degenerate, and actually constitute the polaritonic branches. In total one has eight branches instead of the six ones found in the mechanical case. Triangles are the experimental data

second case, the degree of the secular equation is the double, because of the presence of the denominator in the last term at the right hand side. Thus, in the latter case one meets with a double number of solutions with respect to the mechanical case. Namely, in the dispersion relations there arise further branches, named polaritonic, whose experimental existence was confirmed in the case of semiconductors at the end of the years sixties of the last century, and in the subsequent decade in the case of ionic crystals. Figure 1 reports an example of the dispersion relations obtained through the Dirac-Wheeler-Feynman equations (5) for a Lithium Fluoride model.

5 Conclusion

So we have shown how the Dirac-Wheeler-Feynman classical electrodynamics of charges in bulk allows one to overcome the main objection raised since always against the use of classical Newtonian trajectories in atomic physics. This is due to the Wheeler-Feynman reabsorption property, which is proven in realistic systems and is made plausible in general. Such property apparently also explains relevant macroscopic phenomena such as Kirchhoff's law and reabsorption in plasmas.

With the main objection eliminated, one can proceed along the Einstein classical program in its extreme form involving Newtonian trajectories of charges, in order to recover phenomena that are usually considered inconceivable in a classical framework. In such spirit we reproduced in a remarkably good way the infrared spectra of ionic crystals (through the motions of the ions) [19, 20], and proved (at the moment, only in an essentially qualitative way) the existence of the chemical bond in the simplest case in which it occurs, that of the ion of the H_2 molecule which involves only one electron [21]. In addition, a proof was given of the existence of polaritons in ionic crystals, a phenomenon still unexplained in a quantum framework.

Notice that, while in the case of ionic crystals the considered Born model has a kind of quantum flavor, since it contains a phenomenological inter ionic effective potential that implicitly takes the role of the electrons into account, in the case of the H_2^+ ion no quantum-like element at all is introduced, and one deals with a fully classical purely Coulomb model of point charges. Planck's constant enters through the initial data regarding the angular momentum L_z of the electron about the inter ionic axis. For small L_z the motion is chaotic and the electron falls on one of the protons. Instead, for $L_z \simeq h$ the motion becomes sufficiently ordered for the occurring of a further integral of motion, which plays the role of a suitable effective potential acting on the electron (analogous to the quantum mechanical Born-Oppenheimer one), and a stable ion is formed.

At the moment we are unable to deal with atomic-physics models involving more than one electron, and with magnetic properties. Whether this limitation constitutes an unsurmountable barrier or is just a provisional one, is an open problem that we plan to tackle in the future. The available results (in particular the proof of the existence of polaritons) allow one to conclude at least that the problem of the relations between classical and quantum physics is much subtler than usually believed, and that such relations seem to involve features that are beyond the reach and the aims of semi classical physics.

Acknowledgments We thank Gianenrico Picone for a very careful critical reading of the paper.

References

1. Schilpp, P.A. (ed.): Albert Einstein Philosopher-Scientist. MJF Books, Evanston Ill (1949)
2. Cercignani, C., Galgani, L., Scotti, A.: Zero-point energy in classical non-linear mechanics. Phys. Lett. **38**(A), 403 (1972)
3. Cercignani, C.: On a nonquantum derivation of Planck's distribution law. Found. Phys. Lett. **11**, 180–198 (1998)
4. Galgani,L.: Carlo Cercignani's interests for the foundations of physics. Meccanica **47**, 1723–1735 (2012)
5. Gangemi, F., Gangemi, R., Carati,A., Galgani, L.: The Cercignani conjecture about a classical zero-point energy, and its confirmation for ionic crystals. This conference

6. Wheeler, J.A., Feynman, R.P.: Interaction with the absorber as the mechanism of radiation. Rev. Mod. Phys. **17**, 157–181 (1945)
7. Fokker, A.D.: Ein invarianter Variationssatz für die Bewegung meherer elektrischer Massenteilchen. Z. Phys. **58**, 386–393 (1929)
8. Artsimovich, L.A., Kolb, A.C., Pease, S.: Controlled Thermonuclear Reactions. Gordon and Breach, New York (1964)
9. Bornatici, M., Cano, R., De Barbieri, O., Engelmann, F.: Electron cyclotron emission and absorption in fusion plasmas. Nuclear Fusion **21** (9), 1153–1257 (1983)
10. Carati, A., Zuin, M., Martines, E., Galgani, L.: Loss of magnetic confinement above a critical density in a microscopic model of magnetized plasma. This conference
11. Dirac, P.A.M.: Classical theory of radiating electrons. Proc. R. Soc. A **167**, 148–169 (1938)
12. Carati, A., Galgani, L.: Classical microscopic theory of dispersion, emission and absorption of light in dielectrics. Eur. Phys. J. D **68**, 307 (2014)
13. Carati, A., Galgani, L.: Nonradiating normal modes in a classical many-body model of matter-radiation interaction. Il Nuovo Cimento B **8**, 839–851 (2003)
14. Carati, A., Galgani, L.: A critical remark on Planck's model of black-body. Int. J. Mod. Phys. B **18**, 549–553 (2004)
15. Born, M.: Optik. Springer, Berlin (1933)
16. Marino, M., Carati, A., Galgani, L.: Classical light dispersion theory in a regular lattice. Ann. Phys. **322**, 799–823 (2007)
17. Jaffé, G.: Dispersion und Absorption. In: Handbuch der Experimentalphysik, vol. 19. Akademische Verlagsgesellschaft, Leipzig (1928)
18. Lerose, A., Sanzeni, A., Carati, A., Galgani, L.: Classical microscopic theory of polaritons in ionic crystals. Eur. Phys. J. D **68**, 35 (2014)
19. Carati, A., Galgani, L., Maiocchi, A., Gangemi, F., Gangemi, R.: Classical infrared spectra of ionic crystals and their relevance for statistical mechanics. Physica A **506**, 1–10 (2018)
20. Carati, A., Galgani, L., Gangemi, F., Gangemi, R.: Relaxation times and ergodic properties in a realistic ionic-crystal model, and the modern form of the FPU problem. Physica A **532**, 121911 (2019)
21. Carati, A., Galgani, L., Gangemi, F., Gangemi, R.: Electronic trajectories in atomic physics: The chemical bond in the H_2^+ ion. Chaos **30**, 063109 (2020)
22. Oseen, C.W.: Über die Extinction des Lichtes. Physik. Zeitschr. **17**, 341–343 (1916)

Reabsorption and Density Limit in Magnetized Plasmas Through a First-Principles Toy Model

A. Carati, M. Zuin, E. Martines, and L. Galgani

Abstract In the physics of magnetized plasmas there are problems of principle that apparently still miss a microscopic explanation. A typical case is the reabsorption of cyclotron emission by gyrational electron motions in a magnetic field. On the other hand, it seems that a microscopic explanation of reabsorption is contained in the 1945 paper of Wheeler and Feynman that, however, strangely enough, to our knowledge is never mentioned in this connection. Here we introduce a simple toy model that, on the one hand, contains all the features of the Wheeler–Feynman electrodynamics (which in fact is just the standard electrodynamics, as they repeatedly point out) and, on the other hand, presents such a simple kinematics as to allow for an analytic treatment. In such way reabsorption is proven. As a byproduct, a density limit for magnetic confinement due to a microscopic instability is also deduced.

1 Introduction

Let us briefly recall what is the phenomenon of reabsorption for plasmas in a magnetic field. This is related to the fact that an accelerated charge emits energy with a power proportional to acceleration squared. For example, in the book of Landau it is proposed, as an exercise, to show that in about 10^{-8} sec an electron gyrating around a nucleus would fall on it. Analogously, in a plasma immersed in a magnetic field, the electrons, gyrating around its field lines, should continuously emit energy, coming finally to rest. Consequently magnetic confinement would be impossible, and furthermore a plasma should emit an enormous quantity of energy

A. Carati (✉) · L. Galgani
Department of Mathematics, Università degli Studi di Milano, Milano, Italy
e-mail: andrea.carati@unimi.it; luigi.galgani@unimi.it

M. Zuin · E. Martines
Consorzio RFX (CNR, ENEA, INFN, UNIPD, Acciaierie Venete SpA), Corso Stati Uniti 4, Padova, Italy
e-mail: matteo.zuin@igi.cnr.it; emilio.martinez@igi.cnr.it

P. Barbante et al. (eds.), *From Kinetic Theory to Turbulence Modeling*, Springer INdAM Series 51, https://doi.org/10.1007/978-981-19-6462-6_8

[1] at the cyclotron (or Larmor) frequency.[1] On the other hand, such emission process is proven in the case involving a single charge, whereas the case involving a system of charges is not currently discussed, in this connection, at a microscopic level. However, experimentally, the power radiated from the plasma is by far lower, as the plasma is found to be optically thick at the fundamental harmonic at $\omega_c = eB/m_e$, where ω_c the Larmor gyrational frequency. Evidently this is due to the electromagnetic interaction of any charge with all the other ones, i.e., to the fact that one is dealing with charges in bulk. So, things proceed in such a way that the energy emitted by any single charge is reabsorbed (this is the term used) by the other ones, but the relevant mechanism is apparently unknown. A curious fact is that in our opinion such "cancellation mechanism" is implicitly contained in the Wheeler–Feynman paper of 1945 (see [2], and paper [3], presented at this conference). However, apparently such explanation is ignored.

The first aim of the present paper is to prove that such cancellation occurs in a model that complies with all requirements of standard electrodynamics for a system of charges, but presents a so simplified kinematics as to allow for an analytical discussion.[2]

As a byproduct we will also point out the existence of a microscopic instability due to the repulsive character of the mutual Coulomb forces among the electrons that occurs above a critical value of the electron density. The effects of such repulsion are commonly observed by plasma physicists involved in molecular dynamics simulations. Indeed, in these simulations the initial positions of the electrons are usually extracted randomly in the simulation box, with, however, "*a small region surrounding each particle excluded, to avoid initial explosion*" (see [4]). In the presence of a magnetic field one has instead a competition between the repulsive role of the Coulomb forces, and the "confining" gyrational effect of the field. Such competition was investigated both by the methods of modern perturbation theory [5], and by numerical methods of molecular dynamics [6], and a density threshold for the magnetic confinement was determined, that turns out to scale as the square of the field. Now, the existence of such threshold shows up as a kind of corollary also in the model considered here, and with the same B^2 law. However, while in the mentioned papers the electron dynamics involved only instantaneous Coulomb forces, here the retardation of the mutual forces is taken into account, together with the presence of damping due to radiation emission by the single accelerating charges.

[1] In the case of typical plasmas of fusion interest (with a magnetic field of the order of a few Teslas, electron density of the order of $10^{20} m^{-3}$ electron temperature of a few keV), the radiated power would be $P_e = e^4/(3\pi\epsilon_0 m_e^3 c^3) B^2 n_e T_e$, where ϵ_0 is the vacuum permittivity, c the speed of light, m_e the electron mass, n_e and T_e the electron density and temperature, respectively, and B the magnetic field intensity. The radiated power would thus be larger than $\geq$ 1 MW/m^3) (see [1], section 4.21).

[2] In a context formally analogous to the present one but concerning crystals, the analogue of such cancellation was already observed in paper [7], to which the present work is inspired.

In the next Section the model is introduced and its analytic features are studied. The conclusions follow.

2 The Model, and Its Analytic Solution Through Normal Modes

From the dynamical point of view, we work in the framework of what Wheeler and Feynman call *standard electrodynamics for a system of charges*. This means charges subject to forces that are not only the mutual *retarded* electromagnetic ones (and, in our case, even the Lorentz forces due an external magnetic field) but also to the electromagnetic damping forces acting on each single charge. The latter forces take into account the radiation any single charge emits due its actual motion, irrespective of the motions of the other ones. So, in the nonrelativistic case we are going to consider, the equations of motion are the Newton ones and the damping force is the familiar Abraham–Lorentz force, proportional to the time derivative of acceleration.

The great contribution of Wheeler and Feynman has been to show how, for an infinite system of charges, under rather general conditions a compensation occurs, so that the damping forces are cancelled, i.e., one has reabsorption. Such result is obtained by them in a general framework, in the form of a kind of existence theory. Instead, in the model we are going to consider, which is chosen of such an extremely simple kinematics as to allow for an analytic discussion, such cancellation will be seen to show up in the form of an unexpected analytic identity.

2.1 The Model

First of all, the external magnetic field is taken time-independent and uniform, of a given intensity B, say directed along the z axis of a cartesian coordinate system. The model is aimed at describing only the gyrational motions taking place near a selected field line, say the z axis itself. For the sake of simplicity the electrons are constrained to lie on planes orthogonal to the axis, and moreover a discretization is introduced by considering a lattice of equally spaced planes (say, a distance a apart). Each such plane contains one electron, so that the n-th one has coordinates (x_n, y_n, na), with $n \in Z$. For what concerns the retarded forces, they are taken in the well-known dipole approximation (in particular, the magnetic contribution is neglected). Finally the definition of the model is completed by performing a linearization with respect to the distance of the electrons from the axis, in order that the equation be physically significant only for motions taking place near the chosen axis. In particular the distance $r_{n,m}$ between electrons m and n is approximated as $r_{n,m} = a|n - m|$.

The system of Newton equations of motion for the model is thus

$$
\begin{aligned}
&\ddot{x}_n - \ \omega_c \dot{y}_n \ -\frac{2}{3}\frac{e^2}{4\pi\epsilon_0 m_e c^3}\dddot{x}_n = \\
&-\frac{e^2}{4\pi\epsilon_0 m_e}\sum_{m\neq n}\Big[\frac{x_m(t-r_{nm}/c)}{r_{nm}^3}+\frac{1}{c}\frac{\dot{x}_m(t-r_{nm}/c)}{r_{nm}^2}+\frac{1}{c^2}\frac{\ddot{x}_m(t-r_{nm}/c)}{r_{nm}}\Big] \\
&\ddot{y}_n - \ \omega_c \dot{x}_n \ -\frac{2}{3}\frac{e^2}{4\pi\epsilon_0 m_e c^3}\dddot{y}_n = \\
&+\frac{e^2}{4\pi\epsilon_0 m_e}\sum_{m\neq n}\Big[\frac{y_m(t-r_{nm}/c)}{r_{nm}^3}+\frac{1}{c}\frac{\dot{y}_m(t-r_{nm}/c)}{r_{nm}^2}+\frac{1}{c^2}\frac{\ddot{y}_m(t-r_{nm}/c)}{r_{nm}}\Big]
\end{aligned} \tag{1}
$$

2.1.1 Analytic Treatment Through Normal Modes

Equations (1) constitute an infinite system of linear equations with delay, which is just a simple variant of the system considered in [7]. Our aim is now to investigate the stability properties of the system, as the control parameters of density $n_e = 1/a^3$ and field B (or equivalently ω_c) are varied. Following a completely standard procedure, we compute the normal modes of the system, with the aim of determining the values of the parameters for which the frequencies become complex. It will be seen, however, that, due to the presence of retardation, such apparently innocuous program can be fulfilled only if a suitable identity, that is the counterpart of the Wheeler–Feynman one, turns out to be satisfied.

So we look for traveling wave solutions with wavenumber k and angular frequency ω, i.e., of the form

$$
x_n = A_x e^{i(kan+\omega t)}\ , \ y_n = A_y e^{i(kan+\omega t)}\ . \tag{2}
$$

This leads to a linear system in the unknowns A_x, A_y, from which the dispersion relation between ω and k is found by equating the determinant to zero. Due to retardation, this gives a complex equation, i.e., two real equations in the two unknowns ω and k, namely

$$
(\frac{\omega}{\omega_c})^2 \pm \frac{\omega}{\omega_c} + pF(ka, a\omega/c) = 0\ , \quad p = \frac{\omega_p^2}{\omega_c^2} \tag{3}
$$

$$
\frac{2}{3}\frac{e^2}{4\pi\epsilon_0 m_e c^3}\omega^3 - G(ka, a\omega/c) = 0\ . \tag{4}
$$

Here, ω_p is the familiar plasma frequency defined by

$$\omega_p^2 = \frac{e^2}{\epsilon_0 m_e a^3} = \frac{n_e e^2}{\epsilon_0 m_e}, \tag{5}$$

while F and G, as functions of the variables $\alpha = ka$ and $\beta = a\omega/c$, are defined by

$$F(\alpha, \beta) = \frac{1}{4\pi}[\beta^2 \log(2|\cos(\beta) - \cos(\alpha)|) - f(\alpha, \beta)] \tag{6}$$

$$G(\alpha, \beta) = \beta^2 - g(\alpha, \beta) \tag{7}$$

the functions f and g being the ones already introduced in [7], namely

$$f(\alpha, \beta) = \sum_{n \neq 0} \Big(\frac{(\cos(n\alpha - |n|\beta)}{|n^3|} - \beta \frac{\sin(n\alpha - |n|\beta)}{|n^2|} \Big) \tag{8}$$

$$g(\alpha, \beta) = \sum_{n \neq 0} \Big(\frac{(\sin(n\alpha - |n|\beta)}{|n^3|} + \beta \frac{\cos(n\alpha - |n|\beta)}{|n^2|} \Big). \tag{9}$$

Some details concerning the summation of the series leading to the term $\beta^2 \log(2|\cos(\beta) - \cos(\alpha)|)$ entering the function F are here omitted.

2.1.2 The Wheeler–Feynman Identity as Guaranteeing the Existence of a Dispersion Relation

Now, one meets here with a very relevant question of principle. Indeed, for fixed values of the parameters a and ω_c one has two equations in two unknowns (ω and k), and in general this would not allow for the existence of a dispersion relation, i.e., of a function $\omega = \omega(k)$, for a continuous range of values of k. However, the existence of a dispersion relation is guaranteed by the fact that relation (4) is an identity, as was shown in paper [7] (see section 6) for the analogous one-dimensional model of a crystal and later in [8] for a three-dimensional case. In fact, these are just particular cases of the Wheeler–Feynman identity that was proved by them in a general framework. In order to have a feeling of how much such identity is unexpected and a priori unplausible, it suffices to look at the way in which the parameter a enters relation (4). Indeed it shows up in the second term and not in the first one, so that one would not imagine that they may be equal (apart from the sign).

2.1.3 Evidence for the Density-Induced Instability, and Estimate of a Density Limit

Once the existence of a dispersion relation is guaranteed by the fact that the Wheeler–Feynman identity holds, the dispersion relation can be determined by

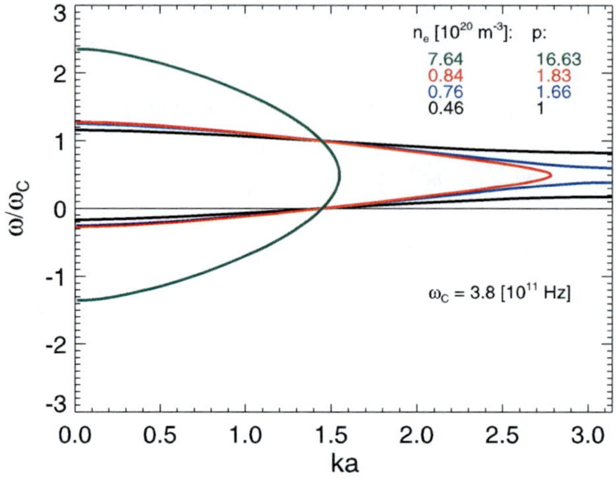

Fig. 1 The dispersion curves, solutions of equation (3) (with the plus sign), in the plane $(ka, \omega/\omega_c)$ for $\omega_c = 3.8 \times 10^{11}$ Hz, and for several values of the parameter $p = \frac{\omega_p^2}{\omega_c^2}$ (or equivalently of the electron density n_e)

standard numerical methods. In the present case one has to solve (3) in the unknown $\omega = \omega(k)$ for any given k, in which $p = \omega_p^2/\omega_c^2$ plays the role of a parameter.

In Fig. 1 the dispersion relations are shown for a cyclotron frequency $\omega_c = 3.8 \times 10^{11}$ Hz, which corresponds to a magnetic field of about 2.1 T (a typical value used for the magnetic confinement of thermonuclear plasmas in the experiments), for several values of the parameter $p = \frac{\omega_p^2}{\omega_c^2}$ (or of the corresponding electron density n_e, at given ω_c).

The most important qualitative result is that normal modes are found to exist (for all k) only below a critical value of p, i.e., below a certain threshold of plasma density. Indeed, starting from low densities, at a certain critical density a bifurcation is seen to occur, characterized by the fact that the curves no more intersect the vertical axis $ka = \pi$. This means that for values of k just below π/a Eq. (3) does not admit a real solution, so that the corresponding frequencies acquire an imaginary part, and the whole system becomes unstable. Numerical computations not reported here show that the characteristic time of the instability is of the order of $2\pi/\omega_c$.

Notice that this phenomenon of the existence of a maximal allowed density is obviously lost if one introduces the continuum approximation, and thus it is a characteristic feature of the discrete structure of matter. Indeed, following [8], the continuum approximation corresponds to deal with wavelengths much larger than the step a, i.e., to assume $k \ll \pi/a$, whereas the existence of a density limit depends on the behavior of the system for $ka \approx \pi$. We have now to determine the bifurcation value of the parameter p. As the bifurcation occurs for $ka = \pi$ and for values of $\omega/\omega_c \leq 1$, i.e., for $a\omega/c \simeq 0$, one can just limit oneself to study equation (3) for a fixed value of the function F, namely $F(\pi, 0)$, so that one is simply reduced to deal with an algebraic equation of second degree. One computes $F(\pi, 0) \simeq 0.14$, and so real values of ω are found to exist only for $p \leq 1.74$. This, recalling the definition of p and using $\epsilon_0\mu_0 = 1/c^2$, gives for the critical density n_e^M the law

$$n_e^M = 1.74\frac{1}{m_e c^2}\frac{B^2}{\mu_0}. \tag{10}$$

This, by the way, is the same law predicted by the modern methods of perturbation theory [5] (apart from the factor 1.5 in place of 1.74) and is also in agreement with the estimate obtained by molecular dynamics numerical simulations [6].

For typical magnetic field values used in the laboratory, the density limit predicted by the model is well in the range of the experimental densities [1]. The instability described here might thus play a significant role in determining the confinement properties of magnetized plasmas. A comparison with the experimental data can be found in paper [5].

It is worth noticing that law (10) has the same form of the Brillouin limit [9], which is known to apply to the case of non-neutral plasmas [10]. The main difference with respect to our procedure is that in the case of the Brillouin limit the mutual electric field acting on electrons is introduced within a mean field approach, whereas here it is computed in the frame of a many-body microscopic theory. Correspondingly, we find that the instability involves normal modes with wavelengths of the order of the mean electron distance, so that it escapes a mean field approach. In particular, such instability is found to occur in neutral plasmas, for which the mean charge density vanishes, and the Brillouin approach cannot be used.

3 Conclusion

In conclusion, we have proved that, in a microscopic toy model of a magnetized plasma, cyclotron emission radiation due to electron gyration is reabsorbed. This occurs in virtue of the Wheeler–Feynman identity, proved in the model, that was proposed by those authors as a general property of matter in bulk.

Just in virtue of such identity, dispersion relations exist, and so it was also possible to draw them for our model, studying their dependence on the parameters, in particular, on density. A relevant result is that a density threshold exists, above which microscopic instabilities occur, and thus magnetization is lost. Such density limit turns out to scale as B^2, in agreement with estimates obtained by modern perturbation methods and by numerical molecular dynamics methods.

A further comment can be added concerning the role the Wheeler–Feynman identity plays in clarifying a general feature of the damping force, namely the analytic form it should have (for example, the Abraham–Lorentz one proportional to the time derivative of acceleration, or some approximation of it). In fact, the identity occurs in the form of a compensation, in which the damping force acting on each charge is exactly cancelled by a part of the electromagnetic forces due to the remaining ones. It is thus clear that the other charges themselves do produce the correct form of the damping force. So, for example, in a relativistic treatment one can show that necessarily the damping force should have exactly the form proposed by Dirac in 1938.

Acknowledgments The authors wish to thank Dr. Nicola Vianello for fruitful discussions.

References

1. Wesson, J.: Tokamaks. Oxford Science Publications, Oxford (2011)
2. Wheeler, J.A., Feynman, R.P.: Interaction with the absorber as the mechanism of radiation. Rev. Mod. Phys. **17**, 157–181 (1945)
3. Carati, A., Galgani, L.: The Einstein classical program, the Wheeler–Feynman identity and Kirchhoff's law. This conference
4. Baalrud, S.D., Daligault, J.: Transport regimes spanning magnetization-coupling phase space. Phys. Rev. E **96**, 043202 (2012)
5. Carati, A., Zuin, M., Marino, M., Maiocchi, A., Martines, E., Galgani, L.: Transition from order to chaos, and density limit, in magnetized plasmas. Chaos **22**, 033124 (2012)
6. Carati, A., Benfenati, F., Maiocchi, A., Zuin, M., Galgani, L.: Chaoticity threshold in magnetized plasmas. Numerical results in the weak coupling regime. Chaos **24**, 013118 (2014)
7. Carati, A., Galgani, L.: Nonradiating normal modes in a classical many-body model of matter-radiation interaction. Nuovo Cim. **118 B**, 839–849 (2003)
8. Marino, M., Carati, A., Galgani, L.: Classical light dispersion theory in a regular lattice. Ann. Phys. **322**, 799–823 (2007)
9. Brillouin, L.: A theorem of Larmor and its importance for electrons in magnetic fields. Phys. Rev. **67**, 260–266 (1945)
10. Davidson, R.C.: Physics of Nonneutral Plasmas. Addison Wesley, Redwood City (1990)

Kinetic Effects in Non-ideal, Two-Phase Shear Flows

Aldo Frezzotti and Henning Struchtrup

Abstract A steady two-phase Couette flow in a liquid film in contact with its vapor phase, through a resolved interface, is studied numerically by Direct Simulation Monte Carlo, based on Enskog–Vlasov kinetic model. Simulations with increasing degree of vapor non-ideality are performed to assess the attenuation of kinetic effects as the vapor phase becomes denser and denser. Deviations of flow properties from hydrodynamic behavior are obtained by direct comparison with a simple hydrodynamic model and by the numerical computation of higher order moments, not present in the hydrodynamic description.

1 Introduction

The paper aims at presenting the results of a mainly numerical study of a simple shear flow in a two-phase fluid, from the kinetic theory point of view. Although focused on a specific and simplified flow geometry, the work finds its motivations in the more general and widely investigated kinetic theory [7] approach to the modeling of flows in which the liquid and vapor phases of the fluid coexist [8, 12, 20]. The vast majority of kinetic theory studies and applications in this research area are based on a mathematically heterogeneous model. The liquid phase bulk and the vapor bulk, far from interfaces, are described by Navier–Stokes–Fourier (NSF) equations. The vapor phase, considered as a dilute gas, is described by the Boltzmann equation [7] in a more or less extended non-hydrodynamic vapor region, in contact with interfaces. Due to the large separation between the space and time scales in the two phases, their coupling is modeled by a phenomenological boundary condition [12, 16, 24]. The latter specifies the distribution function of molecules

A. Frezzotti (✉)
Politecnico di Milano, Dipartimento di Scienze & Tecnologie Aerospaziali, Milano, Italy
e-mail: aldo.frezzotti@polimi.it

H. Struchtrup
University of Victoria, Victoria, BC, Canada
e-mail: struchtr@uvic.ca

P. Barbante et al. (eds.), *From Kinetic Theory to Turbulence Modeling*, Springer INdAM Series 51, https://doi.org/10.1007/978-981-19-6462-6_9

entering the vapor phase because of spontaneous evaporation from the liquid or scattering at the liquid surface. A possible and attractive alternative to the model briefly described above is represented, in principle, by Diffuse Interface Models (DIMs) [2]. DIMs provide a unified description the liquid and vapor phases, on the base of the NSF equations. The vapor-liquid interface structure is explicitly resolved by adding Korteweg's capillary contributions to the Navier–Stokes stress tensor and to the Fourier heat flux [2]. Hence, no condition has to be formulated at the liquid-vapor boundary since the interface is part of the flow field. Unfortunately, when the vapor is dilute DIMs fails to provide an accurate description of flows where a kinetic layer is present [4], as is to be expected. However, their accuracy becomes better and better as the vapor phase gets increasingly denser and non-ideal [13]. Extending DIMs, to give them the capability of describing kinetic layers and, at the same time, non-ideal flows is limited by the difficulty of obtaining a general kinetic description of dense fluids [19]. However, a kinetic extension of the van der Waals fluid has been proposed by Louis de Sobrino in 1967 [9]. In this model, the classical Enskog equation for the dense hard spheres fluid [10] is modified by adding an attractive tail to the repulsive hard sphere potential. To keep the treatment of the interacting particle system within the framework of a closed kinetic equation for the one-particle distribution function [18, 19], the tail contribution is added to Enskog's non-local collision integral in the form of a self-consistent Vlasov force term. The resulting equation, named Enskog–Vlasov (EV) equation, has been used in Ref. [22] to obtain a set of moment equations, based on Grad 13 moments approximation [21]. More recently, an extension to Grad 26 moments [21] has been proposed [23]. However, the formulation of the moment equations is complex and depends on the truncation of the local expansions of non-local terms to a prescribed order [14, 22, 23]. In order to support the development of a proper set of moment equations for EV equation, the latter is solved numerically in a simple test problem, consisting in a two-phase Couette flow. The behavior and generation of the non-hydrodynamic moments in the interface region is investigated, along the liquid-vapor coexistence curve from near ideal to fully non-ideal vapor conditions. The approach to hydrodynamic behavior is assessed by the comparison of the numerical solutions of EV equation with a reduced set of moment equations which represents, in a sense, the DIM approximation of EV equation. The rest of the paper content is subdivided in: Sect. 2, which summarizes the structure and properties of EV equation and the associated moment equations, Sect. 3 devoted to a description of the test problem; the final Sect. 4 describes and discusses the obtained results. Concluding remarks are given in Sect. 5

2 The Enskog–Vlasov Model

Enskog–Vlasov (EV) equation [9] extends the classical Enskog theory of the dense hard spheres fluid [10] by replacing the purely repulsive hard sphere potential with a Sutherland potential which adds a soft attractive tail to the hard sphere contribution.

In its unsteady, spatially one-dimensional form used in this work EV equation takes the following form:

$$\frac{\partial f}{\partial t} + v_x \frac{\partial f}{\partial x} + \frac{F_x(x,t)}{m} \frac{\partial f}{\partial v_x} = C_{hs}(f,f). \tag{1}$$

In Eq. (1), $f(x, \mathbf{v}, t)$ is the one-particle distribution function of molecular velocities $\mathbf{v}$, at the spatial position x and time t.

The collision integral $C_{hs}(f, f)$, on the R.H.S of Eq. (1), describes short range, repulsive molecular interactions as instantaneous, elastic collisions between two spherical molecules of diameter a and mass m. At the time of their impact, molecules spatial positions are $\mathbf{r}$ and $\mathbf{r} - a\hat{\mathbf{k}}$, respectively. Its form:

$$C_{hs}(f,f) = a^2 \int_S \{\chi_{hs}(x, x + ak_x) f(x + ak_x, \mathbf{v}_1^*, t) f(x, \mathbf{v}^*, t) -$$
$$\chi_{hs}(x, x - ak_x) f(x - ak_x, \mathbf{v}_1, t) f(x, \mathbf{v}, t)\} (\mathbf{v}_r \cdot \hat{\mathbf{k}})^+ d\mathbf{v}_1 d^2\hat{\mathbf{k}} \tag{2}$$

is taken from Standard Enskog Theory (SET) [25], in which the uniform equilibrium pair correlation function at contact χ_{hs} is used. The latter is a function of local number density n which is often approximated as:

$$\chi_{hs}(\mathbf{r}, \mathbf{r} \pm a\hat{\mathbf{k}}) = Y\left[n\left(\mathbf{r} \pm a\frac{\hat{\mathbf{k}}}{2}\right)\right], \quad Y(n) = \frac{1}{2}\frac{2-\eta}{(1-\eta)^3}, \quad \eta = \frac{\pi a^3 n}{6} \tag{3}$$

using the relationship between Y and the pressure equation of state of the hard sphere fluid [19], well approximated by the Carnahan and Starling expression [6].

The force term in Eq. (1) is derived by the smooth tail ϕ_t of the molecular interaction potential ϕ:

$$\phi(r) = \begin{cases} +\infty & r < a \\ \phi_t(r) & r \geq a \end{cases} \tag{4}$$

which describes the interaction of two molecules whose centers distance is r. Elementary calculations show that, in a one-dimensional, slab geometry the x component $F_x(x, t)$ of the attractive force has the following expression [14]:

$$F_x(x,t) = -2\pi \left[\phi_t(a) \int_{|y-x| \leq a} (y-x) n(y,t)\, dy + \right.$$
$$\left. \int_{|y-x|>a} (y-x) \phi_t(|y-x|) n(y,t)\, dy \right]. \tag{5}$$

The choice of the tail potential is arbitrary but it affects the thermodynamics of the fluid described by Eq. (1) [14, 15]. It is easily shown that the pressure equation of state of the EV fluid has the form of a generalized van der Waals equation [3]. The critical temperature, T_c, depends on the assumed ϕ_t, whereas the critical number density n_c is determined only by the hard sphere interaction [15].

In this work an algebraic tail in the form:

$$\phi_t(\rho) = -\phi_a \left(\frac{a}{\rho}\right)^{\gamma}, \quad \phi_a, \gamma > 0, \tag{6}$$

has been assumed. The exponent γ has been set equal to 6, to mimic the attractive contribution of the Lennard-Jones $12-6$ potential [17]. The critical parameters for this particular form of the tail potential are [14]:

$$\eta_c = \pi a^3 n_c/6 = 0.13044 \quad T_c^* = \frac{k_B T_c}{\phi_a} = \frac{1}{\alpha_c}\frac{4\gamma}{\gamma-3}, \quad \alpha_c = 10.601. \tag{7}$$

For temperature T_0 below T_c, Eq. (1) admits equilibrium solutions in the form:

$$f_0(x, \mathbf{v}) = n_0(x)\omega_0(\mathbf{v}), \quad \omega_0(\mathbf{v}) = \frac{1}{(2\pi R T_0)^{3/2}} \exp\left(-\frac{\mathbf{v}^2}{2RT_0}\right) \tag{8}$$

being $R = k_B/m$ and k_B the gas and the Boltzmann constants. The non-uniform density of two-phase fluid obeys the equation [14]:

$$k_B T_0 \frac{dn_0}{dx} = n_0(x) F_x(x) + 2\pi a^2 n_0(x) k_B T_0 \int_{-1}^{+1} k_x Y\left[n_0(x - \frac{a}{2}k_x)\right] n_0(x - a k_x)\, dk_x. \tag{9}$$

Table 1 reports about liquid and vapor reduced densities on liquid-vapor equilibrium coexistence curve, for reduced temperatures in the range 0.45–0.70, below T_c.

Table 1 Reduced liquid number density $n_l^* = n_l a^3$, reduced vapor number density $n_v^* = n_v a^3$, reduced pressure $p^* = pa^3/\phi_a$, vapor compressibility $Z_v(T) = p(n,T)/(nk_B T)$, and normalized hard sphere mean free path $\lambda/a = 1/[\sqrt{2}\pi n a^3 Y(n)]$, as a function of reduced equilibrium temperature $T^* = k_B T/\phi_a$.

T^*	T/T_c	n_l^*	n_v^*	p^*	Z_v	λ_v/a
0.45	0.5963	0.77915	5.683e−03	2.4527e−03	0.95909	3.9312e+01
0.50	0.6626	0.7187	1.2213e−02	5.6822e−03	0.9237	1.7988e+01
0.55	0.7288	0.6568	2.3111e−02	1.1154e−02	0.8739	9.4004e+00
0.60	0.7951	0.5917	3.9941e−02	1.9436e−02	0.8093	5.3298e+00
0.65	0.8613	0.5205	6.5517e−02	3.1009e−02	0.7275	3.1434e+00
0.70	0.9276	0.4363	0.1066	4.6261e−02	0.6201	1.8283e+00

As shown by the values of the vapor compressibility Z_v and (nominal) mean free path λ_v, the temperature range spans vapor conditions from near ideal to non-ideal.

2.1 *Simplified Moment Approximation of the Linearized EV Equation*

Let $\psi(\mathbf{v})$ be any velocity dependent molecular property, then multiplying Eq. (1) by $\psi(\mathbf{v})$ and integrating over $\mathbf{v}$ yields the following balance equation:

$$\frac{\partial}{\partial x}\left(J_{\psi}^{(k)}+J_{\psi}^{(c)}\right)-\frac{F_x(x)}{m}\int\frac{\partial\psi}{\partial v_x}f(x,\mathbf{v})\,d\mathbf{v}=S_{\psi}^{(c)}(x). \tag{10}$$

In Eq. (10), $J_{\psi}^{(k)}$ is the kinetic flux, $J_{\psi}^{(c)}$ is the collisional flux, whereas $S_{\psi}^{(c)}(x)$ is the collisional production term. They are defined as follows:

$$J_{\psi}^{(k)}(x)=\int c_x\psi(\mathbf{v})f(x,\mathbf{v})\,d\mathbf{v} \tag{11}$$

$$J_{\psi}^{(c)}(x)=\frac{a^2}{4}\int d\mathbf{v}\,d\mathbf{v}_1\,d^2\mathbf{k}(\mathbf{g}\cdot\mathbf{k})^{+}\int_0^a d\lambda k_x\Delta_{\psi}(\mathbf{v},\mathbf{v}_1,\mathbf{k})Y\left[n\left(x+\left(\lambda-\frac{a}{2}\right)k_x\right)\right]\times$$
$$f(x+\lambda k_x,\mathbf{v})f(x+(\lambda-a)k_x,\mathbf{v}_1) \tag{12}$$

$$S_{\psi}^{(c)}(x)=\frac{a^2}{2}\int d\mathbf{v}\,d\mathbf{v}_1\,d^2\mathbf{k}(\mathbf{g}\cdot\mathbf{k})^{+}\Sigma_{\psi}(\mathbf{v},\mathbf{v}_1,\mathbf{k})Y\left[n\left(x-\frac{a}{2}k_x\right)\right]f(x,\mathbf{v})f(x-ak_x,\mathbf{v}_1), \tag{13}$$

where

$$\Delta_{\psi}(\mathbf{v},\mathbf{v}_1,\mathbf{k})=\left[\psi(\mathbf{v}')-\psi(\mathbf{v})\right]-\left[\psi(\mathbf{v}_1')-\psi(\mathbf{v}_1)\right]$$
$$\Sigma_{\psi}(\mathbf{v},\mathbf{v}_1,\mathbf{k})=\left[\psi(\mathbf{v}')-\psi(\mathbf{v})\right]+\left[\psi(\mathbf{v}_1')-\psi(\mathbf{v}_1)\right].$$

The linearized form of Eq. (10) can be used to obtain a system of closed, linearized moment equations to study isothermal shear flows in the two-phase fluid described below. The distribution function is assumed to have the form:

$$f(x,\mathbf{v})=f_0(x,\mathbf{v})\left[1+u_y(x)\frac{v_y}{RT_0}+\sigma_{xy}(x)\frac{v_xv_y}{\rho_0(x)(RT_0)^2}\right] \tag{14}$$

being $u_y(x)$ the tangential bulk velocity field, $\sigma_{xy}(x)=m\int c_xc_yf(x,\mathbf{v})\,d\mathbf{v}$ the *kinetic* tangential stress tensor component and $\mathbf{c}$ the molecular peculiar velocity. It is to be noted that the above perturbation of equilibrium does not contribute to the density and temperature, assuming that temperature and density perturbations are of second order in the shear rate. The unknowns $u_y(x)$ and $\sigma_{xy}(x)$ are determined by the two following moment equations:

$$\left(1+\frac{4\pi}{15}\frac{a^3}{m}\rho_0 Y_0\right)\sigma_{xy}-\frac{4a^4}{15}\sqrt{\pi R T_0}\frac{\rho_0^2 Y_0}{m}\frac{\partial u_y}{\partial x}=P_{xy} \tag{15}$$

$$-\frac{\sqrt{\pi}a^4}{m}\sqrt{RT_0}\left[\frac{8}{35}\rho_0\frac{\partial}{\partial x}\left(Y_0\frac{\partial\sigma_{xy}}{\partial x}\right)-\frac{88}{105}\sigma_{xy}\frac{\partial}{\partial x}\left(Y_0\frac{\partial\rho_0}{\partial x}\right)-\frac{16}{105}\rho_0\sigma_{xy}\frac{\partial^2 Y_0}{\partial x^2}\right]+$$

$$\left[\rho_0 R T_0\left(1+\frac{4\pi}{15}\frac{a^3}{m}\rho_0 Y_0\right)\right]\frac{\partial u_y}{\partial x}=-\frac{16}{5}\frac{\rho_0\sqrt{\pi R T_0}a^2}{m}Y_0\sigma_{xy}. \tag{16}$$

In Eq. (15), P_{xy} denotes the constant, *total* tangential stress, which reduces to σ_{xy} in the ideal gas limit. Equations (15,16) are obtained by setting $\psi(\mathbf{v}) = m v_y, m v_x v_y$ in Eq. (10) and replacing non-local integrals with local expansions [14, 19] of the unknowns up to the order a^4 [22, 23]. It is worth observing that the above equations reduce to the NS equation for an isothermal shear flow in regions of uniform density but not in the interface region. Although not capable of capturing slip effects in low density vapor phase, their predictions are useful to judge about the influence of higher order moments as vapor density increases, when compared with EV results.

3 Test Problem Formulation

The effects of vapor non-ideality on the deviation of a two-phase shear flow from hydrodynamic behavior have been studied by solving Eq. (1) in the spatial domain

$$\mathcal{D}=\left\{(x,y,z)\in\mathcal{R}^3:-\frac{L_D}{2}<x<\frac{L_D}{2}\right\}$$

of width L_D. A central strip of $\mathcal{D}$, of nominal width $L_{liq} < L_D$, is occupied by an infinite and planar liquid slab, surrounded by two vapor regions of width $(L_D - L_{liq})/2$, symmetrically located with respect to the center of the liquid slab.

The domain is delimited by two parallel, planar, and impermeable walls, where gas-surface interaction is described by Maxwell's model [7], with full accommodation and prescribed wall temperature T_w. The shear flow in the fluid is produced by assigning opposite velocities $U_w\hat{\mathbf{y}}$ and $-U_w\hat{\mathbf{y}}$ to the walls, respectively, located at $x = \frac{L_D}{2}$ and $x = -\frac{L_D}{2}$, being $\hat{\mathbf{y}}$ the unit vector parallel to the walls themselves. It is to be observed that the energy dissipated by the shear motion would cause a continuous increase of the fluid temperature, the effect being proportional to the square of the shear rate. On the other hand, the comparison of EV solutions with the linearized moment equations, described in Sect. 2.1, requires that the temperature is constant and uniform, thus keeping the density as close as possible to its unperturbed equilibrium profile. Energy dissipation has been contrasted by thermostatting a central strip of the liquid slab to the prescribed temperature $T_0 < T_c$ by a Gaussian thermostat [1]. Moreover, walls common temperatures T_w have been also assigned the value T_0 and the shear rate kept as small as permitted by the statistical noise to signal ratio of the numerical method. Equation (1) has been solved numerically by the Direct Simulation Monte Carlo (DSMC) developed in Refs. [5, 11, 14].

4 Numerical Results

The flow geometry and the DSMC particle scheme mentioned in the previous section have been used to obtain the profiles of shear velocity $u_y(x)$, kinetic component of the shear stress $\sigma_{xy}(x)$, transversal component of the heat flux $q_y(x)$ and of the moments m_{xxy}, R_{xy} appearing in the Grad 26 description. The latter two quantities are, respectively, the xxy and xy components of the tensors:

$$m_{i_1i_2i_3} = m \int c_{\langle i_1} c_{i_2} c_{i_3\rangle} f(\mathbf{v}) d\mathbf{v} \quad R_{i_1i_2} = m \int \mathbf{c}^2 c_{\langle i_1} c_{i_2\rangle} f(\mathbf{v}) d\mathbf{v} - 7RT\sigma_{\langle i_1 i_2\rangle} \tag{17}$$

being $\sigma_{\langle i_1 i_2\rangle}$ the stress tensor deviator. The brackets denote a symmetric and traceless tensor. Once again, it is to be noted that $q_y(x)$, m_{xxy}, and R_{xy} do not appear in the hydrodynamic limit of the considered flow. Equation (1) has been solved by a DSMC particle scheme [5], based on reduced units, being a, m, and $a\sqrt{\frac{m}{\phi_\sigma}}$ the length, mass, and time units, respectively. Accordingly, velocities are normalized to $\sqrt{\frac{\phi_a}{m}}$, temperature to the reference value $T_{ref} = \frac{\phi_a}{k_B}$, whereas the reduced number density $a^3 n$ replaces n. In order to obtain a sample size in the vapor region sufficient to compute the small deviations from local equilibrium, simulations with about 24 million particles have been performed for each value of T_0. Spatial grid size has been set equal to $1/10$ and the time step Δt in the range 2–5$\times 10^{-3}$.

4.1 *Ideal Vapor Phase and Comparison with BE*

Before discussing the behavior of moments in non-ideal vapor conditions, it is interesting to compare the results of EV simulations with ideal gas simulations, based on the Boltzmann equation for the hard sphere gas, when $T_0 = 0.45$. In this case, the vapor phase is nearly ideal, as shown by the vapor equilibrium properties listen in the first line of Table 1. The results of an EV DSMC simulation, which includes the liquid, interface, and vapor region, are compared to a companion BE DSMC simulation of the vapor region, only. In the latter, the liquid-vapor boundary has been set $3a$ to the right of the center of the liquid-vapor interface, whose position is determined by the maximum of the density gradient modulus, as shown in Fig. 1a, which zooms on the interface region to show the beginning of the vapor phase where

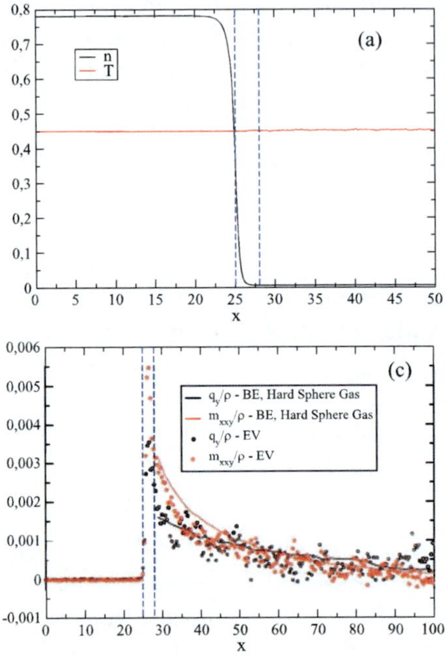

Fig. 1 $T_0 = 0.45$. EV moments profiles in the dilute vapor regime and comparison with BE. Vertical dashed lines indicate the interface center position and the beginning of the vapor region. (**a**) Reduced density and temperature profiles. (**b**) Shear velocity profiles. (**c**) Profiles of third order moments, $q_y(x)$ and m_{xxy}. (**d**) Profiles of fourth order moment R_{xy}

BE description applies. At the liquid surface, fully diffusive boundary condition has been applied to represent scattering from the liquid layer in absence of evaporation or condensation. The small liquid surface velocity has been taken into account in the Maxwellian representing scattered molecules. At the right domain boundary, both simulations (EV and BE) have assumed fully diffuse scattering from a wall at temperature 0.45 and sliding with velocity 0.2 along $\hat{\mathbf{y}}$ direction. On the scale of molecular diameter, the extent of vapor gap is about 177 which corresponds to a Knudsen number of about 0.22. As shown in Fig. 1b, the assumption of fully diffuse scattering at the liquid-vapor interface leads to a very good agreement between EV and BE based simulations. EV equation fully resolves the interface and shows that, on the molecular scale, the velocity slip actually consists of a steep ramp occupying the vapor side half of the interface. Good agreement is also found for the profiles of the third and fourth order moments q_y, m_{xxy} and R_{xy}, as shown in Figs. 1c,d. It is to be noted that, for EV, both q_y and m_{xxy} exhibit sharp peaks within the interface reaching values almost two times larger than their respective liquid-vapor boundary values. The fourth order moment R_{xy} has also a sharp negative peak inside the interface, followed by a rapid growth to reach the boundary value corresponding to diffuse reflection.

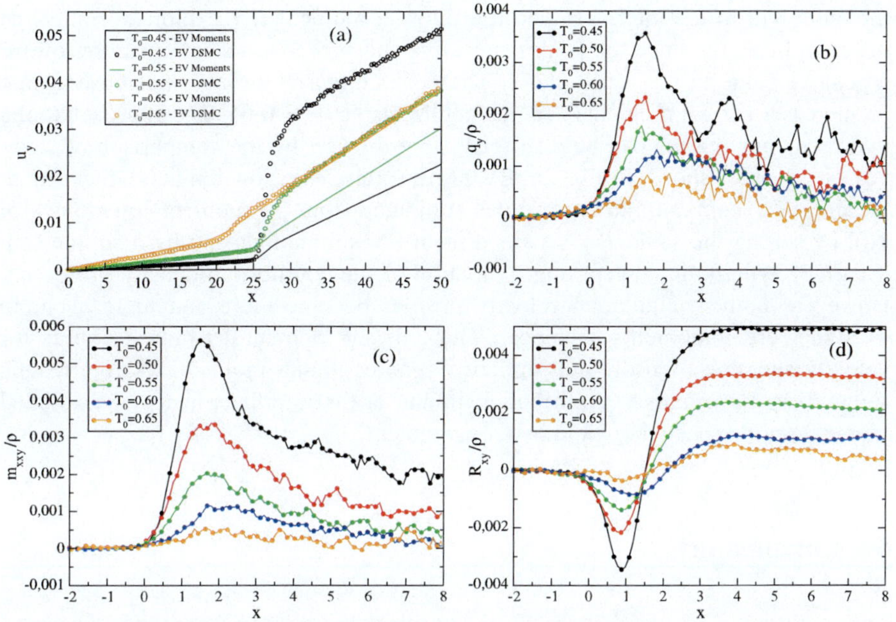

Fig. 2 Flow field properties as function of T_0. (**a**) Comparison of EV and moment approximation of the shear velocity u_y, in the liquid, interface, and vapor regions. (**b–d**) Enlarged views of third and fourth order moments profiles, q_y/ρ, m_{xxy}/ρ, R_{xy}/ρ, in the interface region. $x = 0$ corresponds to the center of the interface

4.2 Non-Ideal Vapor Phase

The behavior of higher order moments, as markers of non-equilibrium flow conditions, has been investigated for the fluid temperatures $T_0 = 0.45, 0.50, 0.55, 0.60, 0.65$, thus obtaining an increasingly denser vapor phase. It was not possible to obtain a stable liquid slab for the value $T_0 = 0.70$, very close to T_c. The wall velocities have been kept equal to ∓ 0.2, the nominal liquid slab thickness, based on the interface centers separation has been kept equal to 50, whereas L_D has been set equal to 200. The choice of the above parameters has to be considered a compromise between the need of keeping the shear driven perturbation small and the necessity of obtaining an acceptable signal to noise ratio in the statistical estimation of moments. Figures 2a–d summarize the results by showing the moments *per particle*, in order to compensate the effect of the strong density variation across the interface. The computed profiles show that higher order moments, absent in the liquid phase, are created in the interface, i.e., within a region a few molecular diameters wide. This behavior is particularly evident in Figs. 2b,c, where the reference frame origin has been placed in the interface center. The third order moments, q_y and m_{xxy}, exhibit a similar behavior (the different noise level depends on the different variance), showing a positive peak in the outer half of the interface. The fourth order moment R_{xy} exhibits a somewhat different behavior since the peak is negative and precedes a steep increase toward a positive value in the vapor phase. The peaks intensity strongly decreases while T_0 increases and all moments become negligible at $T_0 = 0.65$. The approach to the hydrodynamic regime can be also seen more directly by the comparison of shear velocity profiles shown in Fig. 2a. EV profiles (circles) show a much larger slip at the interface, compared to the profiles computed from the moment equations (15, 16), by setting the same P_{xy} obtained from EV simulations and by a simple (but accurate) hyperbolic tangent approximation of the equilibrium density profile n_0. However, moment equations velocity profiles become more and more accurate as higher order moments disappear. The reduced moment set only accounts for velocity and stress tensor, and implicitly ignores higher moments. As the results show, these have marked contribution in and at the interface, hence an extended moment set is expected to yield better agreement.

5 Conclusions

The exploration of kinetic effects in non-ideal fluids, here limited to the study of a simple Couette flow in a two-phase fluid, has shown that deviations from hydrodynamics next to the liquid-vapor interface, typical of dilute vapors, persist in non-ideal flow conditions. The numerical results presented here will help the development and fine tuning of a recently proposed, more extended moment method [23] which is necessary for the full understanding of the considered flow regimes.

References

1. Allen, M., Tildesley, D.: Computer Simulation of Liquids. Clarendon Press, Oxford (1987)
2. Anderson, D.M., McFadden, G.B., Wheeler, A.A.: Diffuse-interface methods in fluid mechanics. Annu. Rev. Fluid Mech. **30**, 139–165 (1998)
3. Balescu, R.: Equilibrium and Nonequilibrium Statistical Mechanics. Wiley, New York (1975)
4. Barbante, P., Frezzotti, A.: A comparison of models for the evaporation of the Lennard-Jones fluid. Eur. J. Mech. B/Fluids **64**, 69–80 (2017)
5. Bruno, D., Frezzotti, A.: Dense gas effects in the Rayleigh-Brillouin scattering spectra of SF6. Chem. Phys. Lett. **731**, 136595 (2019)
6. Carnahan, N., Starling, K.: Equation of state for nonattracting rigid spheres. J. Chem. Phys. **51**, 635 (1969)
7. Cercignani, C.: The Boltzmann Equation and Its Applications. Springer, Berlin (1988)
8. Cercignani, C.: Rarefied Gas Dynamics: From Basic Concepts to Actual Calculations. Cambridge University Press (2000)
9. de Sobrino, L.: On the kinetic theory of a Van der Waals gas. Can. J. Phys. **45**(2), 363–385 (1967)
10. Enskog, D.: Kinetische Theorie der Wärmeleitung, Reibung und Selbstdiffusion in gewissen verdichten Gasen und Flüssigkeiten. Kungl. Svenska Vet.-Ak. Handl. **63**, 3 (1922)

11. Frezzotti, A.: A particle scheme for the numerical solution of the Enskog equation. Phys. Fluids **9**, 1329–1335 (1997)
12. Frezzotti, A.: Boundary conditions at the vapor-liquid interface. Phys. Fluids **23**(3), 030609 (2011)
13. Frezzotti, A., Barbante, P.: Simulation of shock induced vapor condensation flows in the Lennard-Jones fluid by microscopic and continuum models. Phys. Fluids **32**(12) (2020)
14. Frezzotti, A., Gibelli, L., Lorenzani, S.: Mean field kinetic theory description of evaporation of a fluid into vacuum. Phys. Fluids **17**, 012102 (2005)
15. Frezzotti, A., Barbante, P., Gibelli, L.: Direct simulation Monte Carlo applications to liquid-vapor flows. Phys. Fluids **31**(6) (2019)
16. Fujikawa, S., Yano, T., Watanabe, M.: Vapor-Liquid Interfaces, Bubbles and Droplets. Springer (2011)
17. Hansen, J.P., McDonald, I.: Theory of Simple Liquids. Academic Press, London, UK (2006)
18. Karkheck, J., Stell, G.: Mean field kinetic theories. J. Chem. Phys. **75**, 1475–1487 (1981)
19. Resibois, P., DeLeener, M.: Classical Kinetic Theory of Fluids. Wiley, New York (1977)
20. Sone, Y.: Kinetic Theory and Fluid Dynamics. Birkhäuser Publishing, Basel (2002)
21. Struchtrup, H.: Macroscopic Transport Equations for Rarefied Gas Flows—Approximation Methods in Kinetic Theory. Interaction of Mechanics and Mathematics. Springer (2005)
22. Struchtrup, H., Frezzotti, A.: Grad's 13 moments approximation for Enskog-Vlasov equation. In: AIP Conference Proceedings, vol. 2132 (2019)
23. Struchtrup, H., Frezzotti, A.: 26 moment equations for the Enskog-Vlasov equation. J. Fluid Mech. **940** (2022). https://doi.org/10.1017/jfm.2022.98
24. Tsuruta, T., Tanaka, H., Masuoka, T.: Condensation/evaporation coefficient and velocity distribution at liquid-vapor interface. Int. J. Heat Mass Transf. **42**, 4107–4116 (1999)
25. van Beijeren, H., Ernst, M.: The modified Enskog equation. Physica **68**, 437 (1973)

The Cercignani Conjecture About a Classical Zero-Point Energy, and Its Confirmation for Ionic Crystals

F. Gangemi, R. Gangemi, A. Carati, and L. Galgani

Abstract A characteristic feature of Quantum Mechanics is that it predicts the existence of zero-point energy, i.e., of states with a nonvanishing kinetic energy at zero temperature. This is a fact that is experimentally verified and is considered to be inconceivable in a classical frame. In the year 1972 Carlo Cercignani advanced the idea that a classical zero-point energy may be conceived, if one understands the corresponding motions as characterized by being (in the terminology used at those times) of "ordered" rather than of "chaotic" type. Here we illustrate how the Cercignani idea is actually implemented for an ionic crystal model which was already shown to be of such a realistic character as to reproduce in a remarkably good way, in terms of the Newtonian trajectories of the ions, the experimental infrared spectra.

1 Introduction

The existence of "zero-point energy" or ZPE (which in German means zero-temperature energy) was first conceived in 1911 by Planck [1], within an attempt at understanding his blackbody formula in classical terms (as also did Nernst in the year 1916 [2]). Einstein and Stern in 1913 were able to confirm the existence of ZPE in diatomic molecules, making use, in a rather astonishing way, of the experimental data on specific heats [3]. ZPE manifests itself particularly in crystals, as exhibited by the Debye–Waller B-factors, which describe the fluctuations of atoms around their equilibrium positions, measured, e.g., by X-ray diffraction. It is observed that, when absolute temperature T tends to zero, the values of the B-factors don't vanish,

F. Gangemi (✉) · R. Gangemi
DMMT, Università degli Studi di Brescia, Brescia, Italy
e-mail: fabrizio.gangemi@unibs.it; roberto.gangemi@unibs.it

A. Carati · L. Galgani
Department of Mathematics, Università degli Studi di Milano, Milano, Italy
e-mail: andrea.carati@unimi.it; luigi.galgani@unimi.it

P. Barbante et al. (eds.), *From Kinetic Theory to Turbulence Modeling*, Springer INdAM Series 51, https://doi.org/10.1007/978-981-19-6462-6_10

tending instead to finite values [4]. Thus, it is an experimental fact that atoms present a residual mechanical energy at zero temperature.

Since the first paper of Planck, the occurring of a zero-point energy was explicitly associated with the idea that it should correspond to a transition from chaotic to ordered motions. So, the internal energy at temperature T was thought of as the sum of a "thermal energy" and of a residual zero-point energy. In fact such a superposition of two terms came out somehow "automatically" in the frame of quantum mechanics after Heisenberg and Schrödinger. But the idea that zero-point energy should have some kind of ordered character didn't find any explicit implementation in dynamical terms. The only qualitative distinction was of a thermodynamic character, since zero-point energy corresponds to a state of vanishing entropy (which is the "quintessence of the hypothesis of quanta," in Planck's words (see the preface of [5])).

In the year 1972 the idea that zero-point energy may exist in a classical frame, and actually as corresponding to a transition from chaotic to ordered motions, was advanced by Carlo Cercignani, in a joint paper with L. Galgani and A. Scotti [6]. A Fermi–Pasta–Ulam (FPU) model was considered, namely a chain of particles with nearest-neighbor interactions, the latter being taken of Lennard-Jones type with realistic parameters of atomic physics, a fact that implicitly introduces Planck's constant $\hbar$ into the problem.[1]

So one deals with a system of "weakly coupled" harmonic oscillators (the normal modes of the chain, of angular frequencies ω_j), to which a quantum zero-point energy $\sum \hbar\omega_j/2$ can be associated. The Cercignani conjecture was that such a quantum zero-point energy should be of the order of magnitude of the "stochasticity threshold" from chaotic to ordered motions that had been found in the FPU problem.[2] And this was actually checked to be the case.

In the present paper we will illustrate how ZPE naturally appears in a classical model of an ionic crystal, Lithium Fluoride (LiF), which was shown to reproduce with a surprisingly good accuracy experimental measurements of infrared spectra [8, 9]. It is just in terms of such spectra that ZPE comes in. Indeed the spectra are available at many different temperatures ranging from a few K up to values close to the melting point. Now, one of the peaks appearing in these spectra is of an anharmonic origin, since it doesn't correspond to any normal-mode frequency of the system. Thus its relative intensity would be expected to vanish at low temperatures,

---

[1] As pointed out in the previous paper [7], this occurs through the known relation $\sqrt{m\epsilon\sigma} = 2Z\hbar$, where ϵ and σ are the familiar parameters of the Lennard-Jones potential, while m and Z are the mass and the atomic number of the particles.

[2] At those times the question of interest was what Fermi had called his "little discovery," i.e., a lack of attainment of energy equipartition starting from an initial non-equipartition state. In the following years it was found that the Fermi discovery corresponds to a metaequilibrium state, since a final equipartition state is always attained, however small the energy of the system be. On the other hand, a transition from ordered to chaotic motions still persists in a suitable sense. To this point we will come back later.

where the crystal vibrations tend to become harmonic, whereas experimental data show that the peak keeps a significant intensity even at the lowest temperatures.

Tentative interpretations of this phenomenon were suggested in the context of quantum mechanics (see, for example, [10]). We show here that, in a classical model of ionic crystals, such a phenomenon can be interpreted as a manifestation of a state function or equation of state $\varepsilon(T)$, internal energy versus temperature (at fixed room pressure), with the property that $\lim_{T\to 0} \varepsilon(T) \neq 0$, i.e., a ZPE exists.

At the present stage of our investigations the relation between ε and T can only be obtained through an empirical procedure, in which the value of the (specific) internal energy ε corresponding to a given temperature T is chosen so as to give the best fit of the calculated spectrum to the experimental one. As a result of this procedure, the value of internal energy per vibrational mode ε^0 corresponding to $T = 0$ can be estimated to be[3] $\varepsilon^0 \approx 120$ K, which differs by about a factor 2 from the quantum mechanical one. From the description of the procedure used it will appear that it contains an empirical element, a better choice of which might perhaps remove such a discrepancy.

In Sect. 2 the model is presented, in Sect. 3 the results of the numerical simulations for the spectra are illustrated, and are then discussed in Sect. 4 in connection with ZPE. The conclusions follow.

2 The Ionic Crystal Model

The model of LiF we adopt treats each ion as a point charge with a mass equal to the experimental value. Electrons are not explicitly present in the model. However, following a tradition going back to Born, their effect is assumed to be responsible for two features, namely: (1) the value of an "effective charge" e_i for each ion i, which is one of the free parameters in the model, and (2) the production, through their distribution around the ions (in the spirit of Born's adiabatic principle), of a short-range "effective potential" V^{SR} acting among the ions. The pairwise interaction potential among the ions, with a suitable choice for V^{SR}, is thus given by

$$V_{ij}(r) = \frac{e_i e_j}{r} + V_{ij}^{SR}(r), \qquad V_{ij}^{SR}(r) = A_{ij} e^{-B_{ij} r} - \frac{C_{ij}}{r^6} \, . \tag{1}$$

Simulations are performed on a cubic cell of N ions (with N between 512 and 4096) with periodic boundary conditions, so that the system has infinite size. The short-range interactions are calculated for pairs within a cutoff $r \leq 5$ Å, while the familiar Ewald summation method is applied for the Coulomb long-range forces.

[3] Here and in the following we report energy values divided by the Boltzmann constant k_B, so that they are expressed in kelvin.

By means of the Green–Kubo linear response theory applied to dispersion in dielectric media, along the lines of ref. [11] one can show that the electric susceptibility of a system of charges is given by

$$\chi_{ij}(\omega) = \frac{3V}{2\langle K\rangle}\int_0^{+\infty} e^{-i\omega t}\big\langle P_i(t)\dot{P}_j(0)\big\rangle\, dt \,, \tag{2}$$

$$\mathbf{P}(t) = \frac{1}{V}\sum_j e_j\mathbf{x}_j(t) \qquad \text{microscopic polarization,} \tag{3}$$

where $\mathbf{x}_j$ is the position vector of charge e_j, V is the volume of the integration cell, K the kinetic energy, and the notation $\langle\ldots\rangle$ indicates phase averages. The latter are actually replaced by time averages. We perform molecular dynamics (MD) simulations at constant energy and volume,[4] integrating the equations of motion by means of the Verlet integration method with 2 fs integration step. Moreover, in order to enhance phase-space sampling, we average over multiple trajectories with independent initial conditions.

From electric susceptibility one obtains the permittivity tensor

$$\epsilon_{ij}(\omega) = \delta_{ij}\epsilon_\infty + 4\pi\,\chi_{ij}^{(ions)}(\omega) \,, \tag{4}$$

where the constant ϵ_∞ takes into account the electronic contribution and is obtained from experimental data. For isotropic materials such as LiF, one simply has $\epsilon_{ij}(\omega) = \epsilon(\omega)\delta_{ij}$. Once the permittivity is known, the quantities observed in the experiments can be easily calculated: the refractive index is given by

$$n(\omega) = \sqrt{\epsilon(\omega)} \,, \tag{5}$$

while the reflectivity is

$$R(\omega) = \left|\frac{\epsilon(\omega)-1}{\epsilon(\omega)+1}\right| \,. \tag{6}$$

3 The Computed Spectra

A comparison between calculated and measured spectra at room temperature ($T = 295$ K) is shown in Fig. 1. The very good agreement with data is the result of accurate determination of the parameters of the potential, by fitting the

[4] As will be further discussed in the following, the lattice spacing, and thus the density, is suitably chosen at each temperature T, according to the experimental data on thermal expansion at fixed room pressure. In this sense, one may consider that the crystal is simulated at constant pressure.

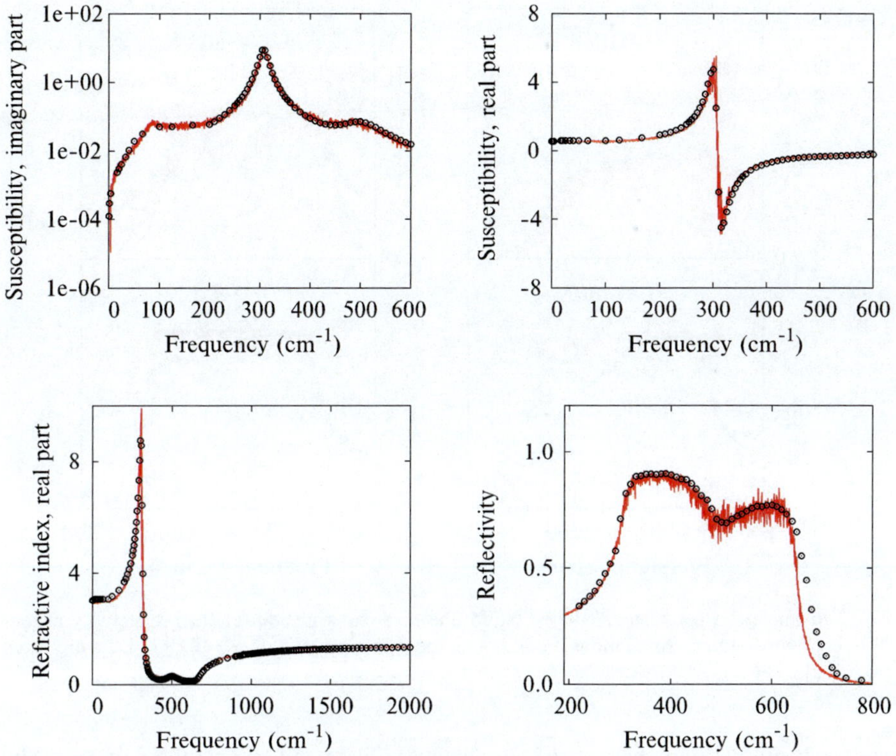

Fig. 1 Comparison between calculated (line) and experimental (circles) values of optical properties at room temperature. Upper left: imaginary part of susceptibility. Upper right: real part of susceptibility. Lower left: refractive index. Lower right: reflectivity. Experimental data are taken from [12] (lower right panel) and [13] (all the other panels)

calculated dispersion relations to the experimental ones. The resulting curves are in better agreement with data than quantum calculations available in the literature [14]. Notice that the fitting was performed within the familiar assumption of MD simulations that $T = 295$ K should correspond to $\varepsilon = 295$ K. On this point we will come back later.

Experimental data on reflectivity are available at many different temperatures, and, in particular, in ref. [12] they are given in a range from 7.5 K to 1060 K. In order to make a comparison with calculated data one has to define what is temperature in a classical model of a system different from a gas. The common prescription used in molecular dynamics studies consists in identifying temperature with a quantity proportional to the average kinetic energy per atom in the system, according to the well known Clausius relation $< K >= \frac{3}{2} k_B T$, assumed to hold at all temperatures, not only for ideal gases but also for other aggregation states. The results shown in Fig. 2, corresponding to four values of temperature, were obtained by applying such a prescription: moreover, since all experimental data are at room pressure,

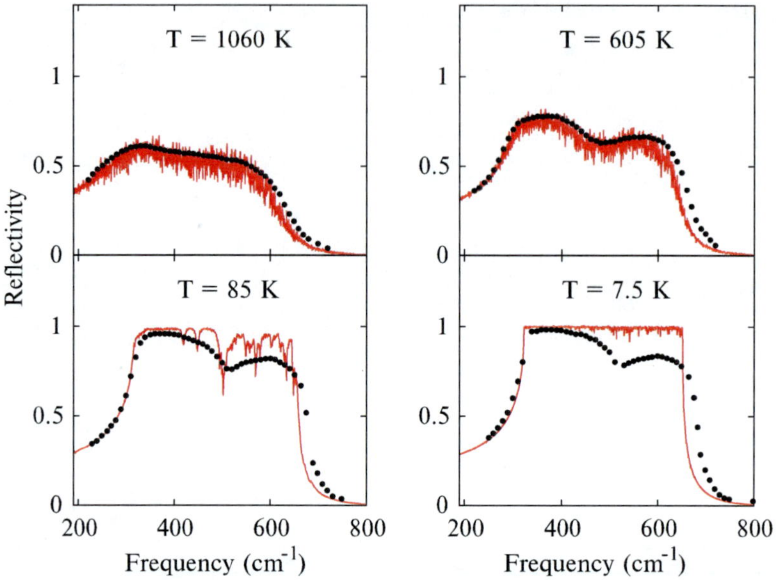

Fig. 2 Comparison between calculated (solid line) and experimental (circles) reflectivity at four different temperatures. Simulations are based on the prescription $< K >= \frac{3}{2}k_B T$. Data are taken from [12]

and each simulation is performed at constant volume and energy, the density of the system was changed at each temperature by adjusting the lattice spacing according to the available experimental data on thermal expansion. The figure shows that the agreement of calculated spectra with experimental data essentially persists at high temperatures, whereas at low temperatures a strong disagreement is found.

A detailed analysis of the LiF reflectivity spectra is needed to better understand these results. The reflectivity curve is determined by susceptibility as shown in Eqs. 4–6, and its features can be interpreted in terms of the peaks of susceptibility, as shown in Fig. 3, where experimental data of both quantities are shown for $T = 295$ K. In the range from 200 to 800 cm^{-1}, where reflectivity data are collected, there are two peaks in the susceptibility curve: one at a frequency $\simeq$ 310 cm^{-1}, of harmonic origin, and one at $\simeq$ 500 cm^{-1}, of anharmonic origin, since no optically active normal mode is found at this frequency. The reflectivity curve consists of a rising part at low frequencies that reaches its maximum approximately at the position of the first peak of susceptibility, followed by a plateau. The height of this plateau is related to the width of the susceptibility peak: the higher the width, the lower is the plateau, while for a vanishing width the plateau tends to 1. At the position of the second susceptibility peak one has a local minimum in the reflectivity plot and then almost a second plateau, lower than the first one. The height of this plateau is related to the width of the second susceptibility peak in the same way as discussed for the previous one. Now, the second peak in susceptibility should disappear when

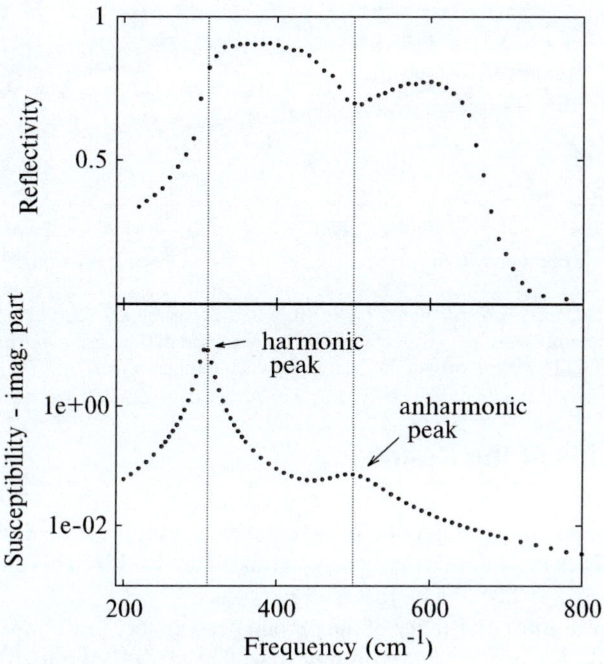

Fig. 3 Analysis of the main features of experimental data on reflectivity (upper panel) and on the imaginary part of susceptibility (lower panel)

the vibrational energy of the crystal vanishes, since the anharmonic interactions between modes vanish. This is what is actually observed in the calculated curve in the lower right panel of Fig. 2, but not in the experimental data. This fact suggests that the specific energy of the crystal doesn't vanish when temperature tends to zero. Indeed a good agreement between calculated and experimental spectra is restored if the values of internal energy of the crystal are suitably chosen in an empirical way, abandoning the Clausius relation between kinetic energy and temperature. The plots shown in Fig. 4 were obtained by means of simulations with average energy per mode $\varepsilon = 180$ K in the case of $T = 85$ K and $\varepsilon = 125$ K in the case of $T = 7.5$ K.[5]

[5] It should be noticed that the parameterization of the potential was done by assuming $\epsilon = 295$ K at $T = 295$ K, which is not consistent with the values of ε obtained at lower T. A more realistic curve $\varepsilon(T)$ should be obtained by a different suitable choice of the ε value corresponding to $T = 295$ K. This we leave for possible future work.

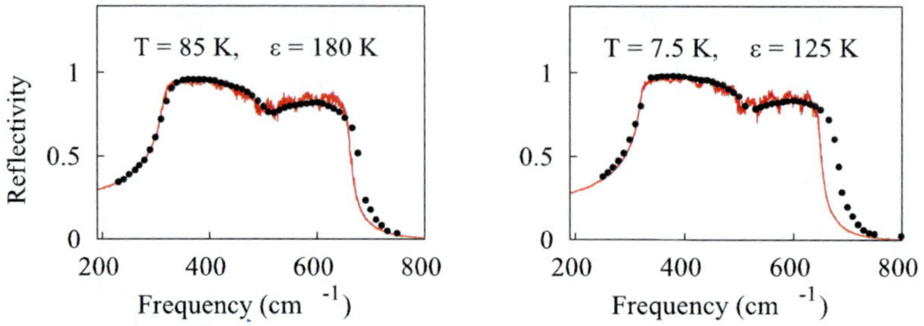

Fig. 4 Comparison between calculated (solid line) and experimental (circles) reflectivity at the two lowest temperatures of Fig. 2. Calculated spectra are fitted in this case to the experimental data by choosing a suitable value ε of specific energy for each temperature T

4 Discussion of the Results

The results shown in Fig. 4 may be interpreted as an empirical determination of the relation $\varepsilon(T)$ between internal energy and temperature for LiF. This clearly indicates that $\lim_{T\to 0} \varepsilon(T) \approx 120$ K, i.e., that a ZPE exists.

In other words, the persistence of the second peak at very low temperatures (down to 7.5 K) could be justified by assuming that, due to ZPE, the total energy of the crystal even at $T = 0$ is sufficiently large to produce significant anharmonic effects as those involved in the generation of the secondary peak.

Although the role of ZPE to explain the behavior of experimental spectra of LiF at low T was invoked by other authors in the framework of quantum calculations, we point out that the model considered here is completely classical as far as the motions of the ions (the only particles entering the model) are concerned. So it is of interest to investigate whether suitable dynamical properties of the system can be found, which reveal the existence of some energy threshold separating two qualitatively different regimes, along the lines of the original Cercignani's conjecture that a transition may occur from chaotic to ordered motions.

A first attempt can be made, using an extremely naive procedure, as follows. In Fig. 5 are reported the projections, on some chosen single normal modes phase planes, of two phase-space trajectories generated by numerical simulations at $\varepsilon =$ 300 K (upper row) and at $\varepsilon = 10$ K (lower row): points are taken at 0.02 ps intervals along the trajectories. One sees that, at 10 K, in most cases the projections exhibit a character that might be considered ordered, whereas they become completely disordered at 300 K. This seems to be in agreement with the hypothesis that the value $\varepsilon \simeq 120$ K of internal energy indicated through the spectra as corresponding to $T = 0$ might correspond to a threshold from partially ordered to completely disordered motions.

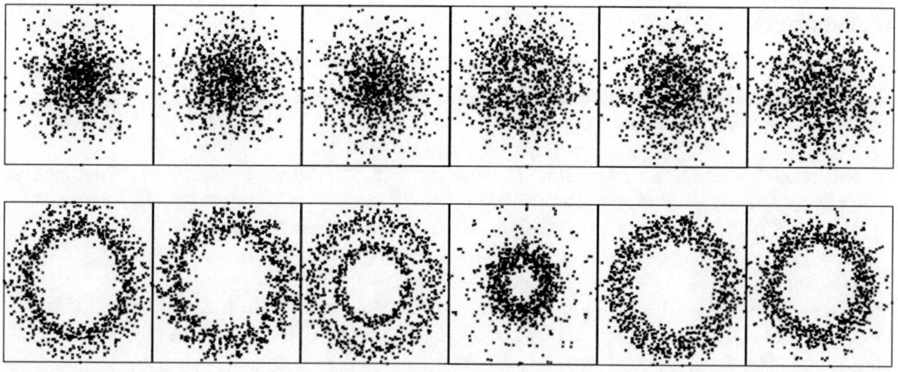

Fig. 5 Projections of simulated phase-space trajectories onto selected normal-mode phase planes for $\varepsilon = 300$ K (upper row) and $\varepsilon = 10$ K (lower row). Points are taken at 0.02 ps intervals from trajectories of 20 ps. Notice that the scales of the different plots are different, being just chosen in order to produce a common size

5 Conclusions

The existence of a zero-point energy (an energy below which motions occur at zero temperature) is considered to be a typical quantum phenomenon, but already in the year 1972 Carlo Cercignani proposed that it may be conceived also in a classical frame. A result supporting such a conjecture was reported here, on the basis of numerical computations performed in a classical Born-like model of ionic crystal, for which the rather astonishing result was already obtained that it reproduces fairly well the infrared spectra in terms of the Newtonian trajectories of the ions. The existence of a zero-point energy was exhibited by extrapolating to zero temperature the state function $\varepsilon(T)$, specific energy versus temperature. In turn, the latter was obtained by the empirical procedure of finding, for each T, the value of ε which gives the best fit of the calculated spectrum to the experimental one. Moreover, such a zero-point energy appears to correspond to a qualitative change in the chaoticity character of the motions, along the lines of the original Cercignani conjecture.

Although it was not yet possible for us to find a rationale for the existence of a zero-point energy within classical statistical thermodynamics,[6] it was for us a gratifying experience to produce the preliminary supporting arguments illustrated above, almost exactly fifty years after the idea of a classical zero-point energy was conceived, in the form of a startling invention, by our dear late friend Carlo.

[6] The key point is to understand why temperature might vanish at all below an energy threshold. Here arguments mimicking the familiar ones of Khinchin should be considered. One should, however, take into account that the microcanonical or the canonical ensembles cannot be used, due to lack of ergodicity in the case of crystals. Indeed the arrangement of the atoms in a crystal is just one among the $N!$ ones a priori available and, as long as a crystal is a crystal, most of the remaining $N! - 1$ arrangements are not dynamically attained.

References

1. Planck, M.: Verh. Deutsche Phys. Ges. **13**, 138 (1911)
2. Nernst, W.: Verh. Deutsche Phys. Ges. **18**, 83 (1916)
3. Einstein, A., Stern, O.: Ann. der Phys. **40**, 551 (1913)
4. Sirdeshmukh, D.B., Sirdeshmukh, L., Subhadra, K.G.: Alkali Halides - A Handbook of Physical Properties. Springer, Berlin (2001)
5. Planck, M.: The Theory of Heat Radiation. Dover, New York (1959)
6. Cercignani, C., Galgani, L., Scotti, A.: Phys. Lett. A **38**, 403 (1972)
7. Galgani, L., Scotti, A.: Phys. Rev. Lett. **28**, 1173 (1972)
8. Gangemi, F., Carati, A., Galgani, L., Gangemi, R., Maiocchi, A.: EPL **110**, 47003 (2015)
9. Carati, A., Galgani, L., Maiocchi, A., Gangemi, F., Gangemi, R.: Physica A **506**, 1 (2018)
10. Lax, M., Burstein, E.: Phys. Rev. **97**, 39 (1955)
11. Carati, A., Galgani, L.: Eur. Phys. J. D **68**, 307 (2014)
12. Jasperse, J., Kahan, A., Plendl, J., Mitra, S.: Phys. Rev. **146**, 526 (1966)
13. Palik, E.: Handbook of Optical Constants of Solids. Academic Press, Amsterdam (1998)
14. Ipatova, I.P., Maradudin, A.A., Wallis, R.F.: Phys. Rev. **155**, 882 (1967)

Turbulence Without Fluctuations

Massimo Germano

Abstract Fluctuations are very important in statistics, but their use in the study of turbulent flows, particularly in the formulation and the analysis of Large Eddy Simulations, is not so essential. In the paper some historical and personal recollections on that are recorded.

1 Introduction

Thirty years ago, in 1992, I was invited by ENEL, the Italian agency for the production and the distribution of electric power, to illustrate the recent progresses in the new emerging computational techniques applied to the Large Eddy Simulation, (LES), of turbulent flows. In Pisa, the morning of May 21, I spoke about the past and the present of LES and the related perspectives of future developments. In the afternoon some participants presented their activity in the field, and Carlo Cercignani was one of them. It was a great day, plenty of suggestions and promises for the future.

It was not my first conversation with Carlo, our friendship started in the sixties of the last century, but it was the first time that we were directly related to some joint project. Both Italians we were used to meet outside, at a congress in the United States or a meeting in Denmark. I was attracted by his way of looking at science, his passionate and disenchanted way of considering the progress, his penetrating and well-disposed judgements of new theories and people. The contacts were rapid but very significative, a dinner in a pub, a little walk in a campus or the casual encounter in a library between one session and another. For me he was like an elder brother, disposed to understand your problems and indulgent with your weaknesses.

At the time of our meeting in Pisa I recall his amused interest on my reports about my personal American Adventure, the joint gestation of the dynamic model with

M. Germano (✉)
Duke University, Department of Civil and Environmental Engineering, Durham, NC, USA
e-mail: mg234@duke.edu

P. Barbante et al. (eds.), *From Kinetic Theory to Turbulence Modeling*, Springer INdAM Series 51, https://doi.org/10.1007/978-981-19-6462-6_11

Stanford. Thirty years ago the interest for LES suddenly had a great impulse due to the new dynamic modeling approach [1] that was presented at the Summer Meeting of the CTR in 1990. This procedure, originally applied to improve dynamically the Smagorinsky model, was very general; Carlo was working with his collaborators, Antonella Abbà and Lorenzo Valdettaro, on anisotropic eddy viscosity models and a possible dynamic improvement was attracting their attention.

For me all that was a great surprise. The basic ingredient of the new modeling approach was a trivial identity between the subgrid stresses at two different filtering resolutions incidentally found in the framework of my speculations on LES. My interest on LES was relatively new, and my main problem at that time was to develop a simple multiscale approach to the analysis of turbulent flows based on a generic hierarchy of filtering operators. The starting point was a curiosity about the fundamentals, in particular, the problem of the average and the analytical formalization of LES. I was attracted by LES mainly from the point of view of a new way to represent turbulent flow, different from the statistical and the spectral ones.

The beginning was discouraging, a good friend of mine, consulted on the subject, was very strong about that: please leave that topic, it is unfitted for you. It requires big computers, you are not an expert in numerical and computational problems and there is little space left for new theoretical work. But I was attracted by some analogies between the classical Leonard formulation of LES [2] and some old ideas of Boussinesq [3] and Reynolds [4], developed in the middle of the last century by Kampé de Fériet and his school [5], concerning the rules of the mean from the algebraic point of view, and that was the starting point. Then I realized that a great obstacle to the formulation of a hierarchical filtering approach was to get rid of the fluctuations, and to remove the fluctuation-conditioned attitude peculiar of the statistical decomposition. At the time of my presentation in Pisa I was in the main of my struggle against the fluctuations, my first paper on that [6] was encountering some difficulties to be published, it was presented to JFM on July 1990 and was finally published on June 1992. That was my main interest and my major concern at that time, and the main argument of my conversation gently supported by Carlo with his particular witty smile that afternoon in Pisa. In the following I would like to recall all that, in the ideal continuity that connect all us in an endless dialogue.

2 Fluctuation-Conditioned Turbulence

Everybody knows that an alternative, rapid, and elegant way to compute the Reynolds stress in turbulent flows is given by

$$R_{ij} = \overline{u_i u_j} - \bar{u}_i \bar{u}_j \tag{1}$$

but for a teacher it is surprisingly scarce the number of students disposed to appreciate this simple expression. Almost all prefer the classical formulation based on the differences, the fluctuations from the mean $u'_i = u_i - \bar{u}_i$

$$R_{ij} = \overline{u'_i u'_j} \tag{2}$$

where the overline stands for the statistical operator, and this fluctuation-conditioned attitude is more evident when applied to the statistical analysis of a filtered representation of a turbulent flow. Let us indicate with the overtilde a generic filtering operator and with $\tilde{u}_i$ a filtered representation of a turbulent flow u_i. We assume as usual that

$$\overline{\tilde{\cdots}} = \overline{\cdots} \quad ; \quad \overline{\tilde{u}_i} = \overline{u_i} \quad ; \quad \overline{\widetilde{u_i u_j}} = \overline{u_i u_j} \tag{3}$$

and we introduce the new fluctuations u''_i and u'''_i defined as

$$u''_i = \tilde{u}_i - \bar{u}_i \;\; ; \;\; u'''_i = u_i - \tilde{u}_i \tag{4}$$

Owing to the identity

$$u'_i = u''_i + u'''_i \tag{5}$$

we can write the relation

$$R_{ij} = \overline{u''_i u''_j} + \overline{u'''_i u'''_j} + \overline{u''_i u'''_j} + \overline{u'''_i u''_j} \tag{6}$$

that if the new supplementary fluctuations u''_i and u'''_i are uncorrelated reduces to the simple result

$$R_{ij} = \overline{u''_i u''_j} + \overline{u'''_i u'''_j} \tag{7}$$

I will not underestimate the importance of these relations. Triple decompositions have been studied by many authors in the past, I will cite one for all the important paper by Reynolds and Hussain [7] where a triple decomposition has been applied to the study of the mechanics of an organized wave in turbulent shear flow. Here I only remark that it is very easy and more general to realize that

$$\overline{u''_i u''_j} = \overline{\tilde{u}_i \tilde{u}_j} - \bar{\tilde{u}}_i \bar{\tilde{u}}_j \tag{8}$$

and that

$$\overline{u'''_i u'''_j} + \overline{u''_i u'''_j} + \overline{u''_i u'''_j} = \overline{\widetilde{u_j u_i}} - \overline{\tilde{u}_i \tilde{u}_j} \tag{9}$$

so that we have the alternative operational decomposition of the Reynolds stress given by

$$R_{ij} = T_{ij} + \overline{\tau}_{ij} \tag{10}$$

where T_{ij} is the resolved stress, explicitly computed

$$T_{ij} = \overline{\tilde{u}_i \tilde{u}_j} - \overline{\tilde{u}}_i \overline{\tilde{u}}_j \tag{11}$$

and τ_{ij} is the subfilter stress, usually modeled

$$\tau_{ij} = \widetilde{u_i u_j} - \tilde{u}_i \tilde{u}_j \tag{12}$$

We remark that the identity (10) is operational and only subjected to the assumption (3), so that can be also applied to two generic filtering operators, and that was the new multiscale identity that I was illustrating that day in Pisa. Here I would like to recall how important in his genesis was for me to remove the fluctuation-conditioned attitude at that time dominant in the field. A fundamental starting point was an old paper published in French in the Italian *Rendiconti del Seminario Matematico e Fisico di Milano* [5]. In that paper Kampé de Fériet poses the following question: *Quelles sont les propriétés de la moyenne, nécessaires et suffisantes pour que les équations de Reynolds se déduisent rigoureusement des équations de Navier en en prenant la moyenne?*, which are the properties of the mean that mathematically justify the RANS equations. The arguments and the developments are very interesting, but mainly two conclusions deserve in our opinion our principal attention. The first is related to the fundamental property of the statistical fluctuations, $\overline{u'_i} = 0$, their nullity when averaged, that is proved unessential: *si on attache donc une telle importance à* $\overline{u'_i} = 0$ *c'est parce qu'on la juge impliquée intuitivement dans le concept de la moyenne, mais nullement parce qu'elle intervient dans les calculs réellement effectués*[1]. The second is that the RANS equations are very simply and directly derived by the Navier-Stokes equations *si l'effet du mouvement de l'agitation turbulent sur le mouvement moyen ne s'exprime plus par le tenseur de Reynolds,* $\overline{u'_i u'_j}$, *mais par un nouveau tenseur* $R_{ij} = \overline{u_i u_j} - \bar{u}_i \bar{u}_j$[2]. More precisely the problem that interested Kampé de Fériet in the fifties of the last century was to better understand algebraically the Reynolds rules of the mean. Given a generic linear and constant preserving filtering operator $\mathcal{F}$ that produces the quantities $\bar{u}_i$ let us introduce the fluctuations $u'_i = u_i - \bar{u}_i$. We

[1] If one attains such importance to this property is that we intuitively think it associated with the concept of mean, but it has no role at all in the computations really done

[2] If the effect of turbulence is not expressed by the Reynolds stress $\overline{u'_i u'_j}$ but by a new tensor $R_{ij} = \overline{u_i u_j} - \bar{u}_i \bar{u}_j$

can write

$$\overline{u_i u_j} = \overline{\bar{u}_i \bar{u}_j} + \overline{u'_i u'_j} + \overline{u'_i \bar{u}_j} + \overline{u'_j \bar{u}_i} \tag{13}$$

and the problem was to find the operational rules that justify the equivalence of this expression with the usual one

$$\overline{u_i u_j} = \bar{u}_i \bar{u}_j + \overline{u'_i u'_j} \tag{14}$$

Obviously this equivalence is satisfied if $\mathcal{F} = \mathcal{E}$, the statistical average, and at that time the interest to this problem was not so appreciated by the scientific community. Let us quote from Monin and Yaglom [8], volume 1, page 209: *However, all these investigations are of a formal mathematical nature and their results do not find direct application in the theory of turbulence. Furthermore, they are not even necessary, since in present day turbulence theory the question of the meaning of averaging is resolved in a completely different manner, and, moreover, in such a way that all the Reynolds conditions are evidently satisfied ....* Averages different from the statistical one, like convolutional averages in space and time, are not so interesting because *these mean values will depend on the form of the weighting function (in particular, when averaging over some time interval or region of space, it will depend on the length of the interval or the form and volume of the region)* and in conclusion *it is desirable in the theory of turbulence to avoid the use of this type of averaging altogether, and to adopt instead some other method of defining the mean value, a method that has simpler properties and is more universal. A convenient definition of this type, which we shall use throughout this book, is found in the probability-theory treatment of the fields of fluid dynamic variables in a turbulent flow as random fields.*

But things were rapidly changing due to a lot of reasons. The discovery of coherent structures and the sensation that a turbulent field at least as regards the large scales, is not so chaotic as presumed was more and more acting on the scientific imagination. The classical double decomposition in ensemble averages and statistical fluctuations was insufficient to capture the multiscale nature of turbulence, and something new had to be implemented. It was clear that in order to go on something new had to be conceived, and following the suggestions of Kampé de Fériet the time of abandoning the fluctuation-conditioned approaches was arrived. It was not so easy, and a fundamental step was to define the Generalized Central Moments.

3 Turbulence Without Fluctuations

Once removed by the help of Kampé de Fériet my fluctuation-conditioned attitude, everything became easier. Explicitly, given a generic velocity field $u_i\ (\mathbf{x}, t, \omega)$, the statistical average is given by

$$\langle u_i \rangle = \int u_i \left(\mathbf{x}', t', \omega\right) p(\omega) d\omega \tag{15}$$

and a generic space average is given by

$$\langle u_i \rangle = \int u_i \left(\mathbf{x}', t', \omega\right) \mathcal{G} \left(\mathbf{x} - \mathbf{x}'\right) d^3\mathbf{x}' \tag{16}$$

where $p(\omega)$ is a probability density function and $\mathcal{G}\left(\mathbf{x} - \mathbf{x}'\right)$ a normalized space filtering kernel, but operationally the filtered, averaged, Reynolds, LES equations are all equivalent. In terms of the generalized Reynolds stress introduced by Kampé de Fériet the Navier-Stokes equations are invariant to the filtering process, and this *averaging invariance* is claimed in my paper [6], but essentially it is due to my connection to the past. To generalize all that was simple. Let us consider a generic filtering operator $\mathcal{F}$ representative of our Large Eddy Simulation, the explicit or implicit *LES filter*, and given the generic turbulent quantities, $a, b, c, \cdots$, let us introduce the Generalized Central Moments (GCM) [6, 9, 10] associated with $\mathcal{F}$

$$\begin{aligned}
\tau_f(a,b) &\equiv \langle ab \rangle_f - \langle a \rangle_f \langle b \rangle_f \\
\tau_f(a,b,c) &\equiv \langle abc \rangle_f - \langle a \rangle_f \tau_f(b,c) - \langle b \rangle_f \tau_f(c,a) - \langle c \rangle_f \tau_f(a,b) - \langle a \rangle_f \langle b \rangle_f \langle c \rangle_f
\end{aligned} \tag{17}$$

Let us now define a *test filter* $\mathcal{G}$ that we only assume linear and constant preserving and let us consider the quantities extracted by this test filter $\mathcal{G}$ when applied to the LES $\mathcal{F}$-filtered quantities. In other words we are interested to the *filter product* $\mathcal{P} = \mathcal{G}\mathcal{F}$. We assume that $\mathcal{G}\mathcal{F} = \mathcal{F}\mathcal{G}$, and our notation is the following

$$\begin{aligned}
\mathcal{P}[\cdots] &\equiv \mathcal{G}[\mathcal{F}[\cdots]] \equiv \mathcal{G}\mathcal{F}[\cdots] \equiv \langle \cdots \rangle_{gf} \equiv \langle \langle \cdots \rangle_f \rangle_g \equiv \langle \cdots \rangle_p \\
&\equiv \mathcal{F}[\mathcal{G}[\cdots]] \equiv \mathcal{F}\mathcal{G}[\cdots] \equiv \langle \cdots \rangle_{fg} \equiv \langle \langle \cdots \rangle_g \rangle_f \equiv \langle \cdots \rangle_p
\end{aligned} \tag{18}$$

It is easy to see that we can write the following identities

$$\begin{aligned}
\tau_p(a,b) &= \tau_g(\langle a \rangle_f, \langle b \rangle_f) + \langle \tau_f(a,b) \rangle_g = \tau_f(\langle a \rangle_g, \langle b \rangle_g) + \langle \tau_g(a,b) \rangle_f \\
\tau_p(a,b,c) &= \tau_g(\langle a \rangle_f, \langle b \rangle_f, \langle c \rangle_f) + \langle \tau_f(a,b,c) \rangle_g \\
&\quad + \tau_g(\langle a \rangle_f, \tau_f(b,c)) + \tau_g(\langle b \rangle_f, \tau_f(c,a)) + \tau_g(\langle c \rangle_f, \tau_f(a,b)) \\
&= \tau_f(\langle a \rangle_g, \langle b \rangle_g, \langle c \rangle_g) + \langle \tau_g(a,b,c) \rangle_f \\
&\quad + \tau_f(\langle a \rangle_g, \tau_g(b,c)) + \tau_f(\langle b \rangle_g, \tau_g(c,a)) + \tau_f(\langle c \rangle_g, \tau_g(a,b))
\end{aligned} \tag{19}$$

where, consistently with the definition of the GCM,

$$
\begin{aligned}
\tau_p(a,b) &\equiv \tau_{gf}(a,b) = \langle ab\rangle_{gf} - \langle a\rangle_{gf}\langle b\rangle_{gf} \\
&\equiv \tau_{fg}(a,b) = \langle ab\rangle_{fg} - \langle a\rangle_{fg}\langle b\rangle_{fg} \\
\tau_p(a,b,c) &\equiv \tau_{gf}(a,b,c) = \langle abc\rangle_{gf} - \langle a\rangle_{gf}\tau_{gf}(b,c) - \langle b\rangle_{gf}\tau_{gf}(c,a) \\
&\quad - \langle c\rangle_{gf}\tau_{gf}(a,b) - \langle a\rangle_{gf}\langle b\rangle_{gf}\langle c\rangle_{gf} \\
&\equiv \tau_{fg}(a,b,c) = \langle abc\rangle_{fg} - \langle a\rangle_{fg}\tau_{fg}(b,c) - \langle b\rangle_{fg}\tau_{fg}(c,a) \\
&\quad - \langle c\rangle_{fg}\tau_{fg}(a,b) - \langle a\rangle_{fg}\langle b\rangle_{fg}\langle c\rangle_{fg} \\
\tau_g(\langle a\rangle_f, \langle b\rangle_f) &\equiv \langle \langle a\rangle_f \langle b\rangle_f\rangle_g - \langle a\rangle_{gf}\langle b\rangle_{gf} \\
\tau_g(\langle a\rangle_f, \tau_f(b,c)) &\equiv \langle \langle a\rangle_f \tau_f(b,c)\rangle_g - \langle a\rangle_{gf}\langle \tau_f(b,c)\rangle_g
\end{aligned} \tag{20}
$$

A particularly important case is $\mathcal{G} = \mathcal{E}$. In this case, if we also assume that $\mathcal{E}\mathcal{F} = \mathcal{E}$, we have

$$
\begin{aligned}
\tau_e(a,b) &= \tau_e(\langle a\rangle_f, \langle b\rangle_f) + \langle \tau_f(a,b)\rangle_e \\
\tau_e(a,b,c) &= \tau_e(\langle a\rangle_f, \langle b\rangle_f, \langle c\rangle_f) + \langle \tau_f(a,b,c)\rangle_e + \tau_e(\langle a\rangle_f, \tau_f(b,c)) \\
&\quad + \tau_e(\langle b\rangle_f, \tau_f(c,a)) + \tau_e(\langle c\rangle_f, \tau_f(a,b))
\end{aligned} \tag{21}
$$

and we remark that the *total* turbulence represented by the statistical central moments $\tau_e(a,b)$ and $\tau_e(a,b,c)$ is operationally decomposed in the *resolved* large scale turbulence $\tau_e(\langle a\rangle_f, \langle b\rangle_f)$ and $\tau_e(\langle a\rangle_f, \langle b\rangle_f, \langle c\rangle_f)$, and in the *subfilter* small scale turbulence given by $\langle \tau_f(a,b)\rangle_e$, $\langle \tau_f(a,b,c)\rangle_e$ and $\tau_e(\langle a\rangle_f, \tau_f(b,c))$.

This multiscale operational technique can be easily extended to an ensemble of filtering operators hierarchically organized

$$
\mathcal{F}\ ,\ \mathcal{G}\mathcal{F}\ ,\ \mathcal{H}\mathcal{G}\mathcal{F}\ ,\ \cdots \tag{22}
$$

and has a lot of different applications. At the time of our meeting in Pisa the main interest was in subgrid modeling, and the first application of this multiscale filtering approach was dedicated to the optimization of the Smagorinsky model. In this case the filter $\mathcal{G}$ is a test filter that applied to the $\mathcal{F}$-filtered quantities should provide a different representation at a different resolution level, useful to compare and optimize the subgrid model at different grid resolution. We refer for more detail on that to [10], and we also remark an interesting review and generalization of this modeling procedure recently provided by Meneveau [11].

Here we would like to also recall the application of this multiscale filtering technique to the analysis of LES data, in particular, the extraction of statistical data from a LES database. Let us first of all consider a constant density turbulent field and let us assume that $\mathcal{EF} = \mathcal{E}$. We explicitly define

$$
\begin{aligned}
R_{ij} &\equiv \tau_e(u_i, u_j) \\
R_{ijk} &\equiv \tau_e(u_i, u_j, u_k) \\
\tau_{ij} &\equiv \tau_f(u_i, u_j) \\
\tau_{ijk} &\equiv \tau_f(u_i, u_j, u_k) \\
T_{ij} &\equiv \tau_e(\langle u_i \rangle_f, \langle u_j \rangle_f) \\
T_{ijk} &\equiv \tau_e(\langle u_i \rangle_f, \langle u_j \rangle_f, \langle u_k \rangle_f)
\end{aligned} \tag{23}
$$

where u_i are the components of the velocity field at a given time and location, and we have

$$
\begin{aligned}
R_{ij} &= \langle \tau_{ij} \rangle_e + T_{ij} \\
R_{ijk} &= \langle \tau_{ijk} \rangle_e + T_{ijk} + \tau_e(\langle u_i \rangle_f, \tau_{jk}) + \tau_e(\langle u_j \rangle_f, \tau_{ki}) + \tau_e(\langle u_k \rangle_f, \tau_{ij})
\end{aligned} \tag{24}
$$

where

$$
\tau_e(\langle u_i \rangle_f, \tau_{jk}) \equiv \langle \langle u_i \rangle_f \tau_{jk} \rangle_e - \langle \langle u_i \rangle_f \rangle_e \langle \langle \tau_{jk} \rangle_f \rangle_e \tag{25}
$$

Let us now only assume that $\mathcal{EF} = \mathcal{FE}$. We define the following additive GCM

$$
\begin{aligned}
\vartheta_{ij} &\equiv \tau_f(\langle u_i \rangle_e, \langle u_j \rangle_e) \\
\vartheta_{ijk} &\equiv \tau_f(\langle u_i \rangle_e, \langle u_j \rangle_e, \langle u_k \rangle_e)
\end{aligned} \tag{26}
$$

and we have finally

$$
\langle R_{ij} \rangle_f + \vartheta_{ij} = \langle \tau_{ij} \rangle_e + T_{ij} \tag{27}
$$

$$
\begin{aligned}
&\langle R_{ijk} \rangle_f + \vartheta_{ijk} + \tau_f(\langle u_i \rangle_e, R_{jk}) + \tau_f(\langle u_j \rangle_e, R_{ki}) + \tau_f(\langle u_k \rangle_e, R_{ij}) = \\
&\langle \tau_{ijk} \rangle_e + T_{ijk} + \tau_e(\langle u_i \rangle_f, \tau_{jk}) + \tau_e(\langle u_j \rangle_f, \tau_{ki}) + \tau_e(\langle u_k \rangle_f, \tau_{ij})
\end{aligned} \tag{28}
$$

Another important application of this fluctuation-free multiscale approach is to Variable Density Turbulent Fields ϱ, u_i. In this case we also remove the mass-weighted averages [12], and in terms of the GCM we explicitly write

$$
\begin{aligned}
R_{\varrho i} &\equiv \tau_e(\varrho, u_i) \\
R_{ij} &\equiv \tau_e(u_i, u_j) \\
R_{\varrho ij} &\equiv \tau_e(\varrho, u_i, u_j) \\
\\
\tau_{\varrho i} &\equiv \tau_f(\varrho, u_i) \\
\tau_{ij} &\equiv \tau_f(u_i, u_j) \\
\tau_{\varrho ij} &\equiv \tau_f(\varrho, u_i, u_j) \\
\\
T_{\varrho i} &\equiv \tau_e(\langle\varrho\rangle_f, \langle u_i\rangle_f) \\
T_{ij} &\equiv \tau_e(\langle u_i\rangle_f, \langle u_j\rangle_f) \\
T_{\varrho ij} &\equiv \tau_e(\langle\varrho\rangle_f, \langle u_i\rangle_f, \langle u_j\rangle_f)
\end{aligned} \tag{29}
$$

where ϱ is the density and u_i are the components of the velocity field at a given time and location. We have

$$
\begin{aligned}
R_{\varrho i} &= \langle\tau_{\varrho i}\rangle_e + T_{\varrho i} \\
R_{ij} &= \langle\tau_{ij}\rangle_e + T_{ij}
\end{aligned} \tag{30}
$$

$$
R_{\varrho ij} = \langle\tau_{\varrho ij}\rangle_e + T_{\varrho ij} + \tau_e(\langle\varrho\rangle_f, \tau_{ij}) + \tau_e(\langle u_i\rangle_f, \tau_{\varrho j}) + \tau_e(\langle u_j\rangle_f, \tau_{\varrho i}) \tag{31}
$$

Let us now only assume that $\mathcal{EF} = \mathcal{FE}$. We define the following additive GCM

$$
\begin{aligned}
\vartheta_{\varrho i} &\equiv \tau_f(\langle\varrho\rangle_e, \langle u_i\rangle_e) \\
\vartheta_{ij} &\equiv \tau_f(\langle u_i\rangle_e, \langle u_j\rangle_e) \\
\vartheta_{\varrho ij} &\equiv \tau_f(\langle\varrho\rangle_e, \langle u_i\rangle_e, \langle u_j\rangle_e)
\end{aligned} \tag{32}
$$

and we have finally

$$
\begin{aligned}
\langle R_{\varrho i}\rangle_f + \vartheta_{\varrho i} &= \langle\tau_{\varrho i}\rangle_e + T_{\varrho i} \\
\langle R_{ij}\rangle_f + \vartheta_{ij} &= \langle\tau_{ij}\rangle_e + T_{ij}
\end{aligned} \tag{33}
$$

and

$$\langle R_{\varrho ij}\rangle_f + \vartheta_{\varrho ij} + \tau_f(\langle\varrho\rangle_e, R_{ij}) + \tau_f(\langle u_i\rangle_e, R_{\varrho j}) + \tau_f(\langle u_j\rangle_e, R_{\varrho i}) =$$
$$\langle\tau_{\varrho ij}\rangle_e + T_{\varrho ij} + \tau_e(\langle\varrho\rangle_f, \tau_{ij}) + \tau_e(\langle u_i\rangle_f, \tau_{\varrho j}) + \tau_e(\langle u_j\rangle_f, \tau_{\varrho i}) \quad (34)$$

These last decompositions of the Reynolds stresses in the case of compressible turbulent flow are new, simple, and fluctuation-free. Moreover they are also mass weighted-free, and the author really hope that they could be appreciated more and more in the future in the analysis of LES databases. We recall finally that some recent applications of the filtering approach concerning the subfilter stress [13], the decomposition of the Reynolds stresses [14, 15], the dynamic modeling of the Shock Driven Turbulent Mixing [16] and the definition of statistical homogeneity indices [17] have been recently reviewed in [18].

4 Conclusions

But let us return to that afternoon in Pisa and to the conversation with Carlo. Everything went on well: two months later my paper was finally published, and in the following years the application of the new dynamic modeling procedure to the anisotropic eddy viscosity model proposed by Carlo, Antonella, and Lorenzo was developed and worked successfully [19]. Fluctuations continue to be applied in LES, but some final lessons deserve to be remarked. The first is related to our starting point, the identity (10). Why this simple and, in our opinion, interesting decomposition of the Reynolds stress was appreciated so lately in turbulence? We remark that in applied statistics the identities (3) and (10) are better known as the *Adam and Eve's laws* [20, 21]: given a partition of the probability space, the total statistical covariance R_{ij} of u_i and u_j is the average $\overline{\tau}_{ij}$ of the partial covariance τ_{ij} plus the statistical covariance T_{ij} of the partial mean values. Science as every human activity is often conditioned by great ideas that in some cases obscure other different possibilities. Obviously the fluctuations remain very important, but not always they are essential, and in the case of LES their importance has been overvalued. Individual deviations from the average are intuitively very appealing, but not always you are obliged to decline them: differences are not so relevant in turbulence.

Another important lesson regards finally our ideal endless dialogue with the past: old papers are often plenty of useful observations and suggestions.

References

1. Germano, M., Piomelli, U., Moin, P., Cabot, W.H.: A dynamic subgrid-scale eddy viscosity model. Phys. Fluids A **3**, 1760–1765 (1991)
2. Leonard, A.: Energy cascade in large-Eddy simulations of turbulent fluid flows. Adv. Geophysics, **18A**, 237–248 (1975)
3. Boussinesq, J.: Essai sur la théorie des eaux courantes, Mém. Savants Etrangers Acad. Sciences **26**, 1–680 (1877). Paris
4. Reynolds, O.: On the dynamical theory of incompressible viscous fluids and the determination of the criterion. Philos. Trans. Roy. Soc. A **186**, 123–164 (1895)
5. Kampé de Fériet, J.: La notion de moyenne dans la théorie de la turbulence. Rend. Sem. Mat. Fis. di Milano **27**, 167–207 (1957)
6. Germano, M.: Turbulence: The filtering approach. J. Fluid Mech. **238**, 325–336 (1992)
7. Reynolds, W.C., Hussain A.K.M.F.: The mechanics of an organized wave in turbulent shear flow. Part 3. Theoretical models and comparisons with experiments. J. Fluid. Mech. **54**, 263–288 (1972)
8. Monin, A.S., Yaglom, A.M.: Statistical Fluid Mechanics. The MIT Press, Cambridge, MA (1971)
9. Germano, M.: A statistical formulation of the dynamic model. Phys. Fluids **8**, 565–570 (1996)
10. Germano, M.: Fundamentals of Large Eddy Simulation. Advanced Turbulent Flows Computations. In: Peyret, R., Krause, E. (eds.) CISM Courses and Lectures, vol. 395, pp. 81–130. Springer (2000)
11. Meneveau, C.: Germano identity-based subgrid-scale modeling: A brief survey of variations on a fertile theme. Phys. Fluids **24**, 121301 (2012)
12. Germano, M.: On the Hybrid RANS-LES of compressible flows. In: Girimaji et al. (eds.) Progress in Hybrid RANS-LES Modelling, pp. 253–263. Springer (2014)
13. Cimarelli, A., Abbà, A., Germano, M.: General formalism for a reduced description and modelling of momentum and energy transfer in turbulence. J. Fluid Mech. **866**, 865–896 (2019)
14. Klein, M., Germano, M.: Decomposition of the Reynolds stress from filtered data. Phys. Rev. Fluids **3**, 114606 (2018)
15. Klein, M., Kasten, C., Germano, M.: Decomposition of Turbulent Fluxes from Filtered Data and Application to Turbulent Premixed Combustion Modelling (2019). Flow, Turbulence and Combustion, Published online 08 May 2019
16. Grinstein, F.F., Saenz, J.A., Rauenzahn, R.M., Germano, M., Israel, D.M.: Dynamic bridging modeling for coarse grained simulations of shock driven turbulent mixing. Comput. Fluids **199**, 104430 (2020)
17. Ferrero, A., Larocca, F., Scovazzi, G., Germano, M.: A numerical study of the spanwise turbulence past a cylinder flow. 17th ETC, 3–6 Sept. 2019, Torino, Italy (2019)
18. Abbà, A, Cimarelli, A., Klein, M., Ferrero, A., Grinstein, F.F., Larocca, F., Saenz, J.A., Scovazzi, G., Germano, M.: The Filtering Approach as a Tool for Modeling and Analyzing Turbulence. In: R. Örlü et al. (eds.) Progress in Turbulence IX, Springer Proceedings in Physics, vol. 267 (2021)
19. Abbà, A., Cercignani, C., Valdettaro, L.: Analysis of subgrid scale models. Comput. Math. Appl. **46**, 521–535 (2003)
20. Rice, J.A.: Mathematical Statistics and Data Analysis, 3rd edn. Thomson (2007)
21. Blitzstein, J.K., Hwang, J.: Introduction to Probability. CRC Press LLC (2014)

An Ellipsoidal-Statistical (ES) Model for a Polyatomic Gas with Temperature-Dependent Specific Heats

Shingo Kosuge and Kazuo Aoki

Abstract In a previous paper by the present authors and H.-W. Kuo [S. Kosuge *et al*, J. Stat. Phys. **177**, 209 (2019)], the ellipsoidal-statistical (ES) model of the Boltzmann equation for a polyatomic gas with constant specific heats (calorically perfect gas), proposed by P. Andries *et al.* [P. Andries *et al.*, Eur. J. Mech. B/Fluids **19**, 813 (2000)], was extended to a polyatomic gas with temperature-dependent specific heats (thermally perfect gas), and the associated Navier–Stokes equations were derived by the Chapman–Enskog procedure. In this paper, the new model, together with the Navier–Stokes equations, is summarized. Then, the form of the appropriate boundary conditions for the latter equations is derived by the analysis of the Knudsen layer.

In memory of Carlo Cercignani (1939–2010)

1 Introduction

The study of nonequilibrium flows of polyatomic gases is an important subject in kinetic theory of gases nowadays in connection with various applications in high-temperature circumstances [1, 2]. However, the original Boltzmann equation, even in the case of a single-component gas, requires detailed information about collision processes of polyatomic molecules, which is not necessarily available. Even if the data are available for a specific gas, the Boltzmann equation becomes very complicated. Therefore, it is generally very hard to use the original Boltzmann

S. Kosuge
Institute for Liberal Arts and Sciences, Kyoto University, Kyoto, Japan
e-mail: kosuge.shingo.6r@kyoto-u.ac.jp

K. Aoki (✉)
Department of Mathematics, National Cheng Kung University, Tainan, Taiwan
e-mail: kazuo.aoki.22v@st.kyoto-u.ac.jp

P. Barbante et al. (eds.), *From Kinetic Theory to Turbulence Modeling*, Springer INdAM Series 51, https://doi.org/10.1007/978-981-19-6462-6_12

equation for practical applications. To bypass this difficulty, various simplified and tractable models of the Boltzmann equation, which satisfy some basic properties of the original Boltzmann equation, have been proposed (see [3–13] and the references therein), and some of them have been successfully applied to practical problems. One of such models is the ellipsoidal-statistical (ES) model proposed by Andries *et al.* [6] and rederived in a systematic way by Brull and Schneider [14].

In a previous paper by the present authors [15], this ES model was successfully applied to analyzing the structure of a plane shock wave in a polyatomic gas with large bulk viscosity. However, the ES model in [6] is for a polyatomic gas with constant specific heats (calorically perfect gas). In reality, however, the specific heats for most gases depend on the temperature even under the condition that the gases are treated as ideal gases. The effect of the temperature dependence of the specific heats becomes particularly important when the temperature varies significantly in the flow field, such as gas flows containing strong shock waves [16].

For this reason, in a paper by the present authors and H.-W Kuo [17], the ES model in [6] has been extended to the case of a polyatomic gas with temperature-dependent specific heats (thermally perfect gas). In the same reference, the compressible Navier–Stokes equations were derived from the new model by the Chapman–Enskog expansion [18–21]. In the present paper, we first summarize the new ES model proposed in [17] for a polyatomic gas with temperature-dependent specific heats (Sect. 2) together with its basic properties (Sect. 3). Then, the associated Navier–Stokes equations are summarized in Sect. 4. Finally, we will show the form of the appropriate boundary conditions for the Navier–Stokes equations that is derived by the analysis of the Knudsen layer (Sect. 5).

2 ES Model for a Gas with Temperature-Dependent Specific Heats

Let us consider a polyatomic rarefied gas. Let t be the time variable, $\boldsymbol{X}$ (or X_i) the position vector in the physical space, $\boldsymbol{\xi}$ (or ξ_i) the molecular velocity, and $\mathcal{E}$ the energy associated with the internal modes per unit mass. We denote the total mass of the gas molecules contained in an infinitesimal volume $d\boldsymbol{X}d\boldsymbol{\xi}d\mathcal{E}$ around a point $(\boldsymbol{X}, \boldsymbol{\xi}, \mathcal{E})$ in the seven-dimensional space consisting of $\boldsymbol{X}$, $\boldsymbol{\xi}$, and $\mathcal{E}$ at time t by

$$f(t, \boldsymbol{X}, \boldsymbol{\xi}, \mathcal{E})d\boldsymbol{X}d\boldsymbol{\xi}d\mathcal{E}. \tag{1}$$

The ES model is the equation governing this $f(t, \boldsymbol{X}, \boldsymbol{\xi}, \mathcal{E})$. In the present paper, we consider the thermally perfect gas, for which the specific heat at constant volume c_{V} and that at constant pressure c_{P} are both functions of the temperature T, and summarize the model Boltzmann equation for such a gas proposed in [17]. It is a straightforward extension of the ES model for a gas with constant c_{V} and c_{P} [6].

We first consider an equilibrium state with a uniform and constant temperature T and assume that the specific heats are constant and the classical equipartition law holds. Then, the internal energy E of the gas per unit mass can be expressed as $E = (3 + \delta_0)RT/2$, where δ_0 is the (constant) number of internal degrees of freedom, and R is the gas constant per unit mass. Now, we consider a more general gas for which the internal energy E is a given (monotonically increasing) function of the temperature T, i.e., $E = E(T)$. Then, we define a function $D(T)$ as $D(T) = 2E(T)/RT - 3$. This $D(T)$ is a *generalized* number of internal degrees of freedom. Note that the relations $c_{\mathrm{v}}(T) = dE(T)/dT$ and $c_{\mathrm{p}}(T) = c_{\mathrm{v}}(T) + R$ hold, so that c_{v} and c_{p} depend on the temperature.

Next, we extend $E(T)$, $D(T)$, $c_{\mathrm{v}}(T)$, and $c_{\mathrm{p}}(T)$ in the nonequilibrium case and use them in the definition of the new model equation, which is described as follows [17]:

$$\frac{\partial f}{\partial t} + \xi_i \frac{\partial f}{\partial X_i} = Q(f), \tag{2}$$

with

$$Q(f) = A_c(T)\rho(\mathcal{G} - f). \tag{3}$$

Here,

$$\mathcal{G} = \frac{\rho \mathcal{E}^{\delta/2-1}}{(2\pi)^{3/2}[\det(\mathsf{T})]^{1/2}(RT_{\mathrm{rel}})^{\delta/2}\Gamma(\delta/2)} \times \exp\left(-\frac{1}{2}(\mathsf{T}^{-1})_{ij}(\xi_i - v_i)(\xi_j - v_j) - \frac{\mathcal{E}}{RT_{\mathrm{rel}}}\right), \tag{4a}$$

$$(\mathsf{T})_{ij} = (1-\theta)[(1-\nu)RT_{\mathrm{tr}}\delta_{ij} + \nu p_{ij}/\rho] + \theta RT\delta_{ij}, \tag{4b}$$

$$\rho = \iint_0^\infty f\, d\mathcal{E} d\boldsymbol{\xi}, \qquad v_i = \frac{1}{\rho}\iint_0^\infty \xi_i f\, d\mathcal{E} d\boldsymbol{\xi}, \tag{4c}$$

$$p_{ij} = \iint_0^\infty (\xi_i - v_i)(\xi_j - v_j) f\, d\mathcal{E} d\boldsymbol{\xi}, \tag{4d}$$

$$T = E^{-1}(e), \qquad \delta = D(T) = 2e/RT - 3, \tag{4e}$$

$$T_{\mathrm{tr}} = 2e_{\mathrm{tr}}/3R, \qquad T_{\mathrm{int}} = 2e_{\mathrm{int}}/R\delta, \qquad T_{\mathrm{rel}} = \theta T + (1-\theta)T_{\mathrm{int}}, \tag{4f}$$

where e, e_{tr}, and e_{int} are defined by

$$e = e_{\mathrm{tr}} + e_{\mathrm{int}}, \quad e_{\mathrm{tr}} = \frac{1}{2\rho}\iint_0^\infty |\boldsymbol{\xi} - \boldsymbol{v}|^2 f\, d\mathcal{E} d\boldsymbol{\xi}, \quad e_{\mathrm{int}} = \frac{1}{\rho}\iint_0^\infty \mathcal{E} f\, d\mathcal{E} d\boldsymbol{\xi}. \tag{5}$$

In Eqs. (2)–(5), ρ is the density, $\boldsymbol{v}$ (or v_i) is the flow velocity, p_{ij} is the stress tensor, e is the internal energy per unit mass, $e_{\rm tr}$ is that associated with the translational motion, $e_{\rm int}$ is that associated with the internal modes, T is the temperature, $T_{\rm tr}$ is the temperature associated with the translational motion, $T_{\rm int}$ is the temperature associated with the energy of the internal modes, $d\boldsymbol{\xi} = d\xi_1 d\xi_2 d\xi_3$, and the domain of integration with respect to $\boldsymbol{\xi}$ is the whole space of $\boldsymbol{\xi}$. The symbol δ_{ij} indicates the Kronecker delta, and $\nu \in [-1/2, 1)$ and $\theta \in [0, 1]$ are parameters. In addition, $A_c(T)$ is a function of T such that $A_c(T)\rho$ is the collision frequency of the gas molecules, $\Gamma(z) = \int_0^\infty s^{z-1} e^{-s} ds$ is the gamma function, T is the 3×3 positive-definite symmetric matrix whose (i, j) component is defined by Eq. (4b), and $\det(\mathsf{T})$ and T^{-1} are, respectively, its determinant and inverse.

Note that all the macroscopic quantities contained in $\mathcal{G}$ are generated from f. To be more specific, (i) ρ, $\boldsymbol{v}$, p_{ij}, $e_{\rm tr}$, $e_{\rm int}$, and e are obtained by Eqs. (4c), (4d), and (5); (ii) T and then δ are determined by Eq. (4e) using the inverse function E^{-1} of the function E; (iii) $T_{\rm tr}$, $T_{\rm int}$, and $T_{\rm rel}$ are determined by Eq. (4f), and then T is established by Eq. (4b). Since $e = e_{\rm tr} + e_{\rm int} = (3T_{\rm tr} + \delta T_{\rm int})R/2$ and also $e = (3 + \delta)RT/2$, we have the relation $T = (3T_{\rm tr} + \delta T_{\rm int})/(3 + \delta)$ [note that δ depends on T: cf. Eq. (4e)].

The pressure p and the heat-flow vector q_i are given by

$$p = R\rho T, \tag{6a}$$

$$q_i = \iint_0^\infty (\xi_i - v_i)\left(\frac{1}{2}|\boldsymbol{\xi} - \boldsymbol{v}|^2 + \mathcal{E}\right) f d\mathcal{E} d\boldsymbol{\xi}, \tag{6b}$$

where Eq. (6a) is the equation of state.

3 Basic Properties

In the following, we summarize the basic properties that follow from the model equation (2) [17]. Some properties are different when the parameter θ is equal to zero.

- *Conservations for* $\theta \neq 0$: For an arbitrary function $f(t, \boldsymbol{X}, \boldsymbol{\xi}, \mathcal{E})$, the following relation holds:

$$\iint_0^\infty \varphi_r Q(f) d\mathcal{E} d\boldsymbol{\xi} = 0, \tag{7}$$

where φ_r $(r = 0, \ldots, 4)$ are the collision invariants, i.e.,

$$\varphi_0 = 1, \qquad \varphi_i = \xi_i \quad (i = 1, 2, 3), \qquad \varphi_4 = \frac{1}{2}|\boldsymbol{\xi}|^2 + \mathcal{E}. \tag{8}$$

- *Conservations for* $\theta = 0$: For an arbitrary function $f(t, \boldsymbol{X}, \boldsymbol{\xi}, \mathcal{E})$, the following relation holds:

$$\iint_0^\infty \varphi_r Q(f) d\mathcal{E} d\boldsymbol{\xi} = 0, \tag{9}$$

where φ_r ($r = 0, \ldots, 5$) are the collision invariants, i.e.,

$$\varphi_0 = 1, \qquad \varphi_i = \xi_i \quad (i = 1, 2, 3), \qquad \varphi_4 = \frac{1}{2}|\boldsymbol{\xi}|^2, \qquad \varphi_5 = \mathcal{E}. \tag{10}$$

- *Equilibrium for* $\theta \neq 0$: The vanishing of the collision term $Q(f) = 0$ is equivalent to the fact that f is the following local equilibrium distribution:

$$f_{\mathrm{eq}} = \frac{\bar{\rho}\mathcal{E}^{\bar{\delta}/2-1}}{(2\pi R\bar{T})^{3/2}(R\bar{T})^{\bar{\delta}/2}\Gamma(\bar{\delta}/2)} \exp\left(-\frac{|\boldsymbol{\xi} - \bar{\boldsymbol{v}}|^2}{2R\bar{T}} - \frac{\mathcal{E}}{R\bar{T}}\right), \tag{11}$$

where $\bar{\rho}$, $\bar{\boldsymbol{v}}$, and $\bar{T}$ are arbitrary functions of t and $\boldsymbol{X}$, and $\bar{\delta} = D(\bar{T})$.
- *Equilibrium for* $\theta = 0$: The vanishing of the collision term $Q(f) = 0$ is equivalent to the fact that f is the following local equilibrium distribution:

$$f_{\mathrm{eq}} = \frac{\bar{\rho}\mathcal{E}^{\bar{\delta}/2-1}}{(2\pi R\bar{T}_{\mathrm{tr}})^{3/2}(R\bar{T}_{\mathrm{int}})^{\bar{\delta}/2}\Gamma(\bar{\delta}/2)} \exp\left(-\frac{|\boldsymbol{\xi} - \bar{\boldsymbol{v}}|^2}{2R\bar{T}_{\mathrm{tr}}} - \frac{\mathcal{E}}{R\bar{T}_{\mathrm{int}}}\right), \tag{12}$$

where $\bar{\rho}$, $\bar{\boldsymbol{v}}$, $\bar{T}_{\mathrm{tr}}$, and $\bar{T}_{\mathrm{int}}$ are arbitrary functions of t and $\boldsymbol{X}$, and $\bar{\delta}$ and $\bar{T}$ are determined by the following coupled equations:

$$\bar{\delta} = D(\bar{T}), \qquad \bar{T} = E^{-1}(3R\bar{T}_{\mathrm{tr}}/2 + \bar{\delta}R\bar{T}_{\mathrm{int}}/2). \tag{13}$$

The solution $(\bar{\delta}, \bar{T})$ of Eq. (13) exists. In particular, it is unique when $\bar{T}_{\mathrm{int}} \leq \bar{T}_{\mathrm{tr}}$.
- *Entropy inequality*: For an arbitrary function $f(t, \boldsymbol{X}, \boldsymbol{\xi}, \mathcal{E})$, the following inequality holds:

$$\iint_0^\infty \left(\ln \frac{f}{\mathcal{E}^{\delta/2-1}}\right) Q(f) d\mathcal{E} d\boldsymbol{\xi} \leq 0, \tag{14}$$

where δ is defined by Eqs. (4e) and (5). The equality sign holds if and only if $f = f_{\mathrm{eq}}$ in Eq. (11) ($\theta \neq 0$) or (12) ($\theta = 0$).
- *H theorem for spatially homogeneous case*: Let $H_\delta(f)$ be defined by

$$H_\delta(f) = \iint_0^\infty f \ln \frac{f}{\mathcal{E}^{\delta/2-1}} d\mathcal{E} d\boldsymbol{\xi}. \tag{15}$$

If f does not depend on $\boldsymbol{X}$, the following inequality holds:

$$dH_\delta/dt \le 0, \tag{16}$$

and the equality sign holds if and only if $f = f_{\mathrm{eq}}$ with constant $\bar{\rho}$, $\bar{\boldsymbol{v}}$, and $\bar{T}$ in Eq. (11) ($\theta \neq 0$) or that with constant $\bar{\rho}$, $\bar{\boldsymbol{v}}$, $\bar{T}_{\mathrm{tr}}$, and $\bar{T}_{\mathrm{int}}$ in Eq. (12) ($\theta = 0$).

- *Mean free path*: The mean free path l_0 of the gas molecules in the equilibrium state at rest at density ρ_0 and temperature T_0 is given by

$$l_0 = \frac{2}{\sqrt{\pi}} \frac{(2RT_0)^{1/2}}{A_c(T_0)\rho_0}, \tag{17}$$

for Eq. (2), since $A_c(T_0)\rho_0$ is the collision frequency at this equilibrium state.

It should be remarked that we were not able to show the H theorem for the spatially inhomogeneous case. It is a drawback of this model, which should be the consequence of the simple and convenient definition of the temperature T and the resulting δ using the given function E [cf. Eq. (4e)].

4 Navier–Stokes Equations

When the mean free path l_0 of the gas molecules in the reference equilibrium state at rest with density ρ_0 and temperature T_0 is much shorter than the characteristic length L of the system, i.e., when the Knudsen number $\mathrm{Kn} = l_0/L$ is small, one can formally derive fluid-dynamic equations by a classical procedure, such as the Chapman–Enskog expansion, from Eq. (2). In this section, restricting ourselves to the case where θ is of the order of unity, we summarize the Navier–Stokes equations derived by the standard Chapman–Enskog procedure. Incidentally, when θ is small, one can derive Navier–Stokes type equations with two temperatures and their appropriate boundary conditions by following the procedure for a polyatomic gas with constant specific heats [22, 23].

The macroscopic quantities ρ, v_i, and T are governed by the following Navier–Stokes equations for a compressible fluid [17]:

$$\frac{\partial \rho}{\partial t} + \frac{\partial \rho v_j}{\partial X_j} = 0, \tag{18a}$$

$$\frac{\partial \rho v_i}{\partial t} + \frac{\partial \rho v_i v_j}{\partial X_j} = -\frac{\partial p}{\partial X_i} + \frac{\partial}{\partial X_j}\left[\mu(T)\left(\frac{\partial v_i}{\partial X_j} + \frac{\partial v_j}{\partial X_i} - \frac{2}{3}\frac{\partial v_k}{\partial X_k}\delta_{ij}\right)\right] + \frac{\partial}{\partial X_i}\left[\mu_b(T)\frac{\partial v_k}{\partial X_k}\right], \tag{18b}$$

$$\frac{\partial}{\partial t}\left[\rho\left(\frac{3+\delta}{2}RT+\frac{1}{2}|\boldsymbol{v}|^2\right)\right]+\frac{\partial}{\partial X_j}\left[\rho v_j\left(\frac{5+\delta}{2}RT+\frac{1}{2}|\boldsymbol{v}|^2\right)\right]$$
$$=\frac{\partial}{\partial X_j}\left[\lambda(T)\frac{\partial T}{\partial X_j}\right]+\frac{\partial}{\partial X_j}\left[\mu(T)v_i\left(\frac{\partial v_i}{\partial X_j}+\frac{\partial v_j}{\partial X_i}-\frac{2}{3}\frac{\partial v_k}{\partial X_k}\delta_{ij}\right)\right]$$
$$+\frac{\partial}{\partial X_j}\left[\mu_b(T)v_j\frac{\partial v_k}{\partial X_k}\right], \tag{18c}$$

with $p = R\rho T$ [cf. Eq. (6a)] and $\delta = D(T)$. Here, the viscosity $\mu(T)$, the bulk viscosity $\mu_b(T)$, and the thermal conductivity $\lambda(T)$ are given by

$$\mu(T)=\mathrm{Pr}\frac{RT}{A_c(T)}, \qquad \mu_b(T)=\frac{1}{\theta}\left[\frac{5}{3}-\gamma(T)\right]\frac{1}{\mathrm{Pr}}\mu(T),$$
$$\lambda(T)=\frac{\gamma(T)}{\gamma(T)-1}R\frac{RT}{A_c(T)}, \qquad \mathrm{Pr}=\frac{1}{1-\nu+\theta\nu}, \tag{19}$$

where $\gamma(T) = c_{\mathrm{p}}(T)/c_{\mathrm{v}}(T) = [c_{\mathrm{v}}(T) + R]/c_{\mathrm{v}}(T)$ is the ratio of the specific heats, and $\mathrm{Pr} = c_{\mathrm{p}}(T)\mu(T)/\lambda(T)$ is the Prandtl number. In addition, the translational and internal temperatures T_{tr} and T_{int} are, respectively, given as

$$T_{\mathrm{tr}}=T-\frac{1}{R}\mu_b(T)\frac{1}{\rho}\frac{\partial v_k}{\partial X_k}, \qquad T_{\mathrm{int}}=T+\frac{3}{R\delta}\mu_b(T)\frac{1}{\rho}\frac{\partial v_k}{\partial X_k}. \tag{20}$$

One of the advantages of the present ES model is that the resulting transport coefficients (19) are simple and explicit, so that the Navier–Stokes equations (18) have a wide applicability for practical flows of a polyatomic gas with temperature-dependent specific heats. The same is true for the so-called higher-order transport coefficients, such as the coefficients occurring in the thermal stresses [21, 24].

5 Boundary Conditions for Navier–Stokes Equations

The Navier–Stokes equations shown in Sect. 4 correspond to the first-order solution of the Chapman–Enskog expansion, that is, the solution that satisfies the model equation (2) formally with the error of the order of Kn^2 in the dimensionless setting in which the collision term is set to be of the order of unity. Here, it should be recalled that the Chapman–Enskog solution is obtained without considering the effect of boundaries. However, in most practical problems, gas flows are in contact with boundaries, where the kinetic boundary condition, i.e., the boundary condition for Eq. (2), is specified. Therefore, the Chapman–Enskog solution should be matched with the kinetic boundary condition on the boundary. Since we are considering the first-order Chapman–Enskog solution, we have to satisfy the kinetic boundary condition also with the error of the order of Kn^2 (in the dimensionless setting). This can be achieved by introducing the Knudsen layer, a thin layer with thickness of the order of the mean free path l_0, adjacent to the boundary. The analysis of the Knudsen layer leads to the appropriate boundary conditions for the compressible Navier–Stokes equations. The Knudsen-layer technique was developed and established in the framework of asymptotic theory of the Boltzmann equation for small Knudsen numbers in a series of papers by Sone (e.g., [25–27]). The reader is referred to [21, 24] for the details.

The Knudsen-layer technique in Sone's asymptotic theory has been adjusted to the Champan–Enskog procedure to derive the boundary conditions for the compressible Navier–Stokes equations in [28] for the Boltzmann equation for a monatomic gas and in [29] for the ES model for a polyatomic gas with constant specific heats. Since the present model is basically the ES model, the Knudsen-layer analysis in [29] can straightforwardly be applied to the present case. In this section, we summarize the resulting form of the boundary conditions.

Before presenting the result, we mention the kinetic boundary condition for Eq. (2). We assume that the gas molecules undergo the Maxwell diffuse-specular reflection on the boundary. Let $\boldsymbol{X}_{\mathrm{w}}$ (or $X_{\mathrm{w}i}$) be a point on the boundary and $\boldsymbol{v}_{\mathrm{w}}$ (or $v_{\mathrm{w}i}$), T_{w}, and $\boldsymbol{n}$ (or n_i) be, respectively, the velocity, the temperature, and the unit normal vector, pointing into the gas, of the boundary at the point $\boldsymbol{X}_{\mathrm{w}}$. The boundary condition for Eq. (2) is written as

$$
\begin{aligned}
f(t, \boldsymbol{X}_{\mathrm{w}}, \boldsymbol{\xi}, \mathcal{E}) &= (1-\alpha)\mathcal{R} f(t, \boldsymbol{X}_{\mathrm{w}}, \boldsymbol{\xi}, \mathcal{E}) \\
&\quad + \alpha \frac{\rho_{\mathrm{w}} \mathcal{E}^{\delta_{\mathrm{w}}/2-1}}{(2\pi R T_{\mathrm{w}})^{3/2} (R T_{\mathrm{w}})^{\delta_{\mathrm{w}}/2} \Gamma(\delta_{\mathrm{w}}/2)} \exp\left(-\frac{|\boldsymbol{\xi} - \boldsymbol{v}_{\mathrm{w}}|^2}{2RT_{\mathrm{w}}} - \frac{\mathcal{E}}{RT_{\mathrm{w}}} \right), \\
&\qquad \text{for } (\boldsymbol{\xi} - \boldsymbol{v}_{\mathrm{w}}) \cdot \boldsymbol{n} > 0,
\end{aligned} \tag{21a}
$$

$$
\rho_{\mathrm{w}} = -\left(\frac{2\pi}{RT_{\mathrm{w}}} \right)^{1/2} \int_{(\boldsymbol{\xi} - \boldsymbol{v}_{\mathrm{w}}) \cdot \boldsymbol{n} < 0} \int_0^\infty (\boldsymbol{\xi} - \boldsymbol{v}_{\mathrm{w}}) \cdot \boldsymbol{n} f(t, \boldsymbol{X}_{\mathrm{w}}, \boldsymbol{\xi}, \mathcal{E}) d\mathcal{E} d\boldsymbol{\xi}, \tag{21b}
$$

where $\delta_{\mathrm{w}} = D(T_{\mathrm{w}})$, $\alpha \in (0, 1]$ is the accommodation coefficient, and $\mathcal{R}$ is the reflection operator defined with an arbitrary function $g(\boldsymbol{\xi})$ by

$$
\mathcal{R} g(\xi_i) = g(\xi_i - 2(\xi_j - v_{\mathrm{w}j}) n_j n_i). \tag{22}
$$

We assume that the shape of the boundary is smooth and $\boldsymbol{v}_{\mathrm{w}}$ and T_{w} change smoothly in t and $\boldsymbol{X}_{\mathrm{w}}$.

The boundary conditions for the Navier–Stokes equations (18) obtained by the Knudsen-layer procedure are in the following form: at $\boldsymbol{X} = \boldsymbol{X}_{\mathrm{w}}$,

$$(v_i - v_{\mathrm{w}i})n_i = 0, \tag{23a}$$

$$(v_i - v_{\mathrm{w}i})t_i = \frac{\sqrt{2}}{R^{1/2}}\frac{1}{\mathrm{Pr}}c_v^I\frac{\mu(T_{\mathrm{w}})}{\rho T_{\mathrm{w}}^{1/2}}\left(\frac{\partial v_i}{\partial X_j} + \frac{\partial v_j}{\partial X_i}\right)n_i t_j + \frac{2}{R}c_T^I\frac{\gamma(T_{\mathrm{w}})-1}{\gamma(T_{\mathrm{w}})}\frac{\lambda(T_{\mathrm{w}})}{\rho T_{\mathrm{w}}}\frac{\partial T}{\partial X_i}t_i, \tag{23b}$$

$$T - T_{\mathrm{w}} = \frac{1}{R}\frac{1}{\mathrm{Pr}}c_v^{II}\frac{\mu(T_{\mathrm{w}})}{\rho}\frac{\partial v_i}{\partial X_j}n_i n_j + \frac{\sqrt{2}}{R^{3/2}}c_T^{II}\frac{\gamma(T_{\mathrm{w}})-1}{\gamma(T_{\mathrm{w}})}\frac{\lambda(T_{\mathrm{w}})}{\rho T_{\mathrm{w}}^{1/2}}\frac{\partial T}{\partial X_i}n_i, \tag{23c}$$

where t_i is an arbitrary unit vector tangent to the boundary at $\boldsymbol{X} = \boldsymbol{X}_{\mathrm{w}}$. The coefficients c_v^I, c_T^I, c_v^{II}, and c_T^{II}, which are the so-called slip coefficients, are determined by solving four half-space problems of the linearized version of the model equation (2). These problems contain the properties of the gas and the boundary through the parameters α, ν, and θ, the boundary temperature T_{w}, and the functions $D(T_{\mathrm{w}})$ and $\gamma(T_{\mathrm{w}})$. Therefore, the coefficients c_v^I, c_T^I, c_v^{II}, and c_T^{II} also depend on these properties. In other words, we have to specify these properties for the gas and the boundary under consideration with the help of available data in the literature and then solve the four problems numerically. As the results, the coefficients c_v^I, c_T^I, c_v^{II}, and c_T^{II} are determined for our gas. Since the space is limited, the explicit forms of the four problems are omitted here. We only mention that they are classical half-space problems; more specifically, the problems of shear slip, thermal creep, temperature jump caused by the normal viscous stress, and temperature jump caused by the normal heat flow. The form of the boundary conditions (23), which is of the type of slip conditions, is basically the same as that for a polyatomic gas with constant specific heats [29]. The difference lies in the fact that in the latter case, the coefficients c_v^I, c_T^I, c_v^{II}, and c_T^{II} do not depend on the boundary temperature T_{w}.

To complete the Navier–Stokes system, we should specify the initial conditions at time $t = 0$, in addition to the boundary conditions (23). In general, the initial condition for the model (2) is given as

$$f(0, \boldsymbol{X}, \boldsymbol{\xi}, \mathcal{E}) = f^{\mathrm{ini}}(\boldsymbol{X}, \boldsymbol{\xi}, \mathcal{E}), \tag{24}$$

where f^{ini} is an arbitrary function. Correspondingly, one may set

$$\rho(0, \boldsymbol{X}) = \rho^{\mathrm{ini}}(\boldsymbol{X}), \qquad \boldsymbol{v}(0, \boldsymbol{X}) = \boldsymbol{v}^{\mathrm{ini}}(\boldsymbol{X}), \qquad T(0, \boldsymbol{X}) = T^{\mathrm{ini}}(\boldsymbol{X}), \tag{25}$$

as the initial conditions for Eq. (18), where ρ^{ini}, $\boldsymbol{v}^{\mathrm{ini}}$, and T^{ini} are the density, flow velocity, and temperature generated from f^{ini} and are assumed to be smooth functions of $\boldsymbol{X}$. One should note that, with the general initial conditions (25), Eq. (18) and the boundary conditions (23) do not give a correct approximation of the solution to Eqs. (2), (21), and (24) at the initial stage $0 < t < O(\text{mean free time})$ (the initial layer). This initial inaccuracy may remain for a long time in some cases, such as the problems with the propagation of strong shock waves generated at the initial stage. However, for many problems in which the boundary has dominant effects, the Navier–Stokes system (18), (23), and (25) may be expected to provide a good approximate solution globally to the solution of the kinetic system (2), (21), and (24) formally with the error of $O(\mathrm{Kn}^2)$ (in the appropriate dimensionless setting), except inside the Knudsen and initial layers.

The occurrence of the initial layer can be avoided by considering a special kinetic initial condition (24) and by imposing some restrictions on $\boldsymbol{v}_{\mathrm{w}}$ and T_{w} in the kinetic boundary condition (21). Suppose that the gas is in the equilibrium state at rest with density ρ_0 and temperature T_0 at $t = 0$, that is,

$$f^{\mathrm{ini}} = f_0 = \frac{\rho_0 \mathcal{E}^{\delta_0/2-1}}{(2\pi R T_0)^{3/2}(R T_0)^{\delta_0/2}\Gamma(\delta_0/2)} \exp\left(-\frac{|\boldsymbol{\xi}|^2}{2RT_0} - \frac{\mathcal{E}}{RT_0}\right), \tag{26}$$

where $\delta_0 = D(T_0)$. Correspondingly, Eq. (25) reduces to

$$\rho(0, \boldsymbol{X}) = \rho_0, \qquad \boldsymbol{v}(0, \boldsymbol{X}) = 0, \qquad T(0, \boldsymbol{X}) = T_0. \tag{27}$$

If we assume that $\boldsymbol{v}_{\mathrm{w}} = 0$ and $T_{\mathrm{w}} = T_0$ in Eq. (21), then f_0 in Eq. (26) is the solution of the problem for all $t\ (> 0)$. Let $\boldsymbol{v}_{\mathrm{w}}$ and T_{w} start changing slowly in position and time after the initial time, with the time scale much longer than the mean free time and with the length scale much longer than the mean free path. In this setting, the initial layer does not appear.

References

1. Nagnibeda, E., Kustova, E.: Non-Equilibrium Reacting Gas Flows: Kinetic Theory of Transport and Relaxation Processes. Springer, Berlin (2009)
2. Boyd, I.D., Schwartzentruber, T.E.: Nonequilibrium Gas Dynamics and Molecular Simulation. Cambridge Univ. Press, Cambridge (2017)
3. Morse, T.F.: Kinetic model for gases with internal degrees of freedom. Phys. Fluids **7**, 159–169 (1964)
4. Holway Jr., L.H.: New statistical models for kinetic theory: Methods of construction. Phys. Fluids **9**, 1658–1673 (1966)
5. Rykov, V.A.: A model kinetic equation for a gas with rotational degrees of freedom. Fluid Dyn. **10**(6), 959–966 (1975)
6. Andries, P., Le Tallec, P., Perlat, J.-P., Perthame, B.: The Gaussian-BGK model of Boltzmann equation with small Prandtl number. Eur. J. Mech. B/Fluids **19**, 813–830 (2000)
7. Rahimi, B., Struchtrup, H.: Capturing non-equilibrium phenomena in rarefied polyatomic gases: A high-order macroscopic model. Phys. Fluids **26**, 052001 (2014)

8. Bisi, M., Spiga, G.: On kinetic models for polyatomic gases and their hydrodynamic limits. Ricerche Mat. **66**, 113–124 (2017)
9. Mathiaud, J., Mieussens, L.: A Fokker-Planck model of the Boltzmann equation with correct Prandtl number for polyatomic gases. J. Stat. Phys. **168**, 1031–1055 (2017)
10. Arima, T., Ruggeri, T., Sugiyama, M.: Rational extended thermodynamics of a rarefied polyatomic gas with molecular relaxation processes. Phys. Rev. E **96**, 042143 (2017)
11. Bisi, M., Ruggeri, T., Spiga, G.: Dynamical pressure in a polyatomic gas: Interplay between kinetic theory and extended thermodynamics. Kin. Rel. Models **11**, 71–95 (2018)
12. Baranger, C., Dauvois, Y., Marois, G., Mathé, J., Mathiaud, J., Mieussens, L.: A BGK model for high temperature rarefied gas flows. Eur. J. Mech. B/Fluids **80**, 1–12 (2020)
13. Dauvois, Y., Mathiaud, J., Mieussens, L.: An ES-BGK model for polyatomic gases in rotational and vibrational nonequilibrium. Eur. J. Mech. B/Fluids **88**, 1–16 (2021)
14. Brull, S., Schneider, J.: On the ellipsoidal statistical model for polyatomic gases. Continuum Mech. Thermodyn. **20**, 489–508 (2009)
15. Kosuge, S., Aoki, K.: Shock-wave structure for a polyatomic gas with large bulk viscosity. Phys. Rev. Fluids **3**, 023401 (2018)
16. Taniguchi, S., Arima, T., Ruggeri, T., Sugiyama, M.: Overshoot of the non-equilibrium temperature in the shock wave structure of a rarefied polyatomic gas subject to the dynamic pressure. Int. J. Non-Linear Mech. **79**, 66–75 (2016)
17. Kosuge, S., Kuo, H.-W., Aoki, K.: A kinetic model for a polyatomic gas with temperature-dependent specific heats and its application to shock-wave structure. J. Stat. Phys. **177**, 209–251 (2019)
18. Chapman, S., Cowling, T.G.: The Mathematical Theory of Non-uniform Gases, 3rd edn. Cambridge Univ. Press, Cambridge (1991)
19. Grad, H.: Principles of the kinetic theory of gases. In: Flügge, S. (ed.) Handbuch der Physik, Band XII, pp. 205–294. Springer, Berlin (1958)
20. Cercignani, C.: The Boltzmann Equation and Its Applications. Springer, Berlin (1988)
21. Sone, Y.: Molecular Gas Dynamics: Theory, Techniques, and Applications. Birkhäuser, Boston (2007); Supplementary Notes and Errata: http://hdl.handle.net/2433/66098
22. Aoki, K., Bisi, M., Groppi, M., Kosuge, S.: Two-temperature Navier–Stokes equations for a polyatomic gas derived from kinetic theory. Phys. Rev. E **102**, 023104 (2020)
23. Kosuge, S., Aoki, K., Bisi, M., Groppi, M., Martalò, G.: Boundary conditions for two-temperature Navier–Stokes equations for a polyatomic gas, Phys. Rev. Fluids **6**, 083401 (2021)
24. Sone, Y.: Kinetic Theory and Fluid Dynamics. Birkhäuser, Boston (2002); Supplementary Notes and Errata: http://hdl.handle.net/2433/66099
25. Sone, Y.: Asymptotic theory of flow of rarefied gas over a smooth boundary I. In: Trilling, L., Wachman, H.Y. (eds.) Rarefied Gas Dynamics, pp. 243–253. Academic, New York (1969)
26. Sone, Y.: Asymptotic theory of flow of rarefied gas over a smooth boundary II. In: Dini, D. (ed.) Rarefied Gas Dynamics, Vol. II, pp. 737–749. Editrice Tecnico Scientifica, Pisa (1971)
27. Sone, Y.: Asymptotic theory of a steady flow of a rarefied gas past bodies for small Knudsen numbers. In: Gatignol, R., Soubbaramayer (eds.) Advances in Kinetic Theory and Continuum Mechanics, pp. 19–31. Springer, Berlin (1991)
28. Aoki, K., Baranger, C., Hattori, M., Kosuge, S., Martalò, G., Mathiaud, J., Mieussens, L.: Slip boundary conditions for the compressible Navier–Stokes equations. J. Stat. Phys. **169**, 744–781 (2017)
29. Hattori, M., Kosuge, S., Aoki, K.: Slip boundary conditions for the compressible Navier–Stokes equations for a polyatomic gas, Phys. Rev. Fluids **3**, 063401 (2018)

Discrete- and Continuous-Time Random Walks in 1D Lévy Random Medium

Marco Lenci

Abstract A Lévy random medium, in a given space, is a random point process where the distances between points, a.k.a. *targets*, are long-tailed. Random walks visiting the targets of a Lévy random medium have been used to model many (physical, ecological, social) phenomena that exhibit superdiffusion as the result of interactions between an agent and a sparse, complex environment.

In this note we consider the simplest non-trivial Lévy random medium, a sequence of points in the real line with i.i.d. long-tailed distances between consecutive targets. A popular example of a continuous-time random walk in this medium is the so-called *Lévy-Lorentz gas*. We give an account of a number of recent theorems on generalizations and variations of such model, in discrete and continuous time.

1 Introduction

In this note we give an overview of recent rigorous results on random walks (RWs) in random medium on the real line. The random medium is given by a point process $\omega = (\omega_k,\ k \in \mathbb{Z}) \subset \mathbb{R}$, where $\omega_0 = 0$ and the distances $\zeta_k := \omega_k - \omega_{k-1}$ between consecutive points are positive i.i.d. random variables with a long tail. By this we mean that the variance of ζ_k is infinite. The points ω_k will be henceforth called *targets*. For reason that will be better clarified below we refer to ω as a *Lévy random medium*.

We consider two types of RWs on $\mathbb{R}$ related to ω. To define them we introduce the auxiliary process $S = (S_n,\ n \in \mathbb{N})$, a $\mathbb{Z}$-valued RW with $S_0 = 0$ and independent increments. We postulate that S is independent of ω and call it the *underlying random walk*. The first process of interest is $Y \equiv Y^\omega := (Y_n,\ n \in \mathbb{N})$, where $Y_n := \omega_{S_n}$. This is the discrete-time RW (DTRW) that "jumps" on the targets of

M. Lenci (✉)
Dipartimento di Matematica, Università di Bologna, Bologna, Italy

Istituto Nazionale di Fisica Nucleare, Sezione di Bologna, Bologna, Italy
e-mail: marco.lenci@unibo.it

P. Barbante et al. (eds.), *From Kinetic Theory to Turbulence Modeling*, Springer INdAM Series 51, https://doi.org/10.1007/978-981-19-6462-6_13

ω as determined by S. For example, if S produces the realization $(0, 2, -3, \ldots)$, the walker Y starts at the origin, then jumps to the second target to the right of 0, then to the third target to the left of 0, etc. The second process of interest is $X \equiv X^\omega := (X(t),\ t \geq 0)$, the continuous-time RW (CTRW) defined as the unit-speed interpolation of Y. This means that the walker X visits all the points Y_n, ordered by n, but "walking" with unit speed rather than jumping. For instance, for S as in the above example, X starts at the origin and moves with velocity 1 until it reaches $Y_1 = \omega_2$, then it instantaneously turns its velocity to -1 and moves until it reaches $Y_2 = \omega_{-3}$, and so on.

In the case where the underlying RW S is simple and symmetric, the process X is generally referred to as the *Lévy-Lorentz gas*, after Barkai, Fleurov, and Klafter introduced it in the physical literature in 2000 [1]. The Lévy-Lorentz gas has been used since as a simple model for a number of phenomena exhibiting superdiffusion, i.e., diffusion at a faster speed than square root of time. They include transport in porous media, disordered optical media (such as *Lévy glasses* [2]), nanowires, etc.; see [1, 3–5] and the references therein.

In the physical literature, DTRWs, respectively, CTRWs, whose distributions of jumps, respectively, inertial stretches, are long-tailed, are often called *Lévy flights*, respectively *Lévy walks*. Lévy flights and walks, in regular or random media, have been employed as models for anomalous diffusion in a wide range of situations, from the physical to the biological and social sciences [6, 7]. A rigorous mathematical treatment of these systems has only been given for the simplest of them, mostly on regular media. In real-world applications, however, the anomalous behavior of a certain diffusing quantity is seldom due to a special law governing the diffusing agent *per se*, but rather to the interaction between the agent and an irregular medium (e.g., a photon in a Lévy glass, a signal in a small-world network, an animal foraging where food is scarce, etc.). Hence the interest in *Lévy media*, namely media that induce superdiffusive behavior, such as the random point process ω defined earlier.

A fair amount of mathematical work on the processes X and Y has been done in recent years by different authors [8–12]. In particular, in a number of cases, limit theorems have been proved for suitable rescalings of either process. In some instances, the convergence of moments has been proved as well. The purpose of this note is to present these results in a concise, unified manner. For reasons of space and self-consistency, we will neglect interesting work by Artuso and collaborators on yet another type of RW related to the Lévy-Lorentz gas, a persistent RW in an *averaged* medium [13, 14].

In Sect. 2 we present results on the DTRW Y and in Sect. 3 on the CTRW X. Understandably, the results depend on the assumptions on the random medium ω and the underlying RW S. Major differences occur depending on whether ζ_k, the distance between two consecutive targets, has infinite variance but finite mean, or infinite mean (and thus infinite variance), so we consider these cases in different subsections. No proofs are given, but references are placed throughout.

1.1 General Notation

Throughout the paper we denote by $\mathbb{P}$ the probability law that governs the whole system, both the random medium ω and the underlying RW S (that latter playing the role of the random dynamics, as it "drives" both X and Y, in a given ω). $\mathbb{P}$ is called the *annealed law* and we denote by $\mathbb{E}$ its expectation. For a fixed ω, the conditional probability $P_\omega := \mathbb{P}(\cdot|\omega)$ is called the *quenched law* relative to the realization ω of the medium. We denote by E_ω its expectation. A limit theorem, such as the CLT or the Invariance Principle, etc., relative to $\mathbb{P}$ is referred to as an *annealed limit theorem*. One speaks instead of a *quenched limit theorem* if the result is proved w.r.t. P_ω, for $\mathbb{P}$-a.e. ω.[1] In what follows, for the most part, we present annealed and quenched results in different subsections.

We indicate with $\xi_n := S_n - S_{n-1}$ ($n \in \mathbb{Z}^+$) the i.i.d. increments of the underlying RW, whose drift is denoted $\nu := \mathbb{E}[\xi_n]$, if it exists in $\mathbb{R} \cup \{\pm\infty\}$. The mean distance between the targets of ω is denoted $\mu := \mathbb{E}[\zeta_k]$. Since $\zeta_k > 0$, μ always exists in $\mathbb{R} \cup \{+\infty\}$. Let us recall that all ξ_n ($n \in \mathbb{Z}^+$) and ζ_k ($k \in \mathbb{Z}$) are independent.

2 Discrete-Time Random Walk

In this section we consider the asymptotic behavior of the DTRW $Y \equiv Y^\omega$, under a number of different assumptions on the distributions of ζ_1 and ξ_1.

2.1 Finite Mean Distance Between Targets, Quenched Theorems

We start with the results of [8] on the quenched version of Y, which only require a very simple condition on the medium, $\mu = \mathbb{E}[\zeta_1] < \infty$, that is, the mean distance between neighboring targets is finite. The assumptions on the underlying RW are instead as follows:

- the increment ξ_1 of S is symmetric, i.e., $\mathbb{P}(\xi_1 = j) = \mathbb{P}(\xi_1 = -j)$, for all $j \in \mathbb{N}$;
- its distribution is unimodal, i.e., $j \mapsto \mathbb{P}(\xi_1 = j)$ is non-increasing for $j \in \mathbb{N}$;
- it has finite variance: $V_\xi := \mathbb{E}[\xi_1^2] < \infty$.

[1] Since we have neither introduced the measurable space $(\Omega, \mathcal{A})$ where $\mathbb{P}$ is defined nor declared that ω are elements of Ω, mathematical formality requires that we define the phrase "$\mathbb{P}$-almost every ω." The counterimages (equivalently, level sets) of the process ω form a partition of Ω. We assume this partition to be *measurable* in the sense of Rohlin [15]. Now, a property is said to hold for $\mathbb{P}$-a.e. ω if the values of ω which do not satisfy the property correspond to elements of the partition whose union has zero $\mathbb{P}$-measure. Incidentally, the existence of such a measurable partition is what guarantees that $\mathbb{P}(\cdot|\omega)$ is well-defined (for $\mathbb{P}$-a.e. ω).

The authors prove a quenched CLT for Y [8, Thm. 1]:

Theorem 1 *Assume the above conditions, most notably $\mu < \infty$. Then, as $n \to \infty$,*

$$\frac{Y_n}{\sqrt{n}} \xrightarrow{\ \mathrm{d}\ } \mathcal{N}(0, \mu^2 V_\xi),$$

w.r.t. P_ω, for $\mathbb{P}$-a.e. ω. Here $\mathcal{N}(0, \mu^2 V_\xi)$ is a Gaussian variable with mean 0 and variance $\mu^2 V_\xi$.

Obviously, a quenched distributional limit theorem with the same limit for a.e. quenched law implies the annealed version of the same theorem:

Corollary 1 *The limit in the statement of Theorem 1 holds w.r.t. $\mathbb{P}$ as well.*

Convergence is known for the quenched moments of $Y_n/\sqrt{n}$ as well, at least of lower order. Let $p := \sup\{q \geq 0 \mid \mathbb{E}[|\xi_1|^q] < \infty\}$. By the assumption on V_ξ, $p \geq 2$. For all $q \in \mathbb{R}^+$, denote by

$$\overline{m}_q := \sqrt{\frac{2^q}{\pi}}\, \Gamma\left(\frac{q+1}{2}\right) \tag{1}$$

the q-th absolute moment of the standard Gaussian $\mathcal{N}(0, 1)$ (here Γ is the usual Gamma function). It is not hard to show that, at least for all $q < p$,

$$\lim_{n\to\infty} \frac{\mathbb{E}[S_n]}{n^{q/2}} = V_\xi^{q/2}\, \overline{m}_q. \tag{2}$$

The following is a reformulation of Theorem 2 of [8].

Theorem 2 *Under the above assumptions and notation, fix $q \in (0, p)$. For a.a. ω,*

$$\lim_{n\to\infty} \frac{E_\omega[Y_n]}{n^{q/2}} = \mu^q\, V_\xi^{q/2}\, \overline{m}_q.$$

Observe that $\mu^q\, V_\xi^{q/2}\, \overline{m}_q$ is the q-th absolute moment of $\mathcal{N}(0, \mu^2 V_\xi)$, so Theorem 2 is consistent with Theorem 1.

2.2 Finite Mean Distance Between Targets, Annealed Theorems

In the next two subsections we report the functional limit theorems of [11]. We refer the reader to [16] for background material on stable laws, Lévy processes, Skorokhod topologies, etc. All distributional convergences in these subsections are meant w.r.t. $\mathbb{P}$, that is, we are considering annealed functional limit theorems.

We assume that ζ_1 is in the normal basin of attraction of a α-stable distribution, with $\alpha \in (1, 2]$. This means that $\mu = \mathbb{E}[\zeta_1] < \infty$ and

$$\frac{1}{n^{1/\alpha}} \sum_{i=1}^{n} (\zeta_i - \mu) \xrightarrow{\mathrm{d}} \widetilde{Z}_1^{(\alpha)}, \tag{3}$$

as $n \to \infty$, for some α-stable variable $\widetilde{Z}_1^{(\alpha)}$ (whose skewness index[2] must then be 0). As for the underlying RW, we assume ξ_1 is in the normal basin of attraction of an β-stable distribution, with $\beta \in (0, 1) \cup (1, 2]$. We must distinguish two cases, depending on whether $\nu = \mathbb{E}[\xi_1]$ exists and differs from 0, or otherwise.

- If $\beta \in (0, 1)$, or $\beta \in (1, 2]$ and $\nu = 0$, we assume that there exists a β-stable variable $W_1^{(\beta)}$ such that, as $n \to \infty$,

$$\frac{1}{n^{1/\beta}} \sum_{i=1}^{n} \xi_i \xrightarrow{\mathrm{d}} W_1^{(\beta)}. \tag{4}$$

- If $\beta \in (1, 2]$ and $\nu \neq 0$, we assume that there exists a β-stable variable $\widetilde{W}_1^{(\beta)}$ such that

$$\frac{1}{n^{1/\beta}} \sum_{i=1}^{n} (\xi_i - \nu) \xrightarrow{\mathrm{d}} \widetilde{W}_1^{(\beta)}. \tag{5}$$

To state the results of this section, we need two spaces of functions with jump discontinuities. In what follows, we denote by $\mathcal{D}^+$ the space of càdlàg[3] functions $\mathbb{R}^+ \longrightarrow \mathbb{R}$ and by $\mathcal{D}$ the space of functions $\mathbb{R} \longrightarrow \mathbb{R}$ whose restriction to $[0, +\infty)$, respectively $(-\infty, 0]$, is càdlàg, respectively càglàd.

Let $(\widetilde{Z}_{\pm}^{(\alpha)}(s),\ s \geq 0)$ be two i.i.d. $\mathcal{D}\alpha$-stable Lévy processes such that $\widetilde{Z}_{\pm}^{(\alpha)}(0) = 0$ and $\widetilde{Z}_{\pm}^{(\alpha)}(1)$ is distributed like $\widetilde{Z}_1^{(\alpha)}$, introduced in (3) (these conditions uniquely determine the distribution of the processes), and set

$$\widetilde{Z}^{(\alpha)}(s) := \begin{cases} Z_{+}^{(\alpha)}(s)\,, & s \geq 0; \\ -Z_{-}^{(\alpha)}(-s)\,, & s < 0. \end{cases} \tag{6}$$

[2] The skewness index is the parameter that, in virtually all textbooks on stable variables (such as [16]) is denoted $\beta \in [-1, 1]$. In this paper β is used for the stability index of ξ_1.

[3] That is, right-continuous with left limits existing everywhere. Càglàd means left-continuous with right limits everywhere.

By construction, every realization $\widetilde{Z}^{(\alpha)}$ belongs to $\mathcal{D}$, and so do the realizations

$$\bar{\omega}^{(n)}(s) := \frac{1}{n} \begin{cases} \omega_{\lfloor ns \rfloor}, & s \geq 0; \\ \omega_{\lceil ns \rceil}, & s < 0, \end{cases} \tag{7}$$

$$\widetilde{\omega}^{(n)}(s) := \frac{1}{n^{1/\alpha}} \begin{cases} \sum_{i=1}^{\lfloor ns \rfloor} (\zeta_i - \mu), & s \geq 0; \\ -\sum_{i=\lceil (n-1)s \rceil}^{0} (\zeta_i - \mu), & s < 0, \end{cases} \tag{8}$$

defining the processes $(\bar{\omega}^{(n)}(s),\ s \in \mathbb{R})$ and $(\widetilde{\omega}^{(n)}(s),\ s \in \mathbb{R})$. The single-variable convergence (3) entails functional convergence of these processes: as $n \to \infty$, $\bar{\omega}^{(n)} \xrightarrow{\text{a.s.}} \mu\,\text{id}$ and $\widetilde{\omega}^{(n)} \xrightarrow{\text{d}} \widetilde{Z}^{(\alpha)}$, relative to the Skorokhod topology J_1. From now on, we will write "in $(\mathcal{D}, J_1)$" for short.

We now introduce continuous-argument processes for the dynamics.

- In the case $\beta \in (0, 1)$, or $\beta \in (1, 2]$ and $\nu = 0$, we denote by $(W^{(\beta)}(t),\ t \geq 0)$ a $\mathcal{D}\beta$-stable Lévy process whose distribution is uniquely determined by the conditions that $W^{(\beta)}(0) = 0$ and $W^{(\beta)}(1)$ be distributed like $W_1^{(\beta)}$; cf. (4). Also define $(\hat{S}^{(n)}(t),\ t \geq 0)$ via

$$\hat{S}^{(n)}(t) := \frac{S_{\lfloor nt \rfloor}}{n^{1/\beta}}. \tag{9}$$

  It follows from (4) that $\hat{S}^{(n)} \xrightarrow{\text{d}} W^{(\beta)}$, in $(\mathcal{D}^+, J_1)$, as $n \to \infty$.
- In the case $\beta \in (1, 2]$ and $\nu \neq 0$, we consider $\widetilde{W}^{(\beta)}$, defined exactly as $W^{(\beta)}$ above but with $\widetilde{W}_1^{(\beta)}$ in place of $W_1^{(\beta)}$; cf. (5). In lieu of (9) we define two processes:

$$\bar{S}^{(n)}(t) := \frac{S_{\lfloor nt \rfloor}}{n}, \qquad \widetilde{S}^{(n)}(t) := \frac{\sum_{i=1}^{\lfloor nt \rfloor} (\xi_i - \nu)}{n^{1/\beta}}. \tag{10}$$

  All these processes take values in $\mathcal{D}^+$. By (5), $\bar{S}^{(n)} \xrightarrow{\text{a.s.}} \nu\,\text{id}$ and $\widetilde{S}^{(n)} \xrightarrow{\text{d}} \widetilde{W}^{(\beta)}$, in $(\mathcal{D}^+, J_1)$, as $n \to \infty$.

The following result extends Theorem 2.3 of [11].

Theorem 3 *Under the above assumptions, in particular, $\alpha \in (1, 2]$, the following convergences hold, w.r.t. $\mathbb{P}$:*

(a) If $\beta \in (0, 1)$, or $\beta \in (1, 2]$ with $\nu = 0$, let $\hat{Y}^{(n)}(t) := \dfrac{Y_{\lfloor nt \rfloor}}{n^{1/\beta}}$, for $t \geq 0$. As $n \to \infty$,

$$\hat{Y}^{(n)} \xrightarrow{\text{d}} \mu\, W^{(\beta)} \qquad \text{in } (\mathcal{D}^+, J_1).$$

(b) *If* $\beta \in (1, 2]$ *with* $v \neq 0$, *let* $\bar{Y}^{(n)}(t) := \dfrac{Y_{\lfloor nt \rfloor}}{n}$, *for* $t \geq 0$. *As* $n \to \infty$,

$$\bar{Y}^{(n)} \xrightarrow{\mathrm{d}} \mu v \,\mathrm{id} \qquad \text{in } (\mathcal{D}^+, J_1).$$

Remark 1 The statement of [11, Thm. 2.3] does not include the cases $\alpha = 2$ and/or $\beta = 2$, because the authors were mostly interested in *bona fide* Lévy media and Lévy flights in them. The proof of the theorem, however, works verbatim if the assumptions are generalized to include ζ_1 and/or ξ_1 in the normal domain of attraction of a 2-stable distribution, i.e., a Gaussian. In this case, of course, $\widetilde{Z}_{\pm}^{(2)}$, $W_1^{(2)}$ and $\widetilde{W}_1^{(2)}$ are Brownian motions. The same remark holds for Theorems 4 and 5 below.

Remark 2 Theorem 3 mostly supersedes Theorem 2.1 of [9] (which is stated for the case where S is simple and symmetric), but not quite, since the hypothesis on ζ_1 there is that $\mathbb{P}(\zeta_1 > x) \approx x^{-p}$, for $x \to +\infty$, with $p \geq 1$. This is weaker than asking that ζ_1 be in the *normal* domain of attraction of an α-stable distribution, with $\alpha \in (1, 2]$. For example, it includes the case $\mathbb{P}(\zeta_1 > x) = Cx^{-2}$, where ζ_1 is in the domain, but not normal domain of attraction of a Gaussian [17, Ex. 5.10]. Also, the assertion of Theorem 3 is of course much stronger than that of Corollary 1, but the hypothesis on the medium for the latter is much weaker: simply $\mu < \infty$.

The case *(b)* of the above theorem is the case where Y has a drift. Understandably, the scaling rate of $Y_{\lfloor n\cdot \rfloor}$ is n (one says that the process is *ballistic*) and the convergence is to a deterministic function. It is therefore natural to study the fluctuations around the deterministic limit.

Theorem 4 *Under the same assumptions and notation as Theorem 3, if* $\alpha, \beta \in (1, 2]$, *the following convergences hold, w.r.t.* $\mathbb{P}$:

(a) *If* $\beta < \alpha$, *let* $\widetilde{Y}^{(n)}(t) := \dfrac{n(\bar{Y}^{(n)}(t) - \mu v t)}{n^{1/\beta}} = \dfrac{Y_{\lfloor nt \rfloor} - n\mu v t}{n^{1/\beta}}$. *As* $n \to \infty$,

$$\widetilde{Y}^{(n)} \xrightarrow{\mathrm{d}} \mu \widetilde{W}^{(\beta)} \qquad \text{in } (\mathcal{D}^+, J_1).$$

(b) *If* $\beta > \alpha$, *let* $\widetilde{Y}^{(n)}(t) := \dfrac{n(\bar{Y}^{(n)}(t) - \mu v t)}{n^{1/\alpha}} = \dfrac{Y_{\lfloor nt \rfloor} - n\mu v t}{n^{1/\alpha}}$. *As* $n \to \infty$,

$$\widetilde{Y}^{(n)} \xrightarrow{\mathrm{d}} \mathrm{sgn}(v)\, |v|^{1/\alpha}\, \widetilde{Z}_{+}^{(\alpha)} \qquad \text{in } (\mathcal{D}^+, J_2).$$

(c) If $\beta = \alpha$*, let* $\widetilde{Y}^{(n)}(t) := \dfrac{n(\bar{Y}^{(n)}(t) - \mu\nu t)}{n^{1/\beta}} = \dfrac{Y_{\lfloor nt \rfloor} - n\mu\nu t}{n^{1/\beta}}$*. As* $n \to \infty$,

$$\widetilde{Y}^{(n)} \xrightarrow{\mathrm{d}} \mu \widetilde{W}^{(\beta)} + \mathrm{sgn}(\nu)\, |\nu|^{1/\alpha}\, \widetilde{Z}_+^{(\alpha)} \qquad \text{in } (\mathcal{D}^+, J_2),$$

where $\widetilde{W}^{(\beta)}$ *and* $\widetilde{Z}_+^{(\alpha)}$ *are two independent processes, defined as in (5) ff.*

Remark 3 The limits in *(b)* and *(c)* involve the unusual Skorokhod topology J_2 [16, §11.5]. But this is the strongest amongst the classical Skorokhod topologies relative to which such limits hold, cf. Remark 2.11 of [11]. However, see Remark A.2 in the same paper.

2.3 Infinite Mean Distance Between Targets, Annealed Theorems

In this subsection we assume that ζ_1 is in the normal basin of attraction of a α-stable distribution with $\alpha \in (0, 1)$. Since $\mathbb{E}[\zeta_1] = \infty$, this means that, for $n \to \infty$,

$$\frac{1}{n^{1/\alpha}} \sum_{i=1}^{n} \zeta_i \xrightarrow{\mathrm{d}} Z_1^{(\alpha)}, \tag{11}$$

for some α-stable variable $Z_1^{(\alpha)}$ (whose skewness index is 1, since $\zeta_i > 0$). Out of $Z_1^{(\alpha)}$, we construct continuous-argument processes $(Z_\pm^{(\alpha)}(s),\ s \geq 0)$ and $(Z^{(\alpha)}(s),\ s \in \mathbb{R})$ in complete analogy with the previous case; cf. (6). $Z^{(\alpha)}$ takes values in $\mathcal{D}$, and the same is true for

$$\hat{\omega}^{(n)}(s) := \frac{1}{n^{1/\alpha}} \begin{cases} \omega_{\lfloor ns \rfloor}, & s \geq 0, \\ \omega_{\lceil ns \rceil}, & s < 0. \end{cases} \tag{12}$$

It is a basic fact that, as $n \to \infty$, $\hat{\omega}^{(n)} \xrightarrow{\mathrm{d}} Z^{(\alpha)}$, in $(\mathcal{D}, J_1)$.

As the underlying RW, we maintain the same assumptions and notation as in Sect. 2.2, recalling that the fundamental assumption is that ξ_1 is in the normal basin of attraction of an β-stable distribution, with $\beta \in (0, 1) \cup (1, 2]$. Once again, ν denotes the expectation of ξ_1, when defined.

The following theorem comprises and extends Theorems 2.1 and 2.2 of [11]:

Theorem 5 *Under the above assumptions, in particular,* $\alpha \in (0, 1)$*, the following convergences hold, w.r.t.* $\mathbb{P}$*:*

(a) *If* $\beta \in (0,1)$, *or* $\beta \in (1,2]$ *with* $\nu = 0$, *let* $\hat{Y}^{(n)}(t) := \dfrac{Y_{\lfloor nt \rfloor}}{n^{1/\alpha\beta}}$, *for* $t \geq 0$. *As* $n \to \infty$, *the finite-dimensional distributions of* $\hat{Y}^{(n)}$ *converge to those of* $Z^{(\alpha)} \circ W^{(\beta)}$. *This means that, for all* $m \in \mathbb{Z}^+$ *and* $t_1, \ldots, t_m \in \mathbb{R}^+$,

$$\left(\hat{Y}^{(n)}(t_1), \ldots, \hat{Y}^{(n)}(t_m)\right) \xrightarrow{\text{d}} \left(Z^{(\alpha)}(W^{(\beta)}(t_1)), \ldots, Z^{(\alpha)}(W^{(\beta)}(t_m))\right).$$

(b) *If* $\beta \in (1,2]$ *with* $\nu \neq 0$, *let* $\hat{Y}^{(n)}(t) := \dfrac{Y_{\lfloor nt \rfloor}}{n^{1/\alpha}}$, *for* $t \geq 0$. *As* $n \to \infty$,

$$\hat{Y}^{(n)} \xrightarrow{\text{d}} \operatorname{sgn}(\nu)\, |\nu|^{1/\alpha}\, Z_+^{(\alpha)} \qquad \text{in } (\mathcal{D}^+, J_2).$$

Remark 4 The convergence of the finite-dimensional distributions in assertion *(a)* is certainly a weak form of convergence, but it is morally the best one can do, given that $Z^{(\alpha)} \circ W^{(\beta)}$ is not $\mathcal{D}$ with positive probability; see the comments after Theorem 2.2 of [10]. As for assertion *(b)*, the considerations of Remark 3 apply here too.

3 Continuous-Time Random Walk

In this section we deal with the CTRW $X \equiv X^\omega$, again under various assumptions, depending on the papers we report on.

3.1 *Finite Mean Distance Between Targets, Quenched Theorems*

Once more, we start by presenting a result by [8], namely the quenched CLT for X [8, Thm. 1]. The assumptions are the same as in Sect. 2.1 above: the mean distance μ between targets is finite and the underlying RW has symmetric, unimodal, finite-variance increments ξ_n. Let us recall, in particular, the notation $V_\xi := \mathbb{E}[\xi_1^2]$.

Theorem 6 *Under the above assumptions, most notably* $\mu < \infty$, *let* $M_\xi := \mathbb{E}[|\xi_1|]$ *denote the first absolute moment of the underlying RW. Then, as* $t \to \infty$ *and w.r.t.* P_ω, *for a.a.* ω,

$$\frac{X(t)}{\sqrt{t}} \xrightarrow{\text{d}} \mathcal{N}\!\left(0, \mu \frac{V_\xi}{M_\xi}\right).$$

Here, once again, $\mathcal{N}(0,\cdot)$ *is a centered Gaussian variable with the specified variance.*

The annealed CLT follows immediately:

Corollary 2 *The limit in the statement of Theorem 6 holds w.r.t.* $\mathbb{P}$ *as well.*

A recent preprint of Zamparo [12] claims the convergence of *all* quenched moments of $X(t)/\sqrt{t}$, under the additional assumption that the underlying RW is simple and symmetric (implying that X is the *bona fide* Lévy-Lorentz gas). Recall the notation $\overline{m}_q$ for the q-th absolute moment of the standard Gaussian, cf. (1).

Theorem 7 *Assume that* $\mu < \infty$ *and* S *is a simple symmetric RW. Then, for a.a.* ω,

$$\lim_{t\to\infty} \frac{E_\omega[|X(t)|^q]}{t^{q/2}} = \mu^{q/2}\,\overline{m}_q .$$

Remark 5 When S is simple and symmetric, $V_\xi = M_\xi = 1$, so Theorem 7 is consistent with Theorem 6, showing that X is completely diffusive in this case.

Theorem 7 descends from another result of independent interest, concerning the *large deviations* of X, namely events of the type $\{|X(t)| > at\}$, for $a > 0$. Since $X(t)$ is centered and scales like $\sqrt{t}$, the probability of such "ballistic events" is expected to be exceedingly small. In [12, Thm. 2.3] it is proved that this probability vanishes like a stretched exponential. We report such result here:

Theorem 8 *Under the same assumptions as in Theorem 7, there exists* $\kappa > 0$ *such that, for all* $a \in (0, 1]$, *the limit*

$$\limsup_{t\to\infty} \frac{1}{\sqrt{at}} \log P_\omega(|X(t)| > at) \le -\kappa$$

holds for a.e. ω.

3.2 Finite Mean Distance Between Targets, Annealed Theorems

Apart from recalling Corollary 2, which establishes the annealed CLT for X under the assumptions of [8] ($\mu < \infty$ and S has symmetric, unimodal, finite-variance increments, cf. §2.1), in this section we present the results of [12] on the moments and large deviations of the annealed version of X.

The assumptions for this part are stronger than for Theorems 7 and 8. Like before, S must be a simple symmetric RW, but now we also posit that the tail of the distribution of ζ_1 is *regularly varying* with index $-p \le -1$. This means that

$$\tau_\zeta(x) := \mathbb{P}(\zeta_1 > x) = \frac{\ell(x)}{x^p}, \tag{13}$$

where ℓ is a *slowly varying* function at $+\infty$, namely for all $c > 0$,[4]

$$\lim_{x\to+\infty} \frac{\ell(cx)}{\ell(x)} = 1. \tag{14}$$

In order to describe the upcoming theorems in their full power, we need more notation. For $0 < r < 1$, set

$$f_p(r) := \sum_{j=0}^{\lceil (1-r)/2r \rceil - 1} \left(\left(\frac{2j+2}{1+r} \right)^p - \left(\frac{2j}{1-r} \right)^p \right). \tag{15}$$

It can be seen [12, §2.1] that $0 < f_p(r) \le r^{-p}$ and, as $r \to 0^+$,

$$f_p(r) \sim \frac{1}{(p+1)\, r^p}. \tag{16}$$

Here and in the rest of the paper $\sim$ denotes *exact* asymptotic equivalence. This limit shows, in particular, that $\int_0^1 r^{q-1} f_p(r)\, dr$ converges for $q = 2p - 1$ and $p > 1$, or $q > 2p - 1$ and $p \ge 1$.

Theorem 9 *Assume that S is a simple symmetric RW, $\mu < \infty$ and τ_ζ is regularly varying with index $-p \le -1$, cf. (13). For all $q > 0$, recall the notation $\overline{m}_q$ for the q-th moment of the standard Gaussian, cf. (1), and set*

$$d_{\mu,p,q} := \sqrt{\frac{2}{\mu}}\, \frac{\Gamma(q-p+1)}{\Gamma(q-p+3/2)} \int_0^1 r^{q-1} f_p(r)\, dr.$$

Then, as $t \to \infty$,

$$\mathbb{E}[|X(t)|^q] \sim \begin{cases} \overline{m}_q\, \mu^{q/2}\, t^{q/2}, & q < 2p-1, \text{ or } q = p = 1; \\ \overline{m}_q\, \mu^{q/2}\, t^{q/2} + d_{\mu,p,q}\, t^{q+1/2}\, \tau_\zeta(t), & q = 2p-1, \text{ and } p > 1; \\ d_{\mu,p,q}\, t^{q+1/2}\, \tau_\zeta(t), & q > 2p-1. \end{cases}$$

A few words of comment: In the subject of anomalous diffusion, an important quantity to investigate is the *scaling exponent* of the moments,

$$\gamma(q) := \lim_{t\to\infty} \frac{\log \mathbb{E}[|X(t)|^q]}{\log t}, \tag{17}$$

assuming this limit exists at least for a.e. $q > 0$. In many relevant models one observes that $q \mapsto \gamma(q)$ is piecewise linear with two branches, a left one with slope

[4] See [18] for a treatise on regularly varying functions.

$1/2$ and a right one with slope 1. Researchers named this situation *strong anomalous diffusion*,[5] cf. [6, 19]. Theorem 9 shows that this is precisely what happens for the annealed Lévy-Lorentz gas, under the above assumptions. The corner between the two branches occurs at the moment of order $2p - 1 \geq 2$, so the behavior of the second moment is still normal, at least in terms of the leading exponent. Even more interestingly, this picture is very different from that of the corresponding quenched Lévy-Lorentz gas, which is fully diffusive, as seen in Theorem 7.[6]

As for the quenched case, Theorem 9 is based on a large deviation result, which, however, is very different from Theorem 8:

Theorem 10 *Under the same assumptions as in Theorem 9, for $a \in (0, 1]$, let*

$$F_{\mu,p,a} := \frac{1}{\sqrt{2\pi\mu}} \int_a^1 f_p\left(\frac{a}{\eta}\right) \frac{\eta^{-p}}{\sqrt{1-\eta}}\, d\eta.$$

Then, as $t \to \infty$,

$$\mathbb{P}(X(t) > at) = \mathbb{P}(X(t) < -at) \sim F_{\mu,p,a}\sqrt{t}\, \tau_\zeta(t).$$

Moreover, for any $\delta \in (0, 1)$, the lower order terms are uniformly bounded for $a \in [\delta, 1]$.

Remark 6 The first equality of the above assertion is obvious because, by the symmetry of the distributions of ω and S, the annealed distribution of $X(t)$ is the same as that of $-X(t)$.

3.3 *Infinite Mean Distance Between Targets, Annealed Theorems*

The last and hardest case is that of the CTRW X in a medium ω with $\mu = \infty$. At least to this author's knowledge, only a limit theorem seems to be available, that of Bianchi *et al*, recently appeared in [10]. We present it after some preparatory material.

---

[5] Though different authors use different terminologies, not always compatible with each other, or even fully self-consistent.

[6] This does not mean that Theorems 7 and 9 are incompatible: what is happening here is that $t^{-q/2} E_\omega[|X(t)|^q]$ converges to the suitable limit for a.a. ω, but, at least for large q, the convergence rate depends heavily on ω. Moreover, the convergence is not monotonic in t. Mathematically speaking, the convergence is neither dominated nor monotonic, so one cannot interchange the limit in t and the integration on ω, to obtain the limit of the annealed moments from that of the quenched moments.

First off, the assumption on the medium is that ζ_1 is in the normal domain of attraction of an α-stable positive variable, with $\alpha \in (0, 1)$. As far ω is concerned, this is the same assumption as in Sect. 2.3, so we use the same notation introduced there, in particular, for the processes $(Z_{\pm}^{(\alpha)}(s),\ s \geq 0)$ and $(Z^{(\alpha)}(s),\ s \in \mathbb{R})$. The underlying random walk S is assumed to be centered and such that $\mathbb{E}[|\xi_1|^q] < \infty$, for some $q > 2/\alpha$. This implies, in particular, that $V_\xi = \mathbb{E}[|\xi_1|^2] < \infty$. So this is a special case of the assumptions on S of Sect. 2.3 (which were the same as in Sect. 2.2).

All the preliminary results seen earlier then apply, in particular, for $n \to \infty$, $\hat{\omega}^{(n)} \xrightarrow{\text{d}} Z^{(\alpha)}$, in $(\mathcal{D}, J_1)$, cf. (12), and $\hat{S}^{(n)} := n^{-1/2}\, S_{\lfloor n\cdot \rfloor} \xrightarrow{\text{d}} W^{(2)}$, in $(\mathcal{D}^+, J_1)$. Here $W^{(2)}$ is a Brownian motion such that $W^{(2)}(t)$ has mean 0 and variance $V_\xi t$. As clarified in Sect. 2.3, the processes $Z_{\pm}^{(\alpha)}$ and $W^{(2)}$ are independent. Recalling the notation $M_\xi := \mathbb{E}[|\xi_1|]$, let $\Delta := (\Delta(t), t \geq 0)$ be defined by

$$\Delta(t) := M_\xi \left(\int_0^\infty L_t(x)\, dZ_+^{(\alpha)}(x) + \int_0^\infty L_t(-x)\, dZ_-^{(\alpha)}(x) \right), \tag{18}$$

where, for all $x \in \mathbb{R}$, $L_t(x) := \#\big(W^{(2)}|_{[0,t]}\big)^{-1}(x)$. In other words, $L_t(x)$ is the local time of the Brownian motion $W^{(2)}$ in x, up to time t. As a function of x, L_t is compactly supported and almost surely continuous, thus the above r.h.s. is well-defined. Since L_t is also strictly increasing in t, Δ is almost surely continuous and strictly increasing. Processes like Δ are called *Kesten-Spitzer processes* and arise in the context of *RW in random scenery* [20], which is one of the technical ingredients of Theorem 2.1 of [10] which we now present.

Theorem 11 *Under the above assumptions, in particular, $\alpha \in (0, 1)$, $v = 0$, and ξ_1 has a finite absolute moment of order $q > 2/\alpha$, let $\hat{X}^{(n)}(t) := \dfrac{X(nt)}{n^{1/(\alpha+1)}}$, for $t \geq 0$. Then the annealed finite-dimensional distributions of $\hat{X}^{(n)}$ converge to those of $Z^{(\alpha)} \circ W^{(2)} \circ \Delta^{-1}$. This means that, for all $m \in \mathbb{Z}^+$ and $t_1, \ldots, t_m \in \mathbb{R}^+$,*

$$\big(\hat{X}^{(n)}(t_1), \ldots, \hat{X}^{(n)}(t_m)\big) \xrightarrow{\text{d}} \big(Z^{(\alpha)}(W^{(2)}(\Delta^{-1}(t_1))), \ldots, Z^{(\alpha)}(W^{(2)}(\Delta^{-1}(t_m)))\big),$$

as $n \to \infty$, relative to $\mathbb{P}$.

Remark 7 The process $Z^{(\alpha)} \circ W^{(2)} \circ \Delta^{-1}$ is not a.s. càdlàg, so the same considerations and reference as in Remark 4 apply here.

4 A Brief Discussion on Perspectives

Just by looking at the titles of the previous subsections, one notices that no quenched theorems were given for the case of infinite mean distance between targets. This is the main shortcoming of the current mathematical description of the processes X and Y. Technically speaking, the problem is that, without the condition $\mu < \infty$, one does not have a strong law of large numbers for the variables ζ_k. This is after all the simplest form of a quenched result and provides the scaling of $k \mapsto \omega_k$, as $|k| \to \infty$, for *each* realization ω of the medium, apart from a negligible set of exceptions. How to prove quenched limit theorems without this basic ingredient is not clear to me at the moment.

An open question of a different nature is that of devising a good model of Lévy-Lorentz gas in dimension $d \geq 2$. Here 'good' means that it should have the following features, in one form or another:

- The random medium should be *homogeneous*, in the sense that the distribution of the relative positions of two or more targets should not depend on the absolute position of any of the targets involved.[7]
- The distances between targets should be heavy-tailed.[8]
- The law of the random medium should be rotation-invariant, at least for a subgroup of rotations, e.g., the coordinate directions. In other words, the model should be *isotropic*, unless it has a clear reason not to be.
- The transition probabilities from one target to the next should not depend on their distance, only on the "degree of accessibility" of the new target. For example, the next target might always be the nearest one along a random (isotropically selected) direction. The meaning of this condition is that it should be the medium, not the walker, to decide how long the next inertial stretch will be.

Even with all these features, a model might not be very interesting. Here is an example of a feasible, yet not very instructive model. Let $(\omega'_{k_1},\, k_1 \in \mathbb{Z})$ and $(\omega''_{k_2},\, k_2 \in \mathbb{Z})$ be two i.i.d. point processes in $\mathbb{R}$, as introduced in Sect. 1. For $k = (k_1, k_2) \in \mathbb{Z}^2$, set $\omega_k := (\omega'_{k_1}, \omega''_{k_2})$. This defines a random medium $\omega = (\omega_k,\, k \in \mathbb{Z}^2)$ in $\mathbb{R}^2$. An independent, $\mathbb{Z}^2$-*valued*, underlying RW $(S_n,\, n \in \mathbb{N})$ is given, whereby we introduce the DTRW $Y := (Y_n := \omega_{S_n},\, n \in \mathbb{N})$. The CTRW $X := (X(t),\, t \geq 0)$ is then defined as the unit-speed interpolation of Y.

---

[7] The reader who feels this condition is not well-defined is right, see footnote below.

[8] This condition is ill-defined in the same way as the previous condition was. It would be well-defined if the points of the random medium were labeled in a consistent way, so that it would make sense to consider, say, the distribution of the distance between ω_k and ω_ℓ (here k and ℓ are generic indices, not necessarily in $\mathbb{Z}$). But no labeling is assumed on the random medium, as it is not easy to think of a general, physically relevant way to label the points of a d-dimensional point process, for $d \geq 2$.

Obviously, the process X is simply the direct sum, in a very natural sense, of two independent, orthogonal, 1D CTRWs X' and X''. Its properties are thus (for the most part) easily derived from those of X' and X'', as presented in Sect. 3.

Introducing and investigating more relevant, and truly d-dimensional, flights and walks in Lévy random medium will be the subject of future work.

Acknowledgments I am indebted to all my coauthors in publications [8, 10, 11]: G. Bet, A. Bianchi, G. Cristadoro, M. Ligabò, E. Magnanini, F. Pène, and S. Stivanello. I acknowledge partial support by the PRIN Grant 2017S35EHN *"Regular and stochastic behaviour in dynamical systems"* (MUR, Italy). This work is also part of my activity within the Gruppo Nazionale di Fisica Matematica (INdAM, Italy).

My research on RWs in Lévy random media is my contribution to the joint UniBo-UniFi-UniPd project *"Stochastic dynamics in disordered media and applications in the sciences"*. While the paper was in preparation, the Scientific Coordinator of the project in Florence, my dear friend and excellent mathematician Francesca Romana Nardi, passed away, leaving in all of us who were lucky enough to know her a vast sense of emptiness. This paper, while honoring the legacy of Carlo Cercignani, is dedicated to her memory.

References

1. Barkai, E., Fleurov, V., Klafter, J.: Phys. Rev. E **61**, 1164–1169 (2000)
2. Barthelemy, P., Bertolotti, J., Wiersma, D.S.: Nature **453**, 495–498 (2008)
3. Akhmerov, A.R., Beenakker, C.W.J., Groth, C.W.: Phys. Rev. B **79**, 024204 (2009)
4. Burioni, R., Caniparoli, L., Vezzani, A.: Phys. Rev. E **81**, 060101(R) (2010)
5. Vezzani, A., Barkai, E., Burioni, R.: Phys. Rev. E **100**, 012108 (2019)
6. Klages, R., Radons, G., Sokolov, I.M. (eds.): Anomalous Transport: Foundations and Applications. Wiley-VCH, Berlin (2008)
7. Zaburdaev, V., Denisov, S., Klafter, J.: Rev. Mod. Phys. **87**, 483–530 (2015)
8. Bianchi, A., Cristadoro, G., Lenci, M., Ligabò, M.: J. Stat. Phys. **163**, 22–40 (2016)
9. Magdziarz, M., Szczotka, W.: Commun. Nonlinear Sci. Numer. Simul. **60**, 100–106 (2018)
10. Bianchi, A., Lenci, M., Pène, F.: Stoch. Process. Their Appl. **130**, 708–732 (2020)
11. Stivanello, S., Bet, G., Bianchi, A., Lenci, M., Magnanini, E.: Electron. J. Probab. **26**, Article no. 57 (2021)
12. Zamparo, M.: Ann. Inst. Henri Poincaré Probab. Stat., online first (2022) https://arxiv.org/abs/2010.09083v2
13. Artuso, R., Cristadoro, G., Onofri, M., Radice, M.: J. Stat. Mech. 083209 (2018)
14. Radice, M., Onofri, M., Artuso, R., Cristadoro, G.: J. Phys. A **53**, 025701 (2020)
15. Rohlin, V.A.: Mat. Sb. **25**(67), 107–150 (1949)
16. Whitt, W.: Stochastic-Process Limits. An Introduction to Stochastic-Process Limits and Their Application to Queues. Springer, New York (2002)
17. Janson, S.: https://arxiv.org/abs/1112.0220
18. Bingham, N.H., Goldie, C.M., Teugels, J.L.: Regular Variation. Cambridge University Press, Cambridge (1987)
19. Castiglione, P., Mazzino, A., Muratore-Ginanneschi, P., Vulpiani, A.: Phys. D **134**, 75–93 (1999)
20. Kesten, H., Spitzer, F.: Z. Wahrscheinlichkeitstheorie Verwandte Geb. **50**, 5–25 (1979)

Mesoscale Modelling of the Tolman Length in Multi-component Systems

Matteo Lulli, Luca Biferale, Giacomo Falcucci, Mauro Sbragaglia, and Xiaowen Shan

Abstract In this paper we analyze the curvature corrections to the surface tension in the context of the Shan-Chen (SC) multi-component Lattice Boltzmann method (LBM). We demonstrate that the same techniques recently applied in the context of the Shan-Chen multi-phase model can be applied to multi-component mixtures. We implement, as a new application, the calculation of the surface of tension radius R_s through the minimization of the generalized surface tension $\sigma[R]$. In turn we are able to estimate the Tolman length, i.e., the first-order coefficient of the curvature expansion of the surface tension $\sigma(R)$, as well as the higher order corrections, i.e., the curvature- and the Gaussian-rigidity coefficients. The SC multi-component model allows to describe both fully symmetric as well as asymmetric interactions among the components. By performing an extensive set of simulations we present a first example of tunable Tolman length in the mesoscopic model, being zero for symmetric interactions and different from zero otherwise. This result paves the way for controlling such interface properties which are paramount in presence of thermal fluctuations. All reported results can be independently reproduced through the "idea.deploy" framework available at https://github.com/lullimat/idea.deploy.

M. Lulli (✉) · X. Shan
Department of Mechanics and Aerospace Engineering, Southern University of Science and Technology, Shenzhen, China
e-mail: lulli@sustech.edu.cn; shanxw@sustech.edu.cn

L. Biferale · M. Sbragaglia
Department of Physics & INFN, University of Rome "Tor Vergata", Rome, Italy
e-mail: biferale@roma2.infn.it; sbragaglia@roma2.infn.it

G. Falcucci
Department of Enterprise Engineering "Mario Lucertini", University of Rome "Tor Vergata", Rome, Italy

John A. Paulson School of Engineering and Applied Physics, Harvard University, Cambridge, MA, USA
e-mail: giacomo.falcucci@uniroma2.it

P. Barbante et al. (eds.), *From Kinetic Theory to Turbulence Modeling*, Springer INdAM Series 51, https://doi.org/10.1007/978-981-19-6462-6_14

List of Symbols

Symbols are reported according to their order of appearance. Bold font symbols refer to tensorial quantities that can be indexed through Greek letters.

n_{J}	Concentration of the J-th fluid component
$\mathbf{u}$	Fluid velocity field
$n_{\mathrm{J,s}}$	Saturation concentration of the J-th fluid component
σ	Surface tension
$n_{\mathrm{J,b}}$	Concentration of the J-th fluid component in the bulk of a J-rich droplet
$n_{\mathrm{J,out}}$	Concentration of the J-th fluid component outside of a J-rich droplet
z_s	Surface of tension position for a flat interface
R	Position of an arbitrary dividing surface for a droplet
$\sigma[R]$	Generalized surface tension
$\left[\frac{\mathrm{d}\sigma}{\mathrm{d}R}\right]$	Notional derivative of the generalized surface tension
$\{A, B\}$	Components labels for a binary mixture
R_s	Position of the surface of tension for a droplet
ΔP	Pressure jump between the inside and outside of a droplet
$\sigma(R)$	Curvature dependent surface tension
σ_0	Flat interface surface tension
δ	Tolman length
$\bar{k}$,k	Curvature- and Gaussian-rigidity coefficients
$f^{(\mathrm{J})}(\mathbf{x}, \boldsymbol{\xi}, t)$	Single-particle distribution function for the J-th component
$\{\mathbf{x}\}$	Set of discrete lattice points
$\boldsymbol{\xi}$	Particle peculiar velocity
t	Time
$\{\boldsymbol{\xi}_i\}$	Discrete velocity set or stencil
$f_i^{(\mathrm{J})}$	i-th population for the J-th component
$n_{\mathrm{J}}\mathbf{u}_{\mathrm{J}}$	J-th component momentum
$F_i^{(\mathrm{J})}$	Forcing term in the Lattice Boltzmann equation for the i-th population of the J-th component
$\Omega_i^{(\mathrm{J})}$	Collision operator for the i-th population of the J-th component
τ_{J}	BGK relaxation time for the J-th component
$f_i^{(\mathrm{eq,J})}$	Discrete equilibrium distribution for the i-th population of the J-th component
c_s^2	Square of the speed of sound
w_i	Weight associated with the discrete velocity $\boldsymbol{\xi}_i$
$\mathbf{F}_{\mathrm{J}}$	Force exerted on the J-th component
G	Inter-component coupling constant
G_{AA}, G_{BB}	Components self-coupling constants
ψ_{J}	Pseudopotential function for self-interactions
W	Weights associated with the forcing directions

G_c	Critical value for phase separation for the self-interactions for the given ψ_J
δG_c	Relative distance from the critical point
δG_{AB}	Asymmetry parameter estimating the relative difference of G_{AA}, G_{BB}
$P^{\mu\nu}$	(Lattice) pressure tensor
P_N	Normal component of the pressure tensor to an interface
p_0	Bulk pressure for a flat interface
P	Bulk pressure
e_2	Second order isotropy coefficient
$\mathbf{q}$	Projector onto the tangential direction to an interface
$\mathbf{n}$	Normal to an interface
r	Radial coordinate
P_{in}, P_{out}	Bulk pressure inside and outside of a droplet
P_J	Pressure jump function
θ	Heaviside step function
L, L_x, L_y, L_z	Linear system size according to the direction
$n_{J,in}$, $n_{J,out}$	Initial concentrations inside and outside a droplet for the J-th component
$n_{J,h}$, $n_{J,l}$	High and low initial concentrations values for the J-th component with a flat interface
x_0, w	Center and width of the initial profile for the flat interface
σ_m, σ_M	Minimum and maximum values of the flat surface tension obtained from the simulations
δ_m, δ_M	Minimum and maximum values of the Tolman length obtained from the simulations

1 Introduction

Multi-component fluids are systems where two or more components, distinguished by their chemical properties, are mutually diffused into each other. The J-th component can be described by the concentration n_J while the flow of the mixture is characterized by a common velocity $\mathbf{u}$. In the appropriate thermodynamic conditions, i.e., temperature and pressure, for values of the concentration above a saturation threshold $n_{J,s}$, it is possible for droplets of the J-th component to form. In particular, this happens when the free-energy gain provided by the formation of a n_J-rich bulk region overcomes the barrier provided by the surface free-energy associated with the interface. The latter contribution is commonly described by the free-energy cost per unit area, i.e., the surface tension σ, which, in the limit of small deformations, allows to describe the mechanic response of the interface as that of an elastic membrane. Only a few configurations are mechanically stable, namely the flat and the spherical interfaces. In settings where the typical scale of the interface is such that thermal fluctuations are negligible it can be useful to adopt a simplified description of the interface as being a discontinuity point for the concentration n_J, i.e., going from the bulk value $n_{J,b} > n_{J,s}$ to a soluble value $n_{J,out} < n_{J,s}$ outside the bulk region. Such a discontinuity can be used to identify the *surface of tension* [1] providing a simple, yet useful, mechanical model of the interface. However, this is a somewhat idealized description since thermal fluctuations naturally induce a finite *interface thickness* which opens the question of the determination of the position of the surface of tension itself. In other words, the average of the concentration profile over thermal fluctuations is a continuous curve rather than a step function. Considering a flat interface, it is possible to determine the position z_s of the surface of tension by means of the pressure tensor [1]. However, when considering spherical interfaces the determination of the surface of tension is more complicated. While in the case of a flat interface the value of z_s does not explicitly enter the definition of the free-energy, for closed interfaces the curvature appears as a new control parameter with the explicit introduction of an arbitrary dividing surface [1–3] whose position with respect to the center of the droplet is denoted by R. Since the position of such interface is arbitrary, i.e., it can either be completely inside or outside the n_J-rich droplet (see Fig. 1a), a natural request is for the free-energy to be independent, i.e., stationary, with respect to arbitrary, or notional [1], changes in R. Starting here we restrict our discussion and results to the case of two different components, hence two concentrations fields n_J with $J \in \{A, B\}$. The stationarity condition reflects on the definition of a *generalized* surface tension $\sigma[R]$, which assumes the shape of a convex function reaching a minimum at the surface of tension R_s. At the latter position the Laplace law applies in the usual form [1–4]. Indeed, it is possible to show [1] that the stationarity of the free-energy at the surface of tension R_s yields

$$\Delta P = \frac{2\sigma[R_s]}{R_s} + \left.\frac{\mathrm{d}\sigma[R]}{\mathrm{d}R}\right|_{R=R_s} = \frac{2\sigma(R_s)}{R_s}. \tag{1}$$

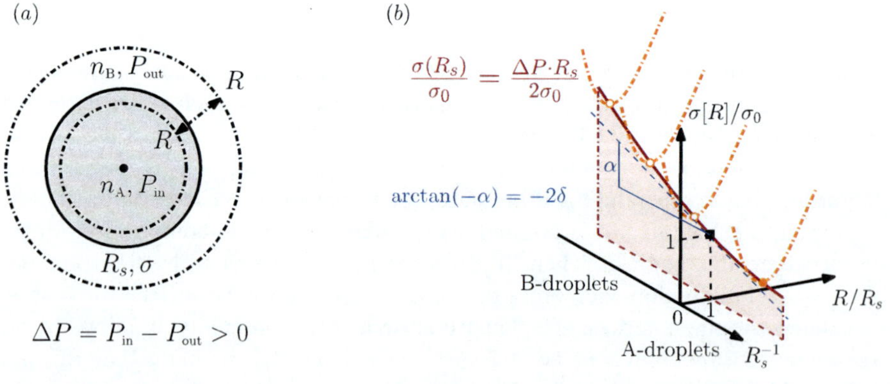

Fig. 1 (**a**) Sketch of a n_A-rich droplet surrounded by the n_B-rich fluid; the position of the *arbitrary dividing surface* is indicated by R while R_s indicates the position of the surface of tension where the surface tension σ acts. (**b**) Sketch for the curvature dependence of the (normalized) surface tension, $\sigma(R_s)/\sigma_0$, reported in brown, as the locus of minima of all the generalized surface tension functions $\sigma[R]/\sigma_0$ reported in dash-dotted orange (cf. with Fig. 2). The horizontal axes express the *physical* dependence on the curvature R_s^{-1} and the *notional* dependence [1] on the (normalized) arbitrary dividing surface position R/R_s

By considering a generic value of $R \neq R_s$ in (1) one obtains the so-called *generalized* Laplace law which explicitly depends on the notional derivative [1] of $\sigma[R]$.

The locus of the minima of $\sigma[R]$ identifies a physical, i.e., non-arbitrary, dependence of the surface tension $\sigma(R)$ on the droplet size at the surface of tension R_s, $\sigma_s = \sigma[R_s] = \sigma(R_s)$. Such a dependence was first examined, for the case of multi-phase systems, in the seminal paper by Tolman [5] (see [6, 7] for reviews). Similar results have been obtained for the case of elastic membranes in Helfrich's work [8] where the curvature dependence is expressed as a power-law expansion in the curvature, i.e., the inverse radius, which at second order reads [4, 9–11]

$$\sigma(R_s) \simeq \sigma_0 \left(1 - \frac{2\delta}{R_s} + \frac{2\bar{k} + k}{R_s^2}\right). \tag{2}$$

The flat interface value σ_0 appears at the leading order, the first-order coefficient δ defines the Tolman length [cf. Fig. 1b] and $\bar{k}$ and k are called curvature- and Gaussian-rigidity coefficients, respectively. It has been shown and studied in the literature [12–14] that the Tolman length provides a measure of the "symmetry" of the interactions under, e.g., a vapor-liquid exchange transformation. Such symmetry can be readily realized (and broken) in mesoscopic multi-component models [15]. In the next Sections we describe how to tune the degree of asymmetry of the interactions which, in turn, will induce a tuning in the Tolman length and the higher order coefficients.

Several works based on the Density Functional Theory (DFT) [10, 11, 16–18] have led to expressions for the coefficients δ, k, and $\bar{k}$, for realistic multi-phase and multi-component systems. From the numerical perspective, simulations have mostly focused on molecular dynamics (MD) for multi-phase [19–22] and multi-component systems [7, 23]. The Tolman length has been measured in nucleation experiments [24], and its role was analyzed both in confined geometries [25] and in colloidal liquids [26]. The curvature dependence of the surface tension is paramount in extending Classical Nucleation Theory (CNT). The standard formulation of CNT relies on the *capillary approximation* by which the nucleation free-energy barrier W linearly depends on the flat interface value of the surface tension σ_0. Nucleation rates are one of the quantities of interest in CNT with an exponential dependence on W, so that, curvature corrections to σ_0 can quantitatively affect the rate to a large extent. Such considerations are valid for both multi-phase and multi-component systems [27], however, much of the literature focuses on multi-phase systems for both theory [28, 29] and experiments [24, 26]. Indeed, at present, the application of CNT to multi-component systems is a most challenging yet very important area of research [27] in which a sound control over curvature corrections can provide a valuable contribution. Indeed, in [30] curvature corrections have been used to eliminate a few important inconsistencies of CNT for the case of a propanol-water mixture, e.g., a negative number of molecules in the critical nucleating cluster, highlighting the relevance of the Tolman length and curvature coefficients. Our present mesoscale modelling bears the possibility of extending these results to the hydrodynamic regime where nucleation rates could be consistently predicted also in heterogeneous stress conditions of relevance in several engineering problems.

In this work, we study the Tolman length and the higher order corrections using a three-dimensional Shan-Chen multi-component [15, 31] lattice Boltzmann method (LBM) [32, 33] by means of an extensive set of hydrostatic simulations. Specifically, we demonstrate that it is possible to tune the value of δ by "breaking," in a controllable way, the symmetry of the system's interactions under the exchange A $\longleftrightarrow$ B, thus making a further step with respect to the results already obtained for the multi-phase case [34] for which the possibility of tuning was left for future works. A strong dependence of the Tolman length on the relative concentrations of the two components has been thoroughly studied in [10] by means of a Square Gradient Theory (SGT) approach which is a first approximation of DFT [35]. In the latter case the variation of the curvature corrections are related to the same physical system, while in this work we investigate a parametrization potentially describing different physical systems. We estimate δ and the combination $2\bar{k} + k$ by leveraging a lattice formulation of the pressure tensor [36] extended to the multi-component case [37]. Using the pressure tensor we can compute $\sigma[R]$ (see Fig. 1b) following a construction stemming by the mechanic equilibrium condition [1] which was detailed in [34]. Most approaches in the field have either leveraged microscopic MD simulations or continuum DFT approaches so that non-equilibrium mesoscopic effects, i.e., hydrodynamics, have mostly been neglected so far. Hence, the present work represents a first step in developing a mesoscale approach for the tuning of the curvature corrections to the surface tension for multi-component systems which can naturally include hydrodynamics. It would also be interesting to investigate the curvature corrections in different models such as the *color gradient* approach [38–40], the *free-energy* approach [41–43] and the *entropic* one [44]. Further, it is important to mention that in the case of multi-phase mixtures, in [45] results compatible with those in [34] have been reported, i.e., same power-law behavior of the Tolman length approaching the critical point. Most importantly, the model in [45] differs from that in [34], for its implementation a for the equation of state, thus providing an important independent validation of the overall approach presented in [34] and extended here. Another interesting perspective is that of studying the relation between curvature corrections and the so-called *near-contact* interactions, e.g., the disjoining pressure that develops when the interfaces of two droplets get close enough and hinders their coalescence. Different lattice Boltzmann models have been devised to correctly capture this kind of interactions [40, 46] and the possibility of an interplay between the Tolman length and the length-scales involved in near-contact interactions will be the focus of future works.

The paper is organized as follows: we describe in Sect. 2 the fundamentals of the LBM formulation adopted in this work; in Sect. 3 we detail the method used to evaluate the position of the surface of tension R_s and how to "break" the symmetry of the interactions in order to achieve a tunable Tolman length; in Sect. 4 we report the results followed by the conclusions in Sect. 5. The simulations source code and a Jupyter notebook to reproduce all the results and figures can be found on GitHub https://github.com/lullimat/idea.deploy.

2 Lattice Boltzmann Model

The lattice Boltzmann method (LBM) allows to simulate the Navier-Stokes dynamics of a multi-component mixture by means of two coupled forced Boltzmann transport equation acting on a discretized phase-space [32, 33]. For each component, the single-particle distribution function $f^{(\mathrm{J})}(\mathbf{x}, \boldsymbol{\xi}, t)$ is defined on the nodes $\{\mathbf{x}\}$ of a three-dimensional lattice at discrete times t. Hence, one defines the *populations* as the single-particle distribution function evaluated at a given discrete velocity $\boldsymbol{\xi}_i$, i.e., $f_i^{(\mathrm{J})}(\mathbf{x}, t) = f^{(\mathrm{J})}(\mathbf{x}, \boldsymbol{\xi}_i, t)$. Remarkably, the convergence to the hydrodynamic limit is very fast even when employing only a few velocity vectors $\{\boldsymbol{\xi}_i\}$ connecting each lattice point to a set neighboring nodes. In this paper we adopt the D3Q19 stencil with nineteen discrete velocity vectors $\boldsymbol{\xi}_i$ with $i = 0, \dots, 18$. The first two moments of the discretized distribution define the component concentration $n_{\mathrm{J}} = \sum_i f_i^{(\mathrm{J})}$ and the momentum density $n_{\mathrm{J}}\mathbf{u}_{\mathrm{J}} = \sum_i f_i^{(\mathrm{J})}\boldsymbol{\xi}_i$, respectively. The lattice transport equation for the J-th component reads

$$f_i^{(\mathrm{J})}\left(\mathbf{x} + \boldsymbol{\xi}_i, t + 1\right) - f_i^{(\mathrm{J})}(\mathbf{x}, t) = \Omega_i^{(\mathrm{J})}(\mathbf{x}, t) + \left(1 - \frac{1}{2\tau_{\mathrm{J}}}\right) F_i^{(\mathrm{J})}(\mathbf{x}, t), \tag{3}$$

where $F_i^{(\mathrm{J})}(\mathbf{x}, t)$ is the forcing term [47] and $\Omega_i^{(\mathrm{J})}(\mathbf{x}, t)$ is the local collision operator conserving mass and momentum, i.e., $\sum_i \Omega_i^{(\mathrm{J})} = \sum_i \boldsymbol{\xi}_i \Omega_i^{(\mathrm{J})} = 0$. Equation (3) is usually interpreted as implementing two separate steps, namely (i) the *streaming* step represented by the left-hand side by which populations freely stream from one lattice node to the other and (ii) the *collision* step represented by the right-hand side which only involves local quantities. The locality of $\Omega_i^{(\mathrm{J})}$ is one of the main features of LBM which renders the approach particularly amenable to parallel implementations [32, 33]. More specifically, the right-hand side of (3) is composed by the Bhatnagar–Gross–Krook (BGK) [48] collision operator

$$\Omega_i^{(\mathrm{J})}(\mathbf{x}, t) = -\frac{1}{\tau_{\mathrm{J}}}\left[f_i^{(\mathrm{J})}(\mathbf{x}, t) - f_i^{(\mathrm{eq,J})}(n_{\mathrm{J}}, \mathbf{u})\right] \tag{4}$$

and by the Guo [47] forcing term

$$F_i^{(\mathrm{J})}(\mathbf{x}, t) = w_i\left[\frac{1}{c_s^2}\xi_i^\beta + \frac{1}{c_s^4}\left(\xi_i^\alpha \xi_i^\beta - c_s^2\delta^{\alpha\beta}\right)u_\alpha\right]F_{\mathrm{J}}^\beta(\mathbf{x}, t), \tag{5}$$

where repeated Greek indices imply summation. This term is used to implement in the LBM the Shan-Chen [15, 31] (SC) force $F_{\mathrm{J}}^\alpha(\mathbf{x}, t)$ responsible for the formation of stable concentration gradients, i.e., interfaces between the two components. The equilibrium populations $f_i^{(\mathrm{eq,J})}$ are obtained as a second order approximation of the Maxwell distribution

$$f_i^{(\mathrm{eq,J})}(n_{\mathrm{J}}, \mathbf{u}) = n_{\mathrm{J}} w_i\left[1 + \frac{1}{c_s^2}\xi_i^\alpha u_\alpha + \frac{1}{2c_s^4}\left(\xi_i^\alpha \xi_i^\beta - c_s^2\delta^{\alpha\beta}\right)u_\alpha u_\beta\right] \tag{6}$$

and the equilibrium fluid velocity is computed according to Guo prescription [32, 47]

$$nu^{\mu} = \sum_{J\in\{A,B\}} \left[\sum_{i=0}^{18} f_i^{(J)} \xi_i^{\mu} + \frac{1}{2} F_J^{\mu} \right]. \tag{7}$$

Several different approaches for multi-component flows [32, 33] have been developed for LBM yielding some of the most successful applications of the method. In this paper we show that the SC multi-component model [15, 31] correctly captures a curvature dependent surface tension while allowing for the tuning of the expansion coefficients, i.e., the Tolman length δ and the combination of the rigidity constants $2\bar{k} + k$. The main feature of the SC model, allowing for the existence of stable gradients of the concentrations $n_J(\mathbf{x}, t)$, is a force computed on the lattice nodes, which, separating the contribution of each component, reads

$$\begin{aligned} F_A^{\mu}(\mathbf{x}) =& - G c_s^2 n_A(\mathbf{x}) \sum_{a=1}^{18} W\left(|\boldsymbol{\xi}_a|^2\right) n_B(\mathbf{x} + \boldsymbol{\xi}_a) \xi_a^{\mu} \\ & - G_{AA} c_s^2 \psi_A(\mathbf{x}) \sum_{a=1}^{18} W\left(|\boldsymbol{\xi}_a|^2\right) \psi_A(\mathbf{x} + \boldsymbol{\xi}_a) \xi_a^{\mu} \\ F_B^{\mu}(\mathbf{x}) =& - G c_s^2 n_B(\mathbf{x}) \sum_{a=1}^{18} W\left(|\boldsymbol{\xi}_a|^2\right) n_A(\mathbf{x} + \boldsymbol{\xi}_a) \xi_a^{\mu} \\ & - G_{BB} c_s^2 \psi_B(\mathbf{x}) \sum_{a=1}^{18} W\left(|\boldsymbol{\xi}_a|^2\right) \psi_B(\mathbf{x} + \boldsymbol{\xi}_a) \xi_a^{\mu}, \end{aligned} \tag{8}$$

where $\psi_J(\mathbf{x}, t) = \psi(n_J(\mathbf{x}, t))$ is the so-called pseudopotential, a local function of the concentration n_J, implicitly depending on space and time, $c_s = 1/\sqrt{3}$ is the speed of sound, G is the inter-component coupling constant while G_{AA} and G_{BB} are the self-coupling constants. If one sets $G = 0$ the two components are completely decoupled and one effectively simulates two parallel multi-phase systems which can display phase separation whenever $G_{AA}, G_{BB} < G_c$ where G_c is the critical coupling constant whose value depends on the choice of $\psi_J(\mathbf{x}, t)$ [15, 31]. A similar approach has been used in [46, 49] for the simulation of emulsions and comparison with experimental results [50, 51]. The vectors $\boldsymbol{\xi}$ are the discrete forcing directions such that their squared lengths are $|\boldsymbol{\xi}_a|^2 = 1, 2$, and $W(1) = 1/6$ and $W(2) = 1/12$ are the weights ensuring 4-th order lattice force isotropy [52, 53]. The set of the forcing vectors $\boldsymbol{\xi}_a$ coincide with that of the lattice velocities $\boldsymbol{\xi}_i$ after excluding the "rest" direction $\boldsymbol{\xi}_0 = (0, 0, 0)$.

Now, a few remarks are in order. From the structure of Eq. (8) above it clearly appears that as long as the $G_{AA} = G_{BB}$ the system is symmetric, or invariant, under the exchange of the two components A $\longleftrightarrow$ B. In order to "break" this symmetry one possibility is that of choosing different self-coupling constants $G_{\mathrm{AA}} \neq G_{\mathrm{BB}}$. In order to do this, we adopt the following parametrization

$$\begin{aligned} G_{\mathrm{AA}} &= G_c \left[1 - \delta G_c \left(1 + \delta G_{\mathrm{AB}}\right)\right] \\ G_{\mathrm{BB}} &= G_c \left[1 - \delta G_c\right], \end{aligned} \tag{9}$$

where $G_c c_s^2 \simeq -2.463$ is the critical value of the self-interaction coupling corresponding to the pseudopotential $\psi_{\mathrm{J}} = \exp(-1/n_{\mathrm{J}})$. It is possible to select other functional forms for ψ_{J} [53, 54]. The present choice is not meant to fulfill a specific requirement and it is only instrumental for the purpose of analyzing the effects on the Tolman length and the rigidity coefficients of switching from symmetric to asymmetric interactions. By setting $\delta G_c = 0.4$ and $\delta G_{\mathrm{AB}} > -1$, we assure that $G_{\mathrm{AA}}, G_{\mathrm{BB}} > G_c$, i.e., the values of the self-coupling constants are above the critical point so that the gradients in the multi-component system are only due to the inter-component interactions. Indeed, the parameter δG_{AB} estimates the degree of asymmetry of the self-interactions, i.e., the ratio $G_{\mathrm{AA}}/G_{\mathrm{BB}}$. One has the following linear relation

$$\delta G_{\mathrm{AB}} \left[\frac{\delta G_c}{1 - \delta G_c}\right] = 1 - \frac{G_{\mathrm{AA}}}{G_{\mathrm{BB}}}, \tag{10}$$

hence, by setting both positive and negative values we can analyze the behavior of the system around the symmetric case $\delta G_{\mathrm{AB}} = 0$.

The SC force defined in Eq. (8) is related to a *lattice* pressure tensor [36, 55–57] that reads

$$\begin{aligned} P^{\mu\nu}(\mathbf{x}) = {} & \left[n_{\mathrm{A}}(\mathbf{x}) + n_{\mathrm{B}}(\mathbf{x})\right] c_s^2 \delta^{\mu\nu} \\ & + \frac{G c_s^2}{2} n_{\mathrm{A}}(\mathbf{x}) \sum_{a=1}^{18} W\left(|\xi_a|^2\right) n_{\mathrm{B}}(\mathbf{x} + \xi_a) \xi_a^{\mu} \xi_a^{\nu} \\ & + \frac{G c_s^2}{2} n_{\mathrm{B}}(\mathbf{x}) \sum_{a=1}^{18} W\left(|\xi_a|^2\right) n_{\mathrm{A}}(\mathbf{x} + \xi_a) \xi_a^{\mu} \xi_a^{\nu} \\ & + \frac{G_{\mathrm{AA}} c_s^2}{2} \psi_{\mathrm{A}}(\mathbf{x}) \sum_{a=1}^{18} W\left(|\xi_a|^2\right) \psi_{\mathrm{A}}(\mathbf{x} + \xi_a) \xi_a^{\mu} \xi_a^{\nu} \\ & + \frac{G_{\mathrm{BB}} c_s^2}{2} \psi_{\mathrm{B}}(\mathbf{x}) \sum_{a=1}^{18} W\left(|\xi_a|^2\right) \psi_{\mathrm{B}}(\mathbf{x} + \xi_a) \xi_a^{\mu} \xi_a^{\nu}. \end{aligned} \tag{11}$$

We wish to highlight that the tensor in the Eq. (11) is such that the flat interface mechanical equilibrium condition, i.e., constant normal component $P_N(x) = p_0$ throughout the interface, is obeyed *on the lattice* with a value of p_0 that is constant to machine precision. This property has allowed for an extremely precise estimation of the coexistence curve in the multi-phase case [36, 54, 56] and it is one of the building blocks for the results presented in this paper. By performing the Taylor expansion of Eq. (11) one obtains, at the leading order, the bulk pressure

$$P(\mathbf{x}) = [n_A(\mathbf{x}) + n_B(\mathbf{x})]\, c_s^2 + G c_s^2 e_2 n_A(\mathbf{x}) n_B(\mathbf{x}) + \frac{G_{AA} c_s^2 e_2}{2} \psi_A^2(\mathbf{x}) + \frac{G_{BB} c_s^2 e_2}{2} \psi_B^2(\mathbf{x}), \tag{12}$$

where $e_2 = \sum_{\mathbf{e}_a} W(|\mathbf{e}_a|^2) e_a^x e_a^x = 1$ [37, 52, 57] for the values of the weights used in this work. The first line represents the ideal gas contribution plus the inter-component interaction contribution while the second line yields the sum of the self-interaction ones. Considering the combination of the ideal and self-interaction parts each component can independently display phase separation whenever $G_{AA}, G_{BB} < G_c$ [15].

The SC model has been widely used to model complex fluids with a non-trivial impact on the study of the interface physics, one may cite heterogeneous cavitation [58] and emulsion rheology physics [59], also in presence of complex boundary conditions [51]. The ability to model and tune the Tolman length and the rigidity coefficients in LBM allows to effectively tackle the study of nucleation and cavitation phenomena in the mesoscale regime for multi-component systems, while providing a computationally efficient tool allowing for a direct bridge with experiments.

3 Method

In Sect. 1 we briefly discussed that in a multi-component system the free-energy needs to be independent on the choice of the position R of an arbitrary dividing spherical surface. Such a condition yields the generalized Laplace law [1–3, 60]

$$\Delta P = \frac{2\sigma[R]}{R} + \left[\frac{d\sigma}{dR}\right] \tag{13}$$

with $\sigma[R]$ the generalized surface tension and its notional derivative [1] $[d\sigma/dR] = \sigma'[R]$ and $\Delta P = P_{in} - P_{out}$, with P_{in} and P_{out} the values of the bulk pressure in the center of the droplet and far away from the interface, respectively (see Fig. 1a). The function $\sigma[R]$ is convex and at its minimum Eq. (13) reduces to the usual Laplace law. The condition $\sigma'[R]|_{R=R_s} = 0$ defines the position of the surface of tension R_s. Hence, comparing Eqs. (2) and (13), it follows that at second order in R_s^{-1} the latter reads

$$\Delta P = \frac{2\sigma_s(R_s)}{R_s} \simeq \frac{2\sigma_0}{R_s}\left(1 - \frac{2\delta}{R_s} + \frac{2\bar{k}+k}{R_s^2}\right). \tag{14}$$

In order to estimate the Tolman length we simulate droplets with n_{A}- and n_{B}-rich bulks for different values of the asymmetry parameter δG_{AB}. Further, we compute the deviations from the Laplace law using the surface of tension radius R_s which is used to determine the droplets sizes. In order to estimate R_s from the simulations we use a construction presented in [1] which only employs the mechanic equilibrium condition $\partial_\mu P^{\mu\nu} = 0$. The same arguments have been adopted in the case of the multi-phase SC model in [34] where they are described in details. Here, we limit our discussion to the most important steps. Let us consider the following decomposition of the pressure tensor

$$P^{\mu\nu} = P_{\mathrm{N}}\delta^{\mu\nu} - (P_{\mathrm{N}} - P_{\mathrm{T}})q^{\mu\nu}, \tag{15}$$

where P_{N} and P_{T} are the (locally) normal and tangential components to the droplet interface, respectively. The projector along the tangential direction is defined as $q^{\mu\nu} = \delta^{\mu\nu} - n^\mu n^\nu$ where n^μ is the normal vector to the interface which is given by the direction of the largest gradient. The mechanic equilibrium condition reads

$$\partial_\mu P^{\mu\nu} = n^\nu n^\mu \partial_\mu P_{\mathrm{N}} + n^\nu \partial_\mu n^\mu \left(P_{\mathrm{N}} - P_{\mathrm{T}}\right) = 0. \tag{16}$$

In three dimensions one has $n^\nu \partial_\mu n^\mu = 2n^\nu/r$, where r is the value of the radial coordinate. Selecting the normal/radial direction to be parallel to the x-axis, i.e., $n^\mu = e_x^\mu$ yields

$$\frac{\mathrm{d}}{\mathrm{d}r} P_{\mathrm{N}}(r) + \frac{2}{r}\left[P_{\mathrm{N}}(r) - P_{\mathrm{T}}(r)\right] = 0. \tag{17}$$

Upon multiplication by r^n followed by some derivatives rearrangements it is possible to obtain a sequence of identities

$$\frac{\mathrm{d}}{\mathrm{d}r}[r^n P_{\mathrm{N}}(r)] = r^{n-1}[(n-2)P_{\mathrm{N}}(r) + 2P_{\mathrm{T}}(r)]. \tag{18}$$

Finally, after introducing the pressure jump function $P_{\mathrm{J}}(r;\,R) = P_{\mathrm{in}} - (P_{\mathrm{in}} - P_{\mathrm{out}})\theta(r-R)$, where $\theta(r-R)$ is the Heaviside function, one can subtract the integral between R_{in} and R_{out} of Eq. (18) and that of $r^n\ P_{\mathrm{J}}(r;\,R)$. After setting $n = 2$ one obtains the following expression for the pressure jump $\Delta P = P_{\text{in}} - P_{\text{out}}$ across the interface

$$\begin{aligned}\Delta P &= \frac{2}{R^2}\int_{R_{\text{in}}}^{R_{\text{out}}} \mathrm{d}r\, r[P_{\mathrm{J}}(r;R) - P_{\mathrm{T}}(r)] \\ &= \frac{2\sigma[R]}{R} + \left[\frac{\mathrm{d}\sigma}{\mathrm{d}R}\right].\end{aligned} \tag{19}$$

It is possible to extract the expressions for $\sigma[R]$ and $[\mathrm{d}\sigma/\mathrm{d}R]$ [1] obtaining

$$\sigma[R] = \int_0^{+\infty} \mathrm{d}r\, \left(\frac{r}{R}\right)^2 [P_{\mathrm{J}}(r;R) - P_{\mathrm{T}}(r)], \tag{20}$$

$$\left[\frac{\mathrm{d}\sigma}{\mathrm{d}R}\right] = -\frac{2}{R^3}\int_0^{+\infty} \mathrm{d}r\, r(r-R)[P_{\mathrm{J}}(r;R) - P_{\mathrm{T}}(r)], \tag{21}$$

where we also considered the limits $R_{\text{in}} \to 0$ and $R_{\text{out}} \to \infty$. In order to estimate the position of the surface of tension R_s, we interpolate the position of the minimum of Eq. (20) after evaluating the expression by means of the SC lattice pressure tensor in Eq. (11), integrating along the x axis so that $P_{\mathrm{N}} = P^{xx}$ and $P_{\mathrm{T}} = P^{yy} = P^{zz}$. The pressure jump across the droplets interfaces ΔP and the position of the surface of tension R_s are the key quantities in our analysis allowing us, by means of hydrostatic simulations of droplets of different sizes, to estimate the curvature dependence of the surface tension as $\sigma(R_s)/\sigma_0 = R_s\,\Delta P/2\sigma_0$. We remark that other choices are possible for the dividing surface, such as the total equimolar interface [10], however, such choices allow the notional derivative [1] in Eq. (13) to play a non-trivial role in the estimation of the coefficients, whereas the surface of tension allows to directly estimate the function $\sigma(R_s)$. Moreover, more than one definition for an equimolar radius is possible, appearing as a further dependence for the rigidity coefficients and not for the Tolman length [10, 16]. All in all, using the surface of tension, as already done in [23], allows for a simpler analysis of the surface tension curvature dependence.

4 Results

The simulations source code can be found on GitHub https://github.com/lullimat/idea.deploy [61–68]. A Jupyter notebook [67] is available from the "idea.deploy" framework to reproduce the results and the plots reported in this paper. The code provided for the multi-component model relies on a straightforward implementation, i.e., not highly optimized, with a set of kernels that can be compiled either in CUDA or OpenCL. This version still does not leverage the automatic code generation already implemented for the multi-phase case which will soon be extended to the multi-component one. In order to give an estimate of the needed simulation time, one needs roughly 5hrs on a Tesla P100 or 1.5hrs on a Tesla A100, for executing the simulations in order to obtain the data for one of the points in Fig. 4 with $L \leq 213$lu for a maximum RAM usage of 8GB per simulation. This is not the full size range presented here which requires 16GB of RAM per simulation. Hence, with the former constraint one would need roughly 5 days on one P100 and 1.5 days on a A100. The largest system size is the most challenging, not only because it requires more resources but also because the actual convergence is slower. The strategy used for the simulations closely follows our previous contribution [34]. Here we report some details for completeness. We simulate three-dimensional droplets in a cubic system of linear size L with periodic boundary conditions using the D3Q19 discrete velocity set with $c_s^2 = 1/3$ [32, 33]. We adopt $\psi_J = \exp(-1/n_J)$ [31] as the pseudopotential function for the self-interaction part in Eq. (8). Other definitions of ψ_J have been used in the literature, however, the present choice is just as suitable for our primary objective, i.e., a first exploration of the tunability of the curvature corrections coefficients. The asymmetry parameter varies in the range $\delta G_{AB} \in \{0, \pm 0.04, \pm 0.08, \pm 0.16, \pm 0.24, \pm 0.32, \pm 0.40, \pm 0.48, \pm 0.56, \pm 0.64, \pm 0.80, \pm 0.88\}$, δG_c is set to 0.4, $G_c c_s^2 \simeq -2.463$ and the inter-component coupling is set to $G c_s^2 = 0.5$ (cf. Eq. (8)). The value of L is chosen to be an odd number so that the center of mass of the system exactly falls on the coordinates of a node. The simulated system sizes are $L \in \{25, 31, 37, 41, 51, 57, 65, 97, 127, 151, 213, 301\}$lu, where "lu" stands for lattice units. The radial concentration fields $n_J(r)$ are initialized to the following profile

$$n_J(r, R) = \frac{1}{2}(n_{J,\text{in}} + n_{J,\text{out}}) - \frac{1}{2}(n_{J,\text{in}} - n_{J,\text{out}}) \tanh(r - R), \tag{22}$$

where the inner $n_{J,\text{in}}$ and outer $n_{J,\text{out}}$ densities are set to the steady-state values obtained from the simulations of a flat interface system and the initial value of the radius is set to maintain a fixed aspect ratio $R = L/4$ for all simulations. The radial coordinate r is computed taking the center of the system as the origin. The values P_{in} and P_{out} are evaluated in the middle of the system $(\lfloor L/2 \rfloor, \lfloor L/2 \rfloor, \lfloor L/2 \rfloor)$ and at the farthest corner $(L-1, L-1, L-1)$, respectively. The outcome of the simulations is analyzed only if all the coordinates of the center of mass lie within a distance of 10^{-3} from the center of the domain. We use two convergence criteria for the simulations, both comparing quantities at a time distance $\delta t = 2^{11}$: (i) we consider the relative variation of the ΔP with respect to the previous configuration, and when the latter is such that $|\Delta P(t) - \Delta P(t + \delta t)|/\Delta P(t) < 10^{-5}$ the simulation is considered as converged; (ii) we consider the magnitude δu of the spatial average of the difference between the components of two velocity fields, $\delta u = L^{-3} \sum_{\mathbf{x}} \sum_{\alpha} |u^{\alpha}(\mathbf{x}, t + \delta t) - u^{\alpha}(\mathbf{x}, t)|$ so that the simulation is considered as converged when $\delta u < 10^{-12}$. Meeting only one of the two criteria is enough to finalize the simulation.

The set of simulations for the flat interface has been performed on a three-dimensional domain of sizes $L_x = 127$, $L_y = L_z = 5$ and the concentration profiles are initialized according to

$$
\begin{aligned}
n_{\mathrm{J}}(x, x_0, w) =& \frac{1}{2}\left(n_{\mathrm{J,h}} + n_{\mathrm{J,l}}\right) \\
& -\frac{1}{2}\left(n_{\mathrm{J,h}} - n_{\mathrm{J,l}}\right)\tanh\left[x - \left(x_0 - \frac{w}{2}\right)\right] \\
& +\frac{1}{2}\left(n_{\mathrm{J,h}} - n_{\mathrm{J,l}}\right)\left\{\tanh\left[x - \left(x_0 + \frac{w}{2}\right)\right] + 1\right\},
\end{aligned}
\tag{23}
$$

where x_0 is the center of the strip and $w = L_x/2$ its width. As a first approximation, in the presence of self-interactions, the flat interface concentrations, $n_{\mathrm{J,h}}$ and $n_{\mathrm{J,l}}$ for the *high* and *low* value, respectively, can be computed using the purely repulsive result of Eq.(43) in [46] with the substitution $\theta(\tau) \to 1$ to take into account Guo's forcing [47]. Such values are then used to initialize the flat interface profile which is simulated until the steady state is reached. The final concentrations are then used to initialize the spherical interface simulations. This procedure proves to be effective in providing a good starting point for the droplets simulations which are able to reach the steady state in a reasonable time avoiding large pressure waves originating by a less precise estimation of the initial concentration values.

As a first result we report in Fig. 2a the data for the rescaled generalized surface tension $\sigma[R]/\sigma_s$, where $\sigma_s = \sigma[R_s] = \sigma(R_s)$ is the value at the minimum, as a function of the normalized position of the arbitrary diving surface R/R_s. It is possible to compare these data with an analytical expression obtained from the integration of the generalized Laplace law: we can rewrite Eq. (13) as $R^2\Delta P = \mathrm{d}\left[R^2\sigma\left[R\right]\right]/\mathrm{d}R$ and integrate from R_s to R and obtain [1] the expression $\frac{\sigma[R]}{\sigma_s} = \frac{1}{3}\left(\frac{R_s}{R}\right)^2 + \frac{2}{3}\frac{R}{R_s}$. The latter one is referred to as "universal" in [13], i.e., not depending on temperature or on the droplet size, mirroring that $\sigma[R]$ depends on the arbitrary value of R. In Fig. 2a we compare the results obtained from the entire set of simulations with the analytical prediction, yielding a good agreement. This result allows us to determine the positions of the surface of tension R_s from the minima of the generalized surface tension curves. Figure 2b displays the points $(R_s^{-1}, \Delta P)$, i.e., the Laplace law, for the symmetric interactions with $\delta G_{\mathrm{AB}} = 0$, and the two most asymmetric cases $\delta G_{\mathrm{AB}} = \pm 0.88$. All curves converge to the slope expected from the flat interface surface tension, i.e., $2\sigma_0$, while sizeable corrections are visible for smaller droplets.

We analyze in further details the surface tension σ_0 as computed from (i) flat interface simulations and (ii) from the $R_s \to \infty$ limit of the data obtained from the droplets simulations. Figure 3a displays the results for different values of δG_{AB} reaching a minimum near the symmetric case $\delta G_{\mathrm{AB}} = 0$ and increasing at the boundary of the interval $\delta G_{\mathrm{AB}} = \pm 0.88$. We report using circles the $R_s \to \infty$ data while those for the flat interface simulations are reported in dashed. In the first case we use parabolic fits of the quantity $\sigma(R_s) = \Delta P \cdot R_s/2$ to estimate the value of σ_0

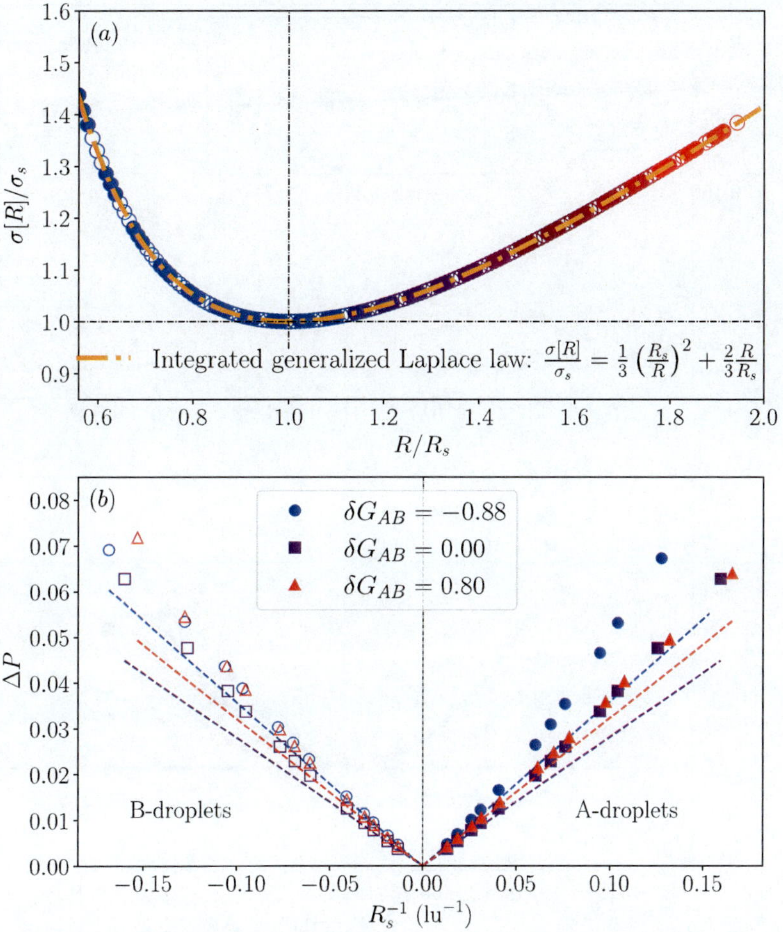

Fig. 2 (**a**) Collapse of the rescaled generalized surface tension $\sigma[R]/R_s$ as a function of the rescaled arbitrary dividing surface position R/R_s for all the simulated n_A- and n_B-rich droplets, related to full and empty circles, respectively, for all the values of δG_{AB}, represented by different colors, and all system sizes L, displaying a good superposition to the expected "universal" behavior. (**b**) Values of the pressure difference ΔP as a function of the curvature (negative for B-rich droplets) at the surface of tension R_s^{-1} for the extrema of the asymmetry parameter $\delta G_{AB} = \pm 0.88$ and the fully symmetric case $\delta G_{AB} = 0$. Corrections to the expected Laplace law (dashed lines) are visible and asymmetric when changing the sign of δG_{AB}. The curvature R_s^{-1} is reported in inverse lattice units, i.e., lu^{-1}

in the $R_s \to \infty$ limit, while for the flat interface we use the mechanical definition of the surface tension

$$\sigma_0 = \int_{L_x/2}^{L_x} \mathrm{d}x[P_\mathrm{N}(x) - P_\mathrm{T}(x)], \tag{24}$$

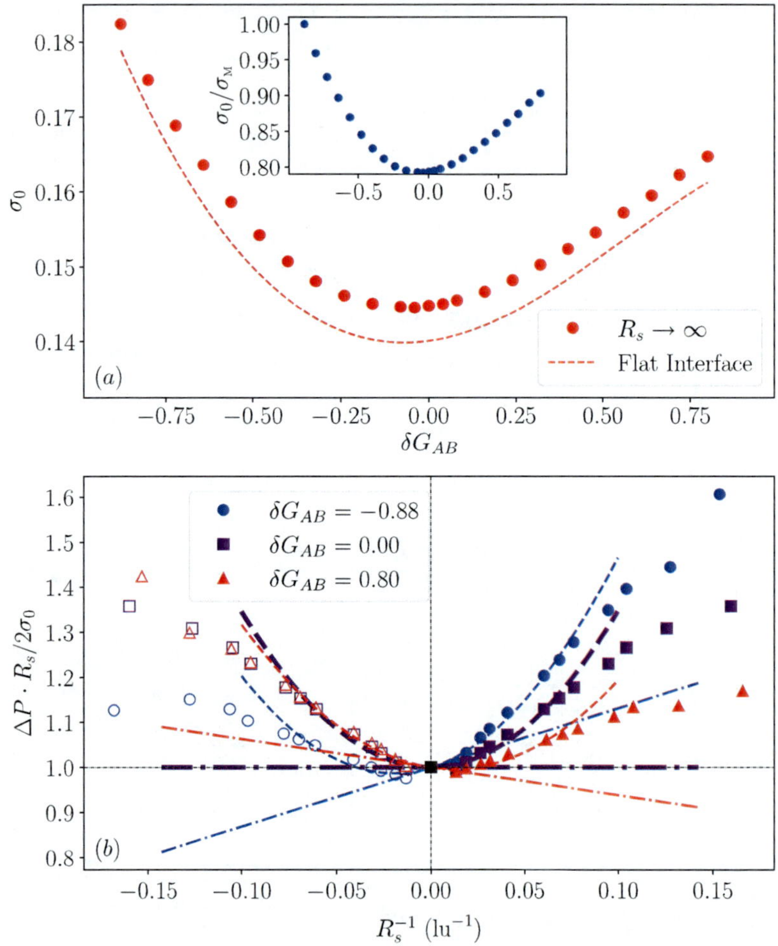

Fig. 3 (**a**) Values of the surface tension as estimated from the droplets simulations in the flat interface limit (symbols) compared to the values computed from flat interface simulations. The discrepancy is at most 3×10^{-2}. The inset displays the variation relative to the maximum which does not exceed 2×10^{-1}. (**b**) Curvature dependence of the surface tension $\sigma(R_s)/\sigma_0 = \Delta P \cdot R_s/2\sigma_0$ for three different choices of the asymmetry parameter $\delta G_{\rm AB}$. Dashed-dotted straight lines indicate the results for the linear term while the dashed parabolas indicate the second order approximation. Thicker lines are used for the symmetric case $\delta G_{\rm AB} = 0$. The curvature R_s^{-1} is reported in inverse lattice units, i.e., lu^{-1}

where $P_{\rm N}(x) = P^{xx}(x)$ and $P_{\rm T}(x) = P^{yy}(x)$ have been obtained from the lattice pressure tensor (11). The relative difference between the two estimates for σ_0 never exceeds 3.7×10^{-2}, while the relative difference between the minimum and the maximum values, $\sigma_{\rm m}$ and $\sigma_{\rm M}$, respectively, as a function of $\delta G_{\rm AB}$ is bounded at $1 - \sigma_{\rm m}/\sigma_{\rm M} \sim 2 \times 10^{-1}$. We continue with the analysis of the curvature corrections to the flat interface value of the surface tension σ_0. In Fig. 3b we report the data for

$\sigma(R_s)/\sigma_0$ as estimated from the droplet simulations through the ratio $\Delta P \cdot R_s/2\sigma_0$ (cf. Eq. (14)). We wish to remark that the values of $\sigma_s = \sigma(R_s)$ estimated from the pressure jump ΔP and those obtained from the minimum of $\sigma[R]$ in Eq. (20) have a relative difference of at most 4×10^{-3}. We only report the symmetric, $\delta G_{\rm AB} = 0$, as well as the most asymmetric cases, $\delta G_{\rm AB} = \pm 0.88$ for ease of reading. We perform quadratic fits in order to estimate the first and second order coefficients. After normalizing the coefficients by the zero-th order one, i.e., σ_0, one obtains -2δ and $2\bar{k} + k$, respectively. The values of σ_0 are reported in Fig. 3. Dash-dotted lines represent the result for the linear coefficient. The thicker line indicates the symmetric case which has a negligible slope, i.e., vanishing Tolman length. On the other hand, the two asymmetric cases display finite slopes of opposite signs, signaling a change in the sign of the Tolman length. We report in dashed lines the results for the fits of the full parabola which give a good approximation in the $R_s^{-1} \to 0$ limit.

Finally, Fig. 4a and b display the Tolman length δ and the rigidity coefficients combination $2\bar{k} + k$, respectively, as a function of the asymmetry parameter $\delta G_{\rm AB}$. As already visible from Fig. 3 the sign of the Tolman length changes with the sign of $\delta G_{\rm AB}$ with an almost monotonic dependence. Moreover, one can notice that the absolute value of δ is larger for negative $\delta G_{\rm AB}$. This can be understood given that for $\delta G_{\rm AB} < 0$ the A-component gets closer to the critical point of the self-interactions. While these are still too weak to induce phase separation, they exert a stronger effect on the interface with respect to $\delta G_{\rm AB} > 0$ branch for which the interface features are set, almost entirely, by the inter-component interactions. A non-trivial competition between inter-component and self-interactions for the formation of the interface is probably responsible for both the presence of a maximum for $2\bar{k}+k$ and the non-monotonic behavior of the first derivative clearly visible around $\delta G_{\rm AB} \simeq -0.75$. A theoretical prediction for both the Tolman length and the higher order curvature coefficients will be paramount to fully capture this competition among interactions. On the other hand, the rigidity coefficients display a non-monotonic behavior, reaching a maximum for $\delta G_{\rm AB} \simeq -0.40$. We notice that the results do not show symmetry under sign exchange for $\delta G_{\rm AB}$ mirroring the asymmetric change in the interactions under exchange of the two components A $\longleftrightarrow$ B. The relative change of δ is around $1 - \delta_{\rm m}/\delta_{\rm M} \sim 1.5$, with $\delta_{\rm m}$ and $\delta_{\rm M}$ the norm of the minimum and the maximum values, respectively. Hence $1 - \delta_{\rm m}/\delta_{\rm M}$ is far larger than $1 - \sigma_{\rm m}/\sigma_{\rm M}$, so that one variation is weakly dependent on the other. A few remarks on the dependence of the results on the simulations parameters are in order. Let us begin from the system size L dependence: limiting the set of simulations to a value $L \leq 213$lu still yields a consistent curve for δ while the estimates for $2\bar{k}+k$ change by roughly 20%. This is due to the fact that the points closest to the flat interface limit $R_s^{-1} \simeq 0$ are the most significant for getting a reliable estimate of the second order coefficient for the curvature corrections. Furthermore, the value of the inter-component coupling G has been chosen so that the largest spurious currents is of order $O(10^{-3}\text{Ma})$, where Ma is the Mach number, which is weak enough not to affect the estimations of ΔP. Moreover, since the magnitude of the spurious currents is correlated to the surface tension [53], the relatively small variation of σ_0 reported

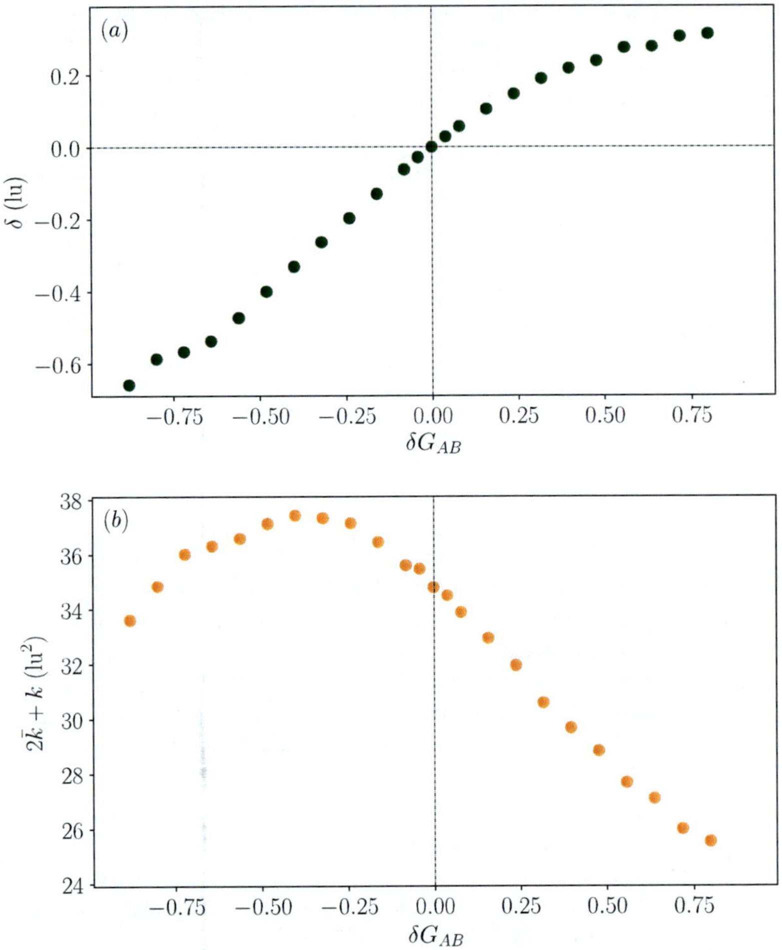

Fig. 4 (**a**) Values of the Tolman length δ as a function of the asymmetry parameter δG_{AB}, vanishing in the symmetric case $\delta G_{\text{AB}} = 0$. (**b**) Values of the combination of the rigidity coefficients $2\bar{k}+k$ as a function of δG_{AB} displaying a non-monotonic behavior. δ, and the curvature coefficients are reported in lattice units, i.e., lu and lu^2

in Fig. 3a assures that the spurious currents consistently stay at the same order in the whole range of δG_{AB}. Finally, in a three-dimensional system, the initial droplet size ratio $L/R = 4$ is large enough to avoid the spurious currents effect to propagate through the periodic boundaries. Future works will aim at completely disentangle the variations of the two quantities making them independent.

5 Conclusions

In the present work we demonstrate, by means of an extensive set of simulations of a two-component system, that (i) the Shan-Chen [15, 31] multi-component model is able to capture the curvature corrections to the surface tension and (ii) it naturally allows for a straightforward method for tuning both the Tolman length and the rigidity coefficients in a wide range of values, while keeping the surface tension in a relatively narrow range. Specifically, this is obtained by tuning the degree of asymmetry [12] of each component self-interaction [46] while keeping the cross-component interaction constant. By this method we demonstrate how the Tolman length, i.e., the first-order curvature correction of the surface tension in the flat interface limit, can be made to vanish in a continuous way by restoring the symmetry of the interaction under exchange of the two components $A \longleftrightarrow B$. The tuning of the Tolman length, especially by means of the tuning of the relative concentrations of the two components, has been thoroughly studied in the context of Density Functional Theory approaches [10]. While those studies focus on the variation of the curvature corrections for the same physical system, here we chose, as a first instance, a parametrization potentially describing different physical systems.

This represents a first step for the tuning of the curvature corrections in order to correctly model different realistic systems. Further studies will address the same results seeking an analytical control also for the multi-phase systems for which recent results [34] have already demonstrated the existence of the Tolman length and its temperature dependence in the Shan-Chen multi-phase model. This research direction holds the potential to allow a more straightforward approach for the study and modelling of nucleation and cavitation problems taking naturally into account the hydrodynamic contributions, while offering, at the same time, a very computationally efficient method capable of dealing with complex and realistic boundary conditions. The simulations source code and a Jupyter notebook to reproduce all the results and figures can be found on GitHub https://github.com/lullimat/idea.deploy.

Acknowledgments The authors wish to thank Øivind Wilhelmsen for useful discussion. Luca Biferale thankfully acknowledges the hospitality from the Department of Mechanics and Aerospace Engineering of Southern University of Science and Technology. This work was supported by National Science Foundation of China Grants 12050410244, 91741101 and 91752204, by Department of Science and Technology of Guangdong Province Grant No. 2019B21203001, Science and Technology Innovation Committee of Shenzhen Grant No. K19325001, and from the European Research Council (ERC) under the European Union's Horizon 2020 research and innovation programme (grant agreement No 882340).

References

1. Rowlinson, J.S., Widom, B.: Molecular Theory of Capillarity. Clarendon, Oxford (1982)
2. Gibbs, J.W.: The Collected Works of J. Willard Gibbs. Yale Univ. Press, New Haven (1948). OCLC: 520226

3. Buff, F.P.: The spherical interface. I. Thermodynamics. J. Chem. Phys. **19**(12), 1591–1594 (1951)
4. Blokhuis, E.M., Bedeaux, D.: Pressure tensor of a spherical interface. J. Chem. Phys. **97**(5), 3576–3586 (1992)
5. Tolman, R.C.: The effect of Droplet size on surface tension. J. Chem. Phys. **17**(3), 333–337 (1949)
6. Malijevsky, A., Jackson, G.: A perspective on the interfacial properties of nanoscopic liquid drops. J. Phys. Condens. Matter **24**, 464121 (2012)
7. Ghoufi, A., Malfreyt, P., Tildesley, D.J.: Computer modelling of the surface tension of the gas–liquid and liquid–liquid interface. Chem. Soc. Rev. **45**(5), 1387–1409 (2016)
8. Helfrich, W.: Elastic properties of lipid bilayers: theory and possible experiments. Zeitschrift für Naturforschung C **28**(11-12), 693–703 (1973)
9. Blokhuis, E.M., Bedeaux, D.: Derivation of microscopic expressions for the rigidity constants of a simple liquid-vapor interface. Phys. A Stat. Mech. Appl. **184**(1-2), 42–70 (1992)
10. Aasen, A., Blokhuis, E.M., Wilhelmsen, Ø.: Tolman lengths and rigidity constants of multicomponent fluids: Fundamental theory and numerical examples. J. Chem. Phys. **148**(20), 204702 (2018)
11. Rehner, P., Aasen, A., Wilhelmsen, Ø.: Tolman lengths and rigidity constants from free-energy functionals–General expressions and comparison of theories. J. Chem. Phys. **151**(24), 244710 (2019)
12. Anisimov, M.A.: Divergence of Tolman's length for a droplet near the critical point. Phys. Rev. Lett. **98**(3), 035702 (2007)
13. Tröster, A., Binder, K.: Positive Tolman length in a lattice gas with three-body interactions. Phys. Rev. Lett. **107**(26), 265701 (2011)
14. Binder, K., Virnau, P.: Overview: Understanding nucleation phenomena from simulations of lattice gas models. J. Chem. Phys. **145**(21), 211701 (2016) https://doi.org/10.1063/1.4959235. http://aip.scitation.org/doi/10.1063/1.4959235
15. Shan, X., Chen, H.: Lattice Boltzmann model for simulating flows with multiple phases and components. Phys. Rev. E **47**(3), 1815–1819 (1993)
16. Boltachev, G.S., Baidakov, V.G., Schmelzer, J.W.: First-order curvature corrections to the surface tension of multicomponent systems. J. Colloid Interf. Sci. **264**(1), 228–236 (2003)
17. Blokhuis, E.M., Van Giessen, A.E.: Density functional theory of a curved liquid-vapour interface: Evaluation of the rigidity constants. J. Phys. Condens. Matter **25**(22), 225003 (2013)
18. Wilhelmsen, Ø., Bedeaux, D., Reguera, D.: Tolman length and rigidity constants of the Lennard-Jones fluid. J. Chem. Phys. **142**(6), 064706 (2015)
19. Nijmeijer, M.J.P., Bruin, C., van Woerkom, A.B., Bakker, A.F., van Leeuwen, J.M.J.: Molecular dynamics of the surface tension of a drop. J. Chem. Phys. **96**(1), 565–576 (1992)
20. van Giessen, A.E., Blokhuis, E.M.: Direct determination of the Tolman length from the bulk pressures of liquid drops via molecular dynamics simulations. J. Chem. Phys. **131**(16), 164705 (2009)
21. Menzl, G., Gonzalez, M.A., Geiger, P., Caupin, F., Abascal, J.L.F., Valeriani, C., Dellago, C.: Molecular mechanism for cavitation in water under tension. Proc. Natl. Acad. Sci. **113**(48), 13582–13587 (2016)
22. Langenbach, K., Heilig, M., Horsch, M., Hasse, H.: Study of homogeneous bubble nucleation in liquid carbon dioxide by a hybrid approach combining molecular dynamics simulation and density gradient theory. J. Chem. Phys. **148**(12), 124702 (2018)
23. Yamamoto, T., Ohnishi, S.: Nano bubbles in liquid of a noble-gas mixture. Phys. Chem. Chem. Phys. **12**(5), 1033–1037 (2010)
24. Bruot, N., Caupin, F.: Curvature dependence of the liquid-vapor surface tension beyond the Tolman approximation. Phys. Rev. Lett. **116**(5), 056102 (2016)
25. Kim, S., Kim, D., Kim, J., An, S., Jhe, W.: Direct evidence for curvature-dependent surface tension in capillary condensation: Kelvin equation at molecular scale. Phys. Rev. X **8**(4), 041046 (2018)

26. Nguyen, V.D., Schoemaker, F.C., Blokhuis, E.M., Schall, P.: Measurement of the curvature-dependent surface tension in nucleating colloidal liquids. Phys. Rev. Lett. **121**(24), 246102 (2018)
27. Kalikmanov, V., Service, S.O.: Nucleation Theory. Lecture Notes in Physics. Springer Netherlands, Dordrecht (2013)
28. Talanquer, V., Oxtoby, D.W.: Density functional analysis of phenomenological theories of gas-liquid nucleation. J. Phys. Chem. **99**(9), 2865–2874 (1995)
29. Tanaka, K.K., Tanaka, H., Angélil, R., Diemand, J.: Simple improvements to classical bubble nucleation models. Phys. Rev. E **92**(2), 022401 (2015)
30. Aasen, A., Reguera, D., Wilhelmsen, Ø.: Curvature corrections remove the inconsistencies of binary classical nucleation theory. Phys. Rev. Lett. **124**(4), 045701 (2020)
31. Shan, X., Chen, H.: Simulation of nonideal gases and liquid-gas phase transitions by the lattice Boltzmann equation. Phys. Rev. E **49**(4), 2941–2948 (1994)
32. Krüger, T., Kusumaatmaja, H., Kuzmin, A., Shardt, O., Silva, G., Viggen, E.M.: The Lattice Boltzmann Method. Springer (2017)
33. Succi, S.: The Lattice Boltzmann Equation: For Complex States of Flowing Matter. Oxford University Press (2018)
34. Lulli, M., Biferale, L., Falcucci, G., Sbragaglia, M., Shan, X.: Mesoscale perspective on the Tolman length. Phys. Rev. E **105**(1) (2022) https://doi.org/10.1103/physreve.105.015301
35. Li, J.S., Wilemski, G.: Temperature dependence of droplet nucleation in a yukawa fluid. J. Chem. Phys. **118**(6), 2845 (2003)
36. Shan, X.: Pressure tensor calculation in a class of nonideal gas lattice Boltzmann models. Phys. Rev. E **77**(6), 066702 (2008)
37. Sbragaglia, M., Belardinelli, D.: Interaction pressure tensor for a class of multicomponent lattice Boltzmann models. Phys. Rev. E **88**(1), 013306 (2013)
38. Gunstensen, A.K., Rothman, D.H., Zaleski, S., Zanetti, G.: Lattice Boltzmann model of immiscible fluids. Phys. Rev. A **43**(8), 4320–4327 (1991)
39. Latva-Kokko, M., Rothman, D.H.: Static contact angle in lattice Boltzmann models of immiscible fluids. Phys. Rev. E **72**(4) (2005). https://doi.org/10.1103/physreve.72.046701
40. Montessori, A., Lauricella, M., Tirelli, N., Succi, S.: Mesoscale modelling of near-contact interactions for complex flowing interfaces. J. Fluid Mech. **872**, 327–347 (2019).
41. Swift, M.R., Orlandini, E., Osborn, W.R., Yeomans, J.M.: Lattice Boltzmann simulations of liquid-gas and binary fluid systems. Phys. Rev. E **54**(5), 5041–5052 (1996)
42. Foglino, M., Morozov, A., Henrich, O., Marenduzzo, D.: Flow of deformable droplets: Discontinuous shear thinning and velocity oscillations. Phys. Rev. Lett. **119**(20) (2017). https://doi.org/10.1103/physrevlett.119.208002
43. Tiribocchi, A., Montessori, A., Aime, S., Milani, M., Lauricella, M., Succi, S., Weitz, D.: Novel nonequilibrium steady states in multiple emulsions. Phys. Fluids **32**(1), 017102 (2020)
44. Chikatamarla, A.M.M.S., Karlin, I.: Entropic lattice Boltzmann method for multiphase flows. Phys. Rev. Lett. **114**(17) (2015). https://doi.org/10.1103/physrevlett.114.174502
45. Hosseini, S.A., Dorschner, B., Karlin, I.V.: Towards a consistent lattice Boltzmann model for two-phase fluid (2021)
46. Benzi, R., Sbragaglia, M., Succi, S., Bernaschi, M., Chibbaro, S.: Mesoscopic lattice Boltzmann modeling of soft-glassy systems: Theory and simulations. J. Chem. Phys. **131**(10), 104903 (2009)
47. Guo, Z., Zheng, C., Shi, B.: Discrete lattice effects on the forcing term in the lattice Boltzmann method. Phys. Rev. E **65**(4), 046308 (2002)
48. Bhatnagar, P.L., Gross, E.P., Krook, M.: A model for collision processes in gases. i. Small amplitude processes in charged and neutral one-component systems. Physical Review **94**(3), 511–525 (1954)
49. Benzi, R., Chibbaro, S., Succi, S.: Mesoscopic lattice Boltzmann modeling of flowing soft systems. Phys. Rev. Lett. **102**(2) (2009). https://doi.org/10.1103/physrevlett.102.026002

50. Derzsi, L., Filippi, D., Mistura, G., Pierno, M., Lulli, M., Sbragaglia, M., Bernaschi, M., Garstecki, P.: Fluidization and wall slip of soft glassy materials by controlled surface roughness. Phys. Rev. E **95**(5) (2017). https://doi.org/10.1103/physreve.95.052602
51. Derzsi, L., Filippi, D., Lulli, M., Mistura, G., Bernaschi, M., Garstecki, P., Sbragaglia, M., Pierno, M.: Wall fluidization in two acts: from stiff to soft roughness. Soft Matter **14**(7), 1088–1093 (2018)
52. Shan, X.: Analysis and reduction of the spurious current in a class of multiphase lattice Boltzmann models. Phys. Rev. E **73**(4), 047701 (2006)
53. Sbragaglia, M., Benzi, R., Biferale, L., Succi, S., Sugiyama, K., Toschi, F.: Generalized lattice Boltzmann method with multirange pseudopotential. Phys. Rev. E **75**(2), 026702 (2007)
54. Sbragaglia, M., Shan, X.: Consistent pseudopotential interactions in lattice Boltzmann models. Phys. Rev. E **84**(3), 036703 (2011)
55. Belardinelli, D., Sbragaglia, M., Biferale, L., Gross, M., Varnik, F.: Fluctuating multicomponent lattice Boltzmann model. Phys. Rev. E **91**(2), 023313 (2015)
56. From, C.S., Sauret, E., Galindo-Torres, S.A., Gu, Y.T.: Interaction pressure tensor on high-order lattice Boltzmann models for nonideal fluids. Phys. Rev. E **99**(6), 063318 (2019)
57. Lulli, M., Biferale, L., Falcucci, G., Sbragaglia, M., Shan, X.: Structure and isotropy of lattice pressure tensors for multirange potentials. Phys. Rev. E **103**(6) (2021). https://doi.org/10.1103/physreve.103.063309
58. Falcucci, G., Jannelli, E., Ubertini, S., Succi, S.: Direct numerical evidence of stress-induced cavitation. J. Fluid Mech. **728**, 362 (2013)
59. Lulli, M., Benzi, R., Sbragaglia, M.: Metastability at the yield-stress transition in soft glasses. Phys. Rev. X **8**(2) (2018). https://doi.org/10.1103/physrevx.8.021031
60. Rowlinson, J.S.: The critical exponent of Tolman's length. J. Phys. A Math. General **17**(6), L357–L360 (1984)
61. Meurer, A., Smith, C.P., Paprocki, M., Čertík, O., Kirpichev, S.B., Rocklin, M., Kumar, A., Ivanov, S., Moore, J.K., Singh, S., Rathnayake, T., Vig, S., Granger, B.E., Muller, R.P., Bonazzi, F., Gupta, H., Vats, S., Johansson, F., Pedregosa, F., Curry, M.J., Terrel, A.R., Roučka, v., Saboo, A., Fernando, I., Kulal, S., Cimrman, R., Scopatz, A.: Sympy: symbolic computing in python. PeerJ Comput. Sci. **3**, e103 (2017)
62. Virtanen, P., Gommers, R., Oliphant, T.E., Haberland, M., Reddy, T., Cournapeau, D., Burovski, E., Peterson, P., Weckesser, W., Bright, J., van der Walt, S.J., Brett, M., Wilson, J., Millman, K.J., Mayorov, N., Nelson, A.R.J., Jones, E., Kern, R., Larson, E., Carey, C.J., Polat, İ., Feng, Y., Moore, E.W., VanderPlas, J., Laxalde, D., Perktold, J., Cimrman, R., Henriksen, I., Quintero, E.A., Harris, C.R., Archibald, A.M., Ribeiro, A.H., Pedregosa, F., van Mulbregt, P., SciPy 1.0 Contributors: SciPy 1.0: Fundamental algorithms for scientific computing in python. Nature Methods **17**, 261–272 (2020)
63. Oliphant, T.E.: A Guide to NumPy, vol. 1. Trelgol Publishing USA (2006)
64. Van Der Walt, S., Colbert, S.C., Varoquaux, G.: The numpy array: a structure for efficient numerical computation. Comput. Sci. Eng. **13**(2), 22 (2011)
65. Pedregosa, F., Varoquaux, G., Gramfort, A., Michel, V., Thirion, B., Grisel, O., Blondel, M., Prettenhofer, P., Weiss, R., Dubourg, V., Vanderplas, J., Passos, A., Cournapeau, D., Brucher, M., Perrot, M., Duchesnay, E.: Scikit-learn: Machine learning in python. J. Mach. Learn. Res. **12**, 2825–2830 (2011)
66. Hunter, J.D.: Matplotlib: A 2d graphics environment. Comput. Sci. Eng. **9**(3), 90–95 (2007)
67. Pérez, F., Granger, B.E.: IPython: a system for interactive scientific computing. Comput. Sci. Eng. **9**(3), 21–29 (2007)
68. Klöckner, A., Pinto, N., Lee, Y., Catanzaro, B., Ivanov, P., Fasih, A.: PyCUDA and PyOpenCL: A scripting-based approach to GPU run-time code generation. Parallel Computing **38**(3), 157–174 (2012)

Kinetic and Macroscopic Epidemic Models in Presence of Multiple Heterogeneous Populations

Andrea Medaglia and Mattia Zanella

Abstract We study the impact of contact heterogeneity on epidemic dynamics. A system characterized by multiple susceptible populations is considered. The description of the spread of an infectious disease is obtained through the study of a system of Boltzmann-type equations for the number densities of social contacts of the introduced compartments. A macroscopic system of equations characterizing observable effects of the epidemic is then derived to assess the impact of contact heterogeneity.

1 Introduction

The recent efforts to design effective non-pharmaceutical measures to mitigate the COVID-19 pandemic were based on the link between social activities and the spreading of a respiratory disease [2]. Several works in mathematical epidemiology characterized the number of contacts of the population taking into account an additional structure that is maintained for the whole dynamics. A classic example is represented by age-structured populations for which realistic contact matrices have been determined, see, e.g., [1, 15, 20]. Nevertheless, recent works highlighted strong changes in contact distribution in the early phases of an epidemic, whose evolution can shape the infection dynamics, see [25]. For these reasons, in [11] it has been proposed in a simple SIR-type compartmentalization a kinetic model to couple the dynamics of an infectious disease with the contact evolution of a system of agents. Based on the observation that people with high number of contacts may by capable to transmit the disease to a large number of people, we introduced a modeling approach in which the social structure of compartments is characterized by the number of contacts. Interestingly enough, in the present setting, models with saturated incidence rates can be easily derived with minimal assumptions [7, 17, 24].

A. Medaglia · M. Zanella (✉)
Department of Mathematics "F. Casorati", University of Pavia, Pavia, Italy
e-mail: andrea.medaglia02@universitadipavia.it; mattia.zanella@unipv.it

P. Barbante et al. (eds.), *From Kinetic Theory to Turbulence Modeling*, Springer INdAM Series 51, https://doi.org/10.1007/978-981-19-6462-6_15

Other recent contributions were centered on the effects of the structure of contacts of agents, we mention in this direction the works [12, 16, 19].

In the present contribution, we concentrate on the influence of contact heterogeneity on the dynamics of the disease in presence of multiple susceptible populations. Each susceptible compartment can be characterized by its mean number of connections. This situation is very common when non-pharmaceutical interventions have different impacts on the population [6] or in presence of sanitary cordon measures, where a portion of the territory results highly affected by the disease. We mention recent contributions in this direction using mobility data [4, 9, 13, 18, 22].

The mathematical tools that we consider are based on kinetic theory for large interacting systems [8, 14, 21] for which we are able to derive the evolution of observable quantities from microscopic, often unobservable, dynamics. In details, we will show how heterogeneity in the contact structure plays a central role in the evolution of an epidemic. In particular, preliminary results will highlight that, in several regimes of parameters, the asymptotic number of recovered can be unexpectedly high in societies with small contact heterogeneity, compared to the ones with high contact heterogeneity. These results are coherent with the recent findings presented in [5].

2 Interplay Between Contact Distribution and Epidemic Dynamics

In this section, we introduce a kinetic model to describe the spreading of an infectious disease depending on an additional variable describing the number of social contacts. Coherently with the modeling approach introduced in the recent works [10], we subdivide the total population into three main compartments: susceptible, who can contract the disease, infected infectious, who can transmit the disease and recovered, corresponding to formerly infected patients that are not infectious. Furthermore, to mimic the early effects of the epidemic, where the collective compliance to reduce the number of daily contacts is often not accepted, we subdivide the susceptible population into two main categories S_+, S_- in relation to their average number of contacts, m_{S_+} and m_{S_-}, respectively.

The contact distribution of the whole population is therefore recovered as

$$f(w,t) = f_{S_+}(w,t) + f_{S_-}(w,t) + f_I(w,t) + f_R(w,t), \qquad \int_{\mathbb{R}_+} f(w,t)dw = 1.$$

Hence, we obtain the mass fractions of population in each compartment and their momentum of order $\alpha > 0$ as

$$J(t) = \int_{\mathbb{R}_+} f_J(w,t)dw, \qquad J(t)m_{\alpha,J}(t) = \int_{\mathbb{R}_+} w^\alpha f_J(w,t)dw.$$

Unambiguously, in the following we will indicate the mean of contact in the compartment J, corresponding to $\alpha = 1$, by m_J.

2.1 Formation of the Contact Distribution

Coherently with [11], we can define a process of contact formation based on microscopic transitions for the variation of contacts of a single agent. At aggregate level, the evolution of the distributions f_J, $J \in \{S_\pm, I, R\}$, can be obtained through a Boltzmann-type equation. As shown in the aforementioned work, to obtain an explicit formulation of the large time distribution, it is possible to derive the following Fokker-Planck-type equation

$$\frac{\partial}{\partial t} f_J(w,t) = \frac{\lambda_J}{2}\partial_w \left[\left(\frac{w}{m_J} - 1\right) f_J\right] + \frac{\sigma_J^2}{2}\partial_w^2 (w f_J(w,t)) \tag{1}$$

with $\lambda_J > 0$, $\sigma_J^2 > 0$ and $m_J > 0$ the mean number of contacts, complemented with no-flux boundary conditions $\frac{\partial}{\partial w}(w f_J(w,t))\Big|_{w=0} = 0, \qquad J = \{S, I, R\}$. The emerging large time contact distributions $f_J^\infty(w)$, $J \in \{S_\pm, I, R\}$ of (1) can be explicitly computed and are of Gamma-type [23] coherently with experimental results in [2]. In particular, if $\mu_J = \lambda_J/\sigma_J^2 > 0$, we have

$$f_J^\infty(w) = \left(\frac{\mu_J}{m_J}\right)^{\mu_J} \frac{1}{\Gamma(\mu_J)} w^{\mu_J - 1} \exp\left\{-\frac{\mu_J}{m_J} w\right\}, \tag{2}$$

whose momenta of order α are

$$\int_{\mathbb{R}_+} w^\alpha f_J^\infty(w) dw = \left(\frac{m_J}{\mu_J}\right)^\alpha \frac{\Gamma(\mu_J + \alpha)}{\Gamma(\mu_J)} = c_{\alpha,J} m_J^\alpha, \tag{3}$$

where $c_{J,\alpha} = \left(\frac{1}{\mu_J}\right)^\alpha \frac{\Gamma(\mu_J + \alpha)}{\Gamma(\mu_J)}$. Since f_J^∞ is a Gamma distribution we also have

$$c_{\alpha+1,J} = \frac{\mu_J + \alpha}{\mu_J} c_{\alpha,J},$$

and $m_{\alpha+1,J} = m_{\alpha,J} \frac{\alpha + \mu_J}{\mu_J} m_J$. We remark that (2) is explicitly dependent on the positive parameter $\mu_J = \lambda_J/\sigma_J^2$ that measures the contact heterogeneity of a population in terms of the variance of the distribution of social contacts. More precisely, small values of μ_J correspond to a larger heterogeneity of the individuals in terms of social contacts. Different models with the defined experimentally consistent equilibrium may be considered, for example, BGK-type operators.

2.2 The Kinetic Model

The resulting system of kinetic equations is given by

$$\begin{aligned}
\partial_t f_{S_+}(w,t) &= -K(f_{S_+}, f_I)(w,t) + \frac{1}{\epsilon} Q_{S_+}(f_{S_+})(w,t) \\
\partial_t f_{S_-}(w,t) &= -K(f_{S_-}, f_I)(w,t) + \frac{1}{\epsilon} Q_{S_-}(f_{S_-})(w,t) \\
\partial_t f_I(w,t) &= K(f_{S_+} + f_{S_-}, f_I)(w,t) - \gamma f_I(w,t) + \frac{1}{\epsilon} Q_I(f_I)(w,t) \\
\partial_t f_R(w,t) &= \gamma f_I(w,t) + \frac{1}{\epsilon} Q_R(f_R)(w,t),
\end{aligned} \tag{4}$$

where $\epsilon > 0$ and $\gamma > 0$ is the recovery rate. The infection transmission is taken into account by the operator

$$K(g, f_I)(w,t) = g(w,t) \int_{\mathbb{R}_+} \kappa(w, w_*) f_I(w_*, t) dw_*, \qquad g = f_{S_+}, f_{S_-},$$

with $\kappa(w, w_*) > 0$ expressing the dependency of the disease transmission by the number of contacts and such that $\kappa(0, y) = \kappa(x, 0) = 0$. In [11] it has been proposed as possible example

$$\kappa(x, y) = \beta x^{\alpha_1} y^{\alpha_2}, \qquad \alpha_1, \alpha_2 > 0.$$

The operators $Q_J(f_J)$, $J \in \{S_\pm, I, R\}$ characterize the thermalization of the distributions $f_J(w,t)$ and as discussed in Sect. 2.1 are given by

$$Q_J(f_J)(w,t) = \frac{\lambda_J}{2} \partial_w \left[\left(\frac{w}{m_J} - 1 \right) f_J \right] + \frac{\sigma_J^2}{2} \partial_w^2 (w f_J(w,t)),$$

that are mass and momentum preserving. From now on we will omit time dependency. Integrating both sides of (4) we get

$$\begin{aligned}
\frac{dS_+}{dt} &= -\beta m_{\alpha_1, S_+} m_{\alpha_2, I} S_+ I \\
\frac{dS_-}{dt} &= -\beta m_{\alpha_1, S_-} m_{\alpha_2, I} S_- I \\
\frac{dI}{dt} &= \beta \left[S_+ m_{\alpha_1, S_+} + S_- m_{\alpha_1, S_-} \right] m_{\alpha_2, I} I - \gamma I \\
\frac{dR}{dt} &= \gamma I
\end{aligned}$$

that is not closed like classical compartmental modeling since it depends on the evolution of local mean values $m_J(t)$. A possible way to obtain a closed system of equations is obtained by resorting to a limit procedure that is classical in statistical physics. Indeed, for $\epsilon \ll 1$ the distribution functions $f_J(w,t)$ collapse to Gamma-type densities with mass fractions $J(t)$ and local mean values $m_J(t)$. After multiplication by w we get

$$\begin{aligned}
\frac{d(S_+ m_{S_+})}{dt} &= -\beta m_{\alpha_1+1,S_+} m_{\alpha_2,I} S_+ I,\\
\frac{d(S_- m_{S_-})}{dt} &= -\beta m_{\alpha_1+1,S_-} m_{\alpha_2,I} S_- I,\\
\frac{d(I m_I)}{dt} &= \beta\left(m_{\alpha_1+1,S_+} S_+ + m_{\alpha_1+1,S_-} S_-\right) m_{\alpha_2,I} I - \gamma m_I I\\
\frac{d(R m_R)}{dt} &= \gamma m_I I.
\end{aligned}$$

Hence, in view of (3) we obtain the following closed system for the evolution of the mass fractions and mean connections in each compartment

$$\begin{aligned}
\frac{dS_+}{dt} &= -\beta c_{\alpha_1,S_+} c_{\alpha_2,I} m_{S_+}^{\alpha_1} m_I^{\alpha_2} S_+ I,\\
\frac{dS_-}{dt} &= -\beta c_{\alpha_1,S_-} c_{\alpha_2,I} m_{S_-}^{\alpha_1} m_I^{\alpha_2} S_- I,\\
\frac{dI}{dt} &= \beta c_{\alpha_2,I}\left[c_{\alpha_1,S_+} S_+ m_{S_+}^{\alpha_1} + c_{\alpha_1,S_-} S_- m_{S_-}^{\alpha_1}\right] m_I^{\alpha_2} I - \gamma I\\
\frac{dR}{dt} &= \gamma I
\end{aligned} \tag{5}$$

and

$$\begin{aligned}
\frac{dm_{S_+}}{dt} &= -\frac{\beta\alpha_1}{\mu_{S_+}} c_{\alpha_1,S_+} c_{\alpha_2,I} m_{S_+}^{\alpha_1+1} m_I^{\alpha_2} I\\
\frac{dm_{S_-}}{dt} &= -\frac{\beta\alpha_1}{\mu_{S_-}} c_{\alpha_1,S_-} c_{\alpha_2,I} m_{S_-}^{\alpha_1+1} m_I^{\alpha_2} I\\
\frac{dm_I}{dt} &= \beta c_{\alpha_2,I} m_I^{\alpha_2}\left\{c_{\alpha_1,S_+} S_+ m_{S_+}^{\alpha_1}\left(\frac{\alpha_1+\mu_{S_+}}{\mu_{S_+}} m_{S_+} - m_I\right)\right.\\
&\quad \left. + c_{\alpha_1,S_-} S_- m_{S_-}^{\alpha_1}\left(\frac{\alpha_1+\mu_{S_-}}{\mu_{S_-}} m_{S_-} - m_I\right)\right\}\\
\frac{dm_R}{dt} &= \gamma(m_I - m_R)\frac{I}{R}.
\end{aligned} \tag{6}$$

We observe that the obtained social SIR model with generalized interaction forces reduces to the one obtained in [10] in the case of a unique susceptible population with the choice $\alpha_1 = \alpha_2 = 1$.

2.3 Saturated Incidence Rate

Fixing $m_I(t) = \tilde{m}_I > 0$ from the first two equations in (6) we get

$$\frac{dm_{S_\pm}}{dt} = -\bar{\beta}_\pm m_{S_\pm}^{\alpha_1+1},$$

being $\bar{\beta}_\pm = \dfrac{\beta\alpha_1}{\mu_{S_\pm}} c_{\alpha_1,S_\pm} c_{\alpha_2,I} \tilde{m}_I^{\alpha_2} I$ complemented with the initial condition $m_{S_\pm}(0)$. The exact solution of the above equation reads

$$m_{S_\pm}(t) = \frac{m_{S_\pm}(0)}{\left(1 + \frac{\beta\alpha_1^2}{\mu_{S_\pm}} m_{S_\pm}^{\alpha_1}(0) c_{\alpha_1,S_\pm} c_{\alpha_2,I} \tilde{m}_I^{\alpha_2} \int_0^t I(s)ds\right)^{1/\alpha_1}}.$$

Hence, with the introduced assumption we get the following set of first order macroscopic equations with saturated incidence rate

$$\begin{aligned}
\frac{dS_+}{dt} &= -\beta c_{\alpha_1,S_+} c_{\alpha_2,I} H_+(t, I(t)) S_+ I, \\
\frac{dS_-}{dt} &= -\beta c_{\alpha_1,S_-} c_{\alpha_2,I} H_-(t, I(t)) S_- I, \\
\frac{dI}{dt} &= \beta c_{\alpha_2,I} (c_{\alpha_1,S_+} H_+(t, I(t)) S_+ + c_{\alpha_1,S_-} H_-(t, I(t)) S_-) I - \gamma I \\
\frac{dR}{dt} &= \gamma I,
\end{aligned} \tag{7}$$

where

$$H_\pm(t, I(t)) = \frac{\tilde{m}_I^{\alpha_2} m_{S_\pm}^{\alpha_1}(0)}{1 + \frac{\beta\alpha_1^2}{\mu_{S_\pm}} m_{S_\pm}^{\alpha_1}(0) c_{\alpha_1,S_\pm} c_{\alpha_2,I} \tilde{m}_I^{\alpha_2} \int_0^t I(s)ds}$$

is a generalization of the classical saturated incidence rate.

To understand the influence of contact heterogeneity we divide the equations for $S_\pm(t)$ in (7) by dR/dt. Hence, in the limit $t \to +\infty$ we have

$$\frac{dS_\pm^\infty}{dR^\infty} = -\xi_\pm \mu_{S_\pm} \frac{S_\pm^\infty}{1 + \alpha_1^2 \xi_\pm R^\infty}, \tag{8}$$

where $\xi_\pm = \dfrac{\beta c_{\alpha_1,S_\pm} c_{\alpha_2,I} \tilde{m}_I^{\alpha_2} m_{S_\pm}^{\alpha_1}(0)}{\gamma \mu_{S_\pm}} > 0$ is a given constant. The solutions of (8) are

$$S_\pm^\infty(R^\infty) = S_\pm(0)\left(1 + \alpha_1^2 \xi_\pm R^\infty\right)^{-\mu_{S_\pm}}, \tag{9}$$

since $S_\pm^\infty(R^\infty = 0) = S_\pm(0)$.

For large times we have

$$S_+^\infty + S_-^\infty + R^\infty = 1.$$

Taking advantage of the explicit solution (9) we can rewrite the last relation as

$$1 - R^\infty = S_+(0)\left(1 + \alpha_1^2 \xi_+ R^\infty\right)^{-\mu_{S_+}} + S_-(0)\left(1 + \alpha_1^2 \xi_- R^\infty\right)^{-\mu_{S_-}}, \tag{10}$$

whose solution defines the dependence of R^∞ by the introduced contact heterogeneity.

3 Numerical Results

In this section, we present several numerical experiments for the system (5)–(6) and the system (7) with saturated incidence rate. In particular, we focus on the relation between the fraction of the recovered at the equilibrium R^∞ and the coefficients $\mu_{S_\pm}$ measuring the heterogeneity of the population of the compartments $S_\pm$ in terms of the variance of the contact distribution. More specifically, small values of $\mu_{S_\pm}$ correspond to a larger heterogeneity of the individuals with respect to the social contact, since $\mu_{S_\pm} = \lambda_{S_\pm}/\sigma_{S_\pm}^2$.

Since R^∞ defined in (10) depends on several parameters defining the initial set-up of the contact distribution, to understand the influence of the contact heterogeneity, we fix the following values

$$m_{S_+}(0) = m_R(0) = m_I(0) = \tilde{m}_I = 15,\ m_{S_-}(0) = 10,$$

$$S_+(0) = 0.68,\ S_-(0) = 0.28,\ I(0) = R(0) = 0.02,$$

$$\alpha_1 = \alpha_2 = 1,\ \gamma = 0.1.$$

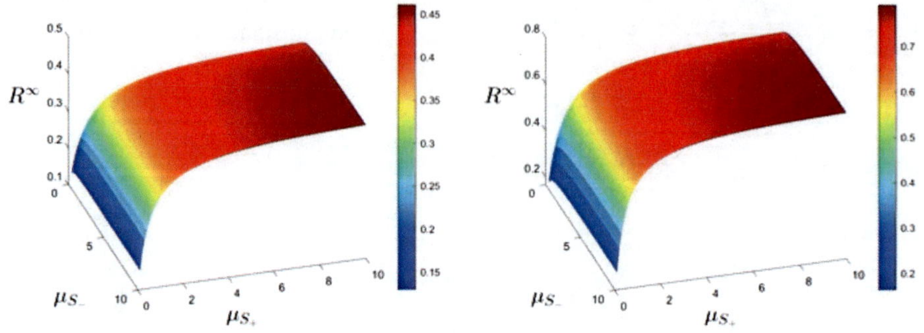

Fig. 1 Numerical solution of equation (10) for R^∞, varying the parameters $\mu_{S_\pm}$. Left: $\mathcal{R}_0 = 1.25$. Right: $\mathcal{R}_0 = 2.5$. The initial data are $m_{S_+}(0) = m_R(0) = m_I(0) = \tilde{m}_I = 15$, $m_{S_-}(0) = 10$, $S_+(0) = 0.68, S_-(0) = 0.28$, $I(0) = R(0) = 0.02$, the other parameters of the model $\alpha_1 = \alpha_2 = 1, \gamma = 0.1$

Therefore, with these choices R^∞ is function of μ_{S_+} and μ_{S_-} with a parametric dependence on β, that is linked to the reproduction number

$$\mathcal{R}_0 = \frac{\beta}{\gamma} m_I(0) \left(S_+(0) m_{S_+}(0) + S_-(0) m_{S_-}(0) \right).$$

The relation between the contact structure of the agents and the spreading of the epidemics has been recently studied in literature [5, 12]. In particular, regarding the COVID-19 pandemic, it has been pointed out that a smaller heterogeneity could be associated with a larger value of the recovered at the equilibrium.

Similarly to [5], we consider first the case $\mathcal{R}_0 = 2.5$. This choice is then compared with the case of an infectious disease characterized by $\mathcal{R}_0 = 1.25$. In details, we solve numerically equation (10) for R^∞, varying $\mu_{S_\pm}$ and taking different values of $\mathcal{R}_0$. As we can observe from Fig. 1, R^∞ is an increasing function of both the coefficients $\mu_{S_\pm}$ regardless of the considered values of the reproduction number $\mathcal{R}_0$.

A rather different behavior can be observed for the *non-saturated* system (5)–(6). In particular, R^∞ exhibits in this case a maximum for small $\mu_{S_\pm}$. This means that a higher heterogeneity is linked to a larger value of the recovered at the equilibrium. As a consequence, we note also that different conditions of the heterogeneity of the social contacts could be associated with the same R^∞, despite they have a distinct time evolution.

As we can easily observe from the left panel of Fig. 2, R^∞ in the $\mathcal{R}_0 = 1.25$ scenario has a maximum for high conditions of heterogeneity and then it decreases as the parameters $\mu_{S_\pm}$ increase. On the contrary, the $\mathcal{R}_0 = 2.5$ case in the right panel of Fig. 2 shows the same trend of the system with saturated incidence rate. In more details, in Fig. 3 we show the time evolution of the system (5)–(6) for $\mathcal{R}_0 = 2.5$. We clearly see that a decreasing contact heterogeneity is associated with bigger fraction of recovered for large times.

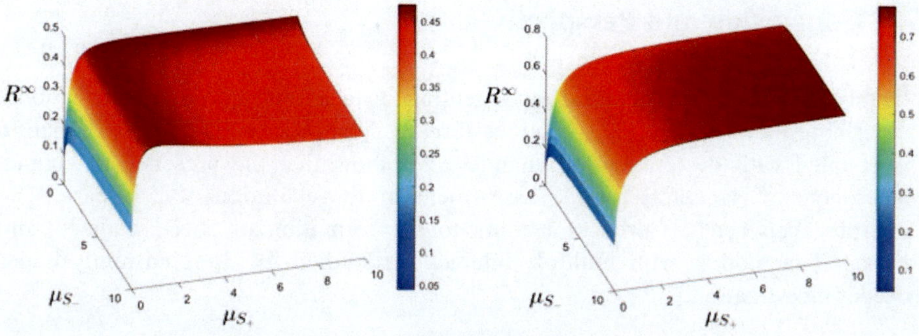

Fig. 2 Fraction of recovered R^∞ versus the parameters $\mu_{S_\pm}$, obtained solving the system (5)–(6). Left: $\mathcal{R}_0 = 1.25$. Right: $\mathcal{R}_0 = 2.5$. The initial data are $m_{S_+}(0) = m_R(0) = m_I(0) = \tilde{m}_I = 15$, $m_{S_-}(0) = 10$, $S_+(0) = 0.68$,$S_-(0) = 0.28$, $I(0) = R(0) = 0.02$, the other parameters of the model $\alpha_1 = \alpha_2 = 1, \gamma = 0.1$

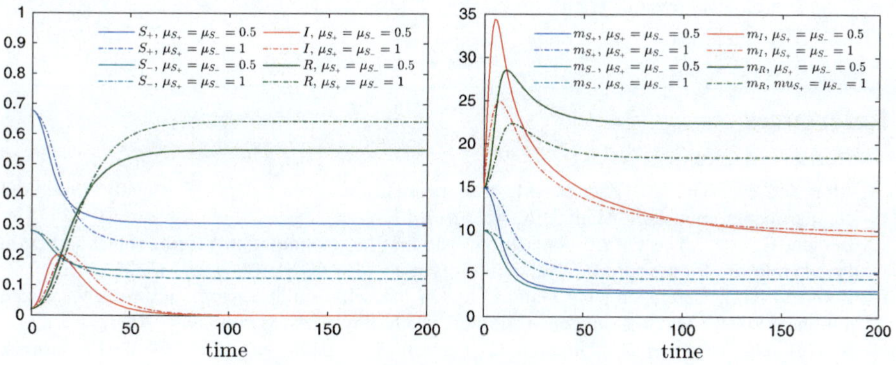

Fig. 3 Evolution of the system (5)–(6) with $\mathcal{R}_0 = 2.5$ for $\mu_{S_+} = \mu_{S_-} = 0.5$ (solid lines) and $\mu_{S_+} = \mu_{S_-} = 1$ (dashed lines). The initial conditions are $m_{S_+}(0) = m_I(0) = m_R(0) = 15$, $m_{S_-}(0) = 10$, $S_+(0) = 0.68$, $S_-(0) = 0.28$ and $I(0) = R(0) = 0.02$. The other parameters are $\alpha_1 = \alpha_2 = 1, \gamma = 0.1$

In the end, we observe that for high values of heterogeneity parameters the model with saturated incidence rate, mimicking non-pharmaceutical protection measures such as a lockdown strategy, exhibits a lower fraction of recovered at the equilibrium than the system (5)–(6). In particular, for small $\mathcal{R}_0$, a fixed mean of contacts in the infected compartment is able to avoid the maximum for small $\mu_{S_\pm}$ shown in the left panel of Fig. 2.

4 Conclusion and Perspectives

In this short note, we focused our attention on a kinetic compartmental model describing the spread of an infectious disease. The process of contact formation is coupled with the epidemic dynamics. We show that the presence of contact heterogeneity is central for the assessment of the evolution of a disease. The interplay between the process leading to the formation of social contacts and Maxwellian models with multiple interactions studied in [3] is currently under deeper investigation.

Acknowledgments This work has been written within the activities of GNFM group of INdAM (National Institute of High Mathematics). The research was partially supported by the Italian Ministry of Education, University and Research (MIUR): Dipartimenti di Eccellenza Program (2018–2022) - Dept. of Mathematics "F.Casorati," University of Pavia. M.Z. acknowledges partial support of MUR-PRIN2020 Project (No. 2020JLWP23) "Integrated mathematical approaches to socio-epidemiological dynamics."

References

1. Albi, G., Pareschi, L., Zanella, M.: Control with uncertain data of socially structured compartmental models. J. Math. Biol. **82**, 63 (2021)
2. Béraud, G. et al.: The French connection: The first large population-based contact survey in France relevant for the spread of infectious diseases. PLoS ONE **10**(7), e0133203 (2015)
3. Bobylev, A.V., Cercignani, C., Gamba, I.: On the self-similar asymptotics for generalized nonlinear kinetic Maxwell models. Commun. Math. Phys. **291**(3), 599–644 (2009)
4. Bertaglia, G., Boscheri, W., Dimarco, G., Pareschi, L.: Spatial spread of COVID-19 outbreak in Italy using multiscale kinetic transport equations with uncertainty. Math. Biosci. Eng. **18**(5), 7028–7059 (2021)
5. Britton, T., Ball, F., Trapman, P.: A mathematical model reveals the influence of population heterogeneity on herd immunity to SARS-CoV-2. Science **369**, 6505 (2020)
6. Buonomo, B., Della Marca, R.: Effects of information-induced behavioural changes during the COVID-19 lockdowns: The case of Italy: COVID-19 lockdowns and behavioral change. R. Soc. Open Sci. **7**, 201635 (2020)
7. Capasso, V., Serio, G.: A generalization of the Kermack-McKendrick deterministic epidemic model. Math. Biosci. **42**, 43–61 (1978)
8. Cercignani, C.: The Boltzmann Equation and its Applications. Springer Series in Applied Mathematical Sciences, vol. 67. Springer, New York, NY (1988)
9. Della Marca, R., Loy, N., Tosin, A.: An SIR-like kinetic model tracking individuals' viral load 2021. Preprint. https://doi.org/10.13140/RG.2.2.32046.02883
10. Dimarco, G., Pareschi, L., Toscani, G., Zanella M.: Wealth distribution under the spread of infectious diseases. Phys. Rev. E **102**, 022303 (2020)
11. Dimarco, G., Perthame, B., Toscani, G., Zanella M.: Kinetic models for epidemic dynamics with social heterogeneity. J. Math. Biol. **83**, 4 (2021)
12. Dolbeault, J., Turinici, G.: Heterogeneous social interactions and the COVID-19 lockdown outcome in a multi-group SEIR model. Math. Model. Nat. Phenom. **15**, Article Number 36 (2020)
13. Dutta, R., Gomes, S., Kalise, D., Pacchiardi, L.: Using mobility data in the design of optimal lockdown strategies for the COVID-19 pandemic. PLoS Comput. Biol. **17**(8), e1009236 (2020)

14. Furioli, G., Pulvirenti, A., Terraneo, E., Toscani, G.: Fokker–Planck equations in the modelling of socio-economic phenomena. Math. Mod. Meth. Appl. Sci. **27**(1), 115–158 (2017)
15. Hethcote, H.W.: Modeling heterogeneous mixing in infectious disease dynamics. In: V. Isham, G.F.H. Medley (eds.) Models for Infectious Human Diseases, pp. 215–238. Cambridge University Press, Cambridge, UK (1996)
16. Lewis, D.: Superspreading drives the COVID pandemic - and could help to tame it. Nature **590**, 544–546 (2021)
17. Liu, X., Stechlinski, P.: Infectious disease models with time-varying parameters and general nonlinear incidence rate. Appl. Math. Model. **36**(5), 1974–1994 (2012)
18. Loy, N., Tosin, A.: A viral load-based model for epidemic spread on spatial networks. Math. Biosci. Eng. **18**(5), 5635–5663 (2021)
19. Nielsen, B.F., Simonsen, L., Sneppen, K.: COVID-19 Superspreading suggests mitigation by social network modulation. Phys. Rev. Lett. **126**, 118301 (2021)
20. Novozhilov, A.S.: On the spread of epidemics in a closed heterogeneous population. Math. Biosci. **215**, 177–185 (2008)
21. Pareschi, L., Toscani, G.: Interacting Multiagent Systems: Kinetic Equations and Monte Carlo Methods. Oxford University Press, Oxford (2013)
22. Salam, P.S.A., Bock, W., Klar, A., Tiwari, S.: Disease contagion models coupled to crowd motion and mesh free simulation. Math. Mod. Meth. Appl. Sci. **31**(6), 1277–1295 (2021)
23. Toscani, G.: Entropy-type inequalities for generalized Gamma densities. Ric. Mat. **70**, 35–50 (2021)
24. Zanella, M., Bardelli, C., Dimarco, G., Deandrea, S., Perotti, P., Azzi, M., Figini, S., Toscani, G.: A data-driven epidemic model with social structure for understanding the COVID-19 infection on a heavily affected Italian Province. Math. Mod. Meth. Appl. Sci. **18**(4), 3384–3403 (2021)
25. Zhang, J., Litvinova, M., Liang, Y., Wang, Y., Wang, W., Zhao, S., Wu, Q., Merler, S., Viboud, C., Vespignani, A.: Changes in contact patterns shape the dynamics of the COVID-19 outbreak in China. Science **368**(6498), 1481–148 (2020)

Electron Transport in Graphene Nanoribbons

Giovanni Nastasi and Vittorio Romano

Abstract Lately, graphene has attracted the attention of several scientists because of its interesting properties. In particular, charge transport in graphene nanoribbons has peculiar effects which reveal very challenging, especially for future generations of electron devices. The possibility to replace traditional semiconductor materials with graphene in the active area of electron devices constitutes the ultimate miniaturization since graphene has the width of a single atom. Here we present an analysis of charge transport in graphene nanoribbons in the framework of semiclassical transport.

1 Introduction

The interest in charge transport in low dimensional structures [1] has recently increased both for the aim to shrink the dimension of the existing electron devices and for the discovery of 2D materials such as graphene and transition metal dichalcogenides. Much effort has been devoted to devise novel graphene field effect transistors [2–4] but the gapless nature of pristine graphene makes it a semimetal and hampers its use in microelectronics. A viable way is to use carbon nanotube or graphene nanoribbons (GNRs) instead of large area graphene. In fact, it has been experimentally observed that GNRs have an energy gap like semiconductors, see, for example, [5], due to quantum confinement effects. Therefore GNRs are potential candidates to replace standard semiconductors, e.g., Si or GaAs, in the active area of electron devices. This should constitute the ultimate miniaturization since graphene has the width of a single atom.

In order to employ GNRs, an accurate analysis of their transport properties is mandatory, in particular, the energy bands, the electron–phonon scatterings, and the edge effects. Here we present an analysis of charge transport in graphene

G. Nastasi (✉) · V. Romano
Dipartimento di Matematica e Informatica, Catania, Italy
e-mail: giovanni.nastasi@unict.it; romano@dmi.unict.it

P. Barbante et al. (eds.), *From Kinetic Theory to Turbulence Modeling*, Springer
INdAM Series 51, https://doi.org/10.1007/978-981-19-6462-6_16

nanoribbons in the framework of semiclassical transport. The plan of the paper is as follows. In Sect. 2 the semiclassical Boltzmann equations for electrons in the valence and conduction bands are introduced along with the structure of the energy band and scattering terms. In Sect. 3 an efficient numerical scheme based on the Discontinuous Galerkin (DG) method is presented. In Sec. 4 models of mobilities are deduced from the numerical simulations of the semiclassical transport equations.

2 Semiclassical Charge Transport in Graphene

The main peculiarity of large area graphene is that the energy bands $\varepsilon(\mathbf{k})$ around the Dirac point have a conical shape [6]

$$\varepsilon_\alpha(\mathbf{k}) = \alpha \hbar v_F \sqrt{k_x^2 + k_y^2}, \tag{1}$$

where v_F is the (constant) Fermi velocity, $\hbar$ the reduced Planck constant and $\mathbf{k} = (k_x, k_y)$ is the wave-vector, we assume to vary in $\mathbb{R}^2$. The parameter α assumes the value 1 for the conduction band and -1 for the valence band. The conduction and valence bands touch at the Dirac points making graphene a gapless semimetal. This feature has important consequences. The most relevant one for the design of a Field Effect Transistor (FET) with the active area made of graphene, the so-called GFET, is the difficulty to get an efficient on-off switch of the current.

Recently, several papers have tackled the problem to model the charge confinement in graphene nanoribbons. In the idealized cases of armchair edges, band-structure calculation predicts that a gap is present; zigzag GNRs are semimetallic instead. In more realistic cases the edge disorder makes the analysis much more involved. The usual approaches are based on a square potential well [1] of width W and lead to the energy subbands

$$\varepsilon_n(\mathbf{k}) = \alpha \hbar v_F \sqrt{k_x^2 + \left(\frac{n\pi}{W}\right)^2}, \quad n = 1, 2, \ldots . \tag{2}$$

However, as shown in [7] by a comparison with DFT calculations, the subband models suffer from a certain discrepancy regarding the density of states (DOS). In the same paper the authors propose a simpler model in good agreement with DFT up to energies of 1 eV, enough for many practical device applications. Such an approach consists of still retaining a single band but changing the dispersion relation as follows

$$\varepsilon_\alpha(\mathbf{k}) = \alpha \hbar v_F \sqrt{k_x^2 + k_y^2 + \left(\frac{\pi}{W}\right)^2}. \tag{3}$$

We set $C = \dfrac{\pi}{W}$ which is related to the energy gap E_g through $E_g = 2\hbar v_F C$. We remark that this model for the energy bands tends to the dispersion relation of a monolayer graphene for large W and it does not require a multi-subband treatment. Moreover, if W is so small that the bandgap is greater than the phonon energy, the inter-band scatterings can be neglected [8, 9].

The semiclassical Boltzmann equations (BEs) in the conduction and valence bands describe the flow of electrons in GNRs with a good accuracy. The electrons in the valleys K and K' are treated as equivalent. Since electrons are confined in the transversal direction, we consider the transport only along the longitudinal direction and, as a consequence, the BEs write

$$\frac{\partial f_\alpha}{\partial t} + (v_\alpha)_x \frac{\partial f_\alpha}{\partial x} - \frac{e}{\hbar} E \frac{\partial f_\alpha}{\partial k_x} = C(f_\alpha, f_{-\alpha}), \qquad (\alpha = \pm 1). \tag{4}$$

The unknowns $f_\alpha = f_\alpha(t, x, \mathbf{k})$ denote the distribution functions of electrons, being t the time, x the position along the longitudinal direction and $\mathbf{k}$ the (2D) wave-vector belonging to the first Brillouin zone which is extended to $\mathbb{R}^2$. We observe that equations are written both for electrons. The alternative descriptions with holes in the valence band will be used for calculating the average quantities only. We have denoted by e the elementary (positive) charge. $E = E(t, x)$ is the component of the electric field along the x-direction. In principle it must be determined via the Poisson equation but in the investigated cases it can be considered as external.

In each energy band the following group velocity $\mathbf{v}_\alpha$ is obtained

$$\mathbf{v}_\alpha = ((v_\alpha)_x, (v_\alpha)_y) = \frac{1}{\hbar} \nabla_{\mathbf{k}}\, \varepsilon_\alpha = \frac{\alpha v_F}{\sqrt{k_x^2 + k_y^2 + C^2}} \mathbf{k}, \tag{5}$$

$\nabla_{\mathbf{k}}$ being the gradient with respect to the vector $\mathbf{k}$. Note that, at variance with the gapless case, the modulus of $\mathbf{v}_\alpha$ is no longer constant but $|\mathbf{v}_\alpha| < v_F$.

The collision terms $C(f_\alpha, f_{-\alpha})$ include both the interaction of electrons with the phonons of graphene and an additional scattering modelling the edge effects.

The general form of the electron–phonon scattering $C^{(\text{el}-\text{ph})}(f_\alpha, f_{-\alpha})$ reads

$$C^{(\text{el}-\text{ph})}(f_\alpha, f_{-\alpha}) = \sum_{\alpha'} \int_{\mathbb{R}^2} S^{(\text{el}-\text{ph})}_{\alpha',\alpha}(\mathbf{k}', \mathbf{k})\, f_{\alpha'}(t, x, \mathbf{k}')\, (1 - f_\alpha(t, x, \mathbf{k}))\, d\mathbf{k}'$$
$$- \int_{\mathbb{R}^2} S^{(\text{el}-\text{ph})}_{\alpha,\alpha'}(\mathbf{k}, \mathbf{k}')\, f_\alpha(t, x, \mathbf{k})\, \big(1 - f_{\alpha'}(t, x, \mathbf{k}')\big)\, d\mathbf{k}',$$

where the total transition rate is the sum of interactions due to acoustic, optical and K phonons of graphene

$$S^{(\text{el-ph})}_{\alpha',\alpha}(\mathbf{k}',\mathbf{k}) = \sum_{\lambda} \left| M^{(\lambda)}(\mathbf{k}',\mathbf{k}) \right|^2 \left[\left(n^{(\lambda)}_{\mathbf{q}} + 1 \right) \delta \left(\varepsilon_{\alpha}(\mathbf{k}) - \varepsilon_{\alpha'}(\mathbf{k}') + \hbar\, \omega^{(\lambda)}_{\mathbf{q}} \right) + n^{(\lambda)}_{\mathbf{q}}\, \delta \left(\varepsilon_{\alpha}(\mathbf{k}) - \varepsilon_{\alpha'}(\mathbf{k}') - \hbar\, \omega^{(\lambda)}_{\mathbf{q}} \right) \right].$$

We have labelled with λ the λth phonon mode and indicated with $\left| M^{(\lambda)}(\mathbf{k}',\mathbf{k}) \right|$ the corresponding matrix element of the scattering due to the phonons of type λ [10, 11]. The symbol δ denotes the Dirac distribution, $\omega^{(\lambda)}_{\mathbf{q}}$ is the the λth phonon frequency, $n^{(\lambda)}_{\mathbf{q}}$ is the Bose–Einstein distribution for the phonon of type λ

$$n^{(\lambda)}_{\mathbf{q}} = \left[\exp\left(\frac{\hbar \omega^{(\lambda)}_{\mathbf{q}}}{k_B T_L} \right) - 1 \right]^{-1},$$

k_B is the Boltzmann constant and T_L is the lattice temperature, assumed constant in this article. The scattering with the acoustic phonons is considered in the elastic approximation

$$2\, n^{(ac)}_{\mathbf{q}} \left| M^{(ac)}(\mathbf{k}',\mathbf{k}) \right|^2 = y \frac{\pi\, D^2_{ac}\, k_B\, T_L}{4\hbar\, \sigma_m\, v_p^2} \left(1 + \cos \vartheta_{\mathbf{k},\mathbf{k}'} \right), \tag{6}$$

where $y = 2/(2\,\pi)^2$, D_{ac} is the acoustic phonon coupling constant, v_p is the sound speed in graphene, σ_m the graphene areal density, and $\vartheta_{\mathbf{k}\,\mathbf{k}'}$ is the convex angle between $\mathbf{k}$ and $\mathbf{k}'$.

The matrix elements of the three relevant optical phonon scatterings, i.e., the longitudinal optical (LO), the transversal optical (TO) and the K phonons, read

$$\left| M^{(O)}(\mathbf{k}',\mathbf{k}) \right|^2 = \left| M^{(LO)}(\mathbf{k}',\mathbf{k}) \right|^2 + \left| M^{(TO)}(\mathbf{k}',\mathbf{k}) \right|^2 = y\, \frac{\pi\, D_O^2}{\sigma_m\, \omega_O} \tag{7}$$

$$\left| M^{(K)}(\mathbf{k}',\mathbf{k}) \right|^2 = y \frac{\pi\, D_K^2}{\sigma_m\, \omega_K} \left(1 - \cos \vartheta_{\mathbf{k},\mathbf{k}'} \right), \tag{8}$$

where D_O is the optical phonon coupling constant, ω_O the optical phonon frequency, D_K is the K-phonon coupling constant and ω_K the K-phonon frequency.

We observe that the acoustic phonon scattering is intra-valley and intra-band while the scattering with optical phonons is intra-valley and can be both intra-band and inter-band. Scattering with optical phonon of type K is inter-valley and can be also both intra-band and inter-band. Regarding the physical parameters of the collision terms see ref. [8].

If the nanoribbon lies on an oxide, e.g., SiO_2, an additional scattering arises with the phonons of the substrate; the main observed effect is a degradation of the mobility [3, 9, 12]. Here, for the sake of simplicity, only suspended nanoribbons will be studied.

A major difference with respect to the large area graphene is the edge effect. Following [13], a way for including the edges roughness is to adopt the Berry-Mondragon (or infinite-mass) boundary conditions. It corresponds to the single Dirac cone approximation and therefore is applicable for smooth enough disorder near the edges. Under the assumption that the type and distributions of scatterers are the same at both edges and requiring that the scattering with rather strong longitudinal momentum transfer is effectively suppressed, the edges roughness gives rise to an additional scattering term $C^{(\text{el}-\text{edg})}(f_\alpha)$ (see [13] for more details) which reads

$$C^{(\text{el}-\text{edg})}(f_\alpha) = \int_{\mathbb{R}^2} S^{(\text{el}-\text{edg})}(\mathbf{k}', \mathbf{k})[f_\alpha(t, x, \mathbf{k}') - f_\alpha(t, x, \mathbf{k})]\, d\mathbf{k}' \tag{9}$$

with

$$S^{(\text{el}-\text{edg})}(\mathbf{k}', \mathbf{k}) = y \frac{\pi N_i V_0^2}{\hbar W} \exp(-2(k_x - k_x')^2 a^2) \delta(\varepsilon_\alpha(\mathbf{k}) - \varepsilon_\alpha(\mathbf{k}')). \tag{10}$$

N_i is the linear density of defects along the graphene edge and the quantity

$$V_0 \exp(-(k_x - k_x')^2 a^2) \tag{11}$$

is the matrix element due to the potential of a single scattering at the edge, V_0 being a constant and a a characteristic range, along the edge direction, so that electron scattering with rather strong longitudinal momentum transfer (along the x-direction), $|k_x - k_x'| > 1/a$, is effectively suppressed. The physical parameters for the edge collision terms are summarized in Table 1.

In the rest of the paper, we will consider a spatially homogeneous graphene nanoribbon (see [14]) under an external constant electric field E. Therefore, the BEs reduce to

$$\frac{\partial f_\alpha}{\partial t} - \frac{e}{\hbar} E \frac{\partial f_\alpha}{\partial k_x} = C(f_\alpha, f_{-\alpha}), \qquad (\alpha = \pm 1), \tag{12}$$

Table 1 Physical parameters for the electron-edge collision term

Parameter	Value
N_i	$10^4\ \text{cm}^{-1}$
V_0	$4.56 \times 10^{-14}\ \text{eV cm}^2$
a	10^{-8} cm

where the total collision term is the sum of the electron–phonon and the electron-edge collision terms

$$C(f_\alpha, f_{-\alpha}) = C^{(\text{el-ph})}(f_\alpha, f_{-\alpha}) + C^{(\text{el-edg})}(f_\alpha). \tag{13}$$

Now $f_\alpha = f_\alpha(t, \mathbf{k})$ and therefore Eq. (12) is an evolutive integro-differential equation which is 2D in $\mathbf{k}$.

3 Numerical Scheme

An efficient numerical scheme for solving the transport equations has been proposed in [8, 15]. It is based on a discontinuous Galerkin scheme (which can be also viewed as a finite volume scheme when a piecewise approximation is considered). From physical argumentations one can reasonably assume that f_1 and $1 - f_{-1}$ tend to zero sufficiently fast as $|\mathbf{k}| \to +\infty$. Let $\Omega \subseteq \mathbb{R}^2$ be a compact domain containing the support of f_1 and $1 - f_{-1}$ for every $t > 0$. We decompose the domain Ω with open sets C_α, $\alpha = 1, 2, \ldots, N$, satisfying

$$C_\alpha \subseteq \Omega \quad \forall \alpha\,, \qquad C_\alpha \cap C_\beta = \emptyset \quad \forall \alpha \neq \beta\,, \qquad \bigcup_{\alpha=1}^{N} \overline{C}_\alpha = \Omega\,.$$

If we approximate the solution of the transport equation in each cell with a piecewise constant function $f_s(t, \mathbf{k})$ with respect to the wave-vector $\mathbf{k}$

$$f_s(t, \mathbf{k}) \approx f_s^\alpha(t), \; \forall \mathbf{k} \in C_\alpha,$$

a set of partial differential equations for the new unknowns $f_s^\alpha(t)$'s is obtained by formally integrating Eq. (4) over each cell C_α

$$\int_{C_\alpha} \frac{\partial f_s}{\partial t}\, d\mathbf{k} - \int_{C_\alpha} \frac{e}{\hbar} E \frac{\partial f_s}{\partial k_x}\, d\mathbf{k} = \int_{C_\alpha} C(f_s, f_{-s})\, d\mathbf{k}\,. \tag{14}$$

For the first integral in Eq. (14) we have

$$\int_{C_\alpha} \frac{\partial f_s}{\partial t}(t, \mathbf{k})\, d\mathbf{k} \approx \text{meas}(C_\alpha)\, \frac{\mathrm{d} f_s^\alpha(t)}{\mathrm{d}t}$$

with meas(C_α) the measure (area) of the cell C_α. The drift term is discretized by the divergence theorem with a reconstruction of the fluxes based on a Min-Mod slope limiter (for details see [15]).

Regarding the collision terms, if we introduce the coefficients,

$$A^{\alpha,\beta} = \int_{C_\alpha} \left[\int_{C_\beta} S(\mathbf{k}, \mathbf{k}')\, d\mathbf{k}' \right] d\mathbf{k}\,, \tag{15}$$

one has (see [8])

$$\int_{C_\alpha} C(f_s, f_{-s})\, d\mathbf{k} \approx \sum_{s'} \sum_{\beta=1}^{N} \left[A^{\beta,\alpha} \left(1 - f_s^\alpha(t)\right) f_{s'}^\beta(t) - A^{\alpha,\beta} f_s^\alpha(t) (1 - f_{s'}^\beta(t)) \right].$$

As numerical $\mathbf{k}$-space, we take the circle $|\mathbf{k}| \leq k_{max}$, where k_{max} is a fixed maximum value chosen so that f_1 and $1 - f_{-1}$ are negligible if $|\mathbf{k}| > k_{max}$. The parameter k_{max} is determined by numerical experiments.

In order to remove apparent singularities arising from the source term, modified polar coordinates $\mathbf{k} = \sqrt{p}\,(\cos\vartheta, \sin\vartheta)$ are adopted. Therefore the generic cell C_α is defined by the inequalities

$$0 \leq p_\alpha < p < p_\alpha + \Delta p \leq k_{max}^2 \quad \text{and} \quad 0 \leq \vartheta_\alpha < \vartheta < \vartheta_\alpha + \Delta\vartheta \leq 2\pi,$$

where Δp and $\Delta\vartheta$ are constant, while the dispersion relation (3) reads

$$\varepsilon(\mathbf{k}) = \pm\varepsilon(p) = \pm\hbar v_F \sqrt{p + C^2}. \tag{16}$$

Regarding intra and inter-band electron–phonon scatterings, the computation of the coefficients (15) requires the evaluation of integrals of the kind

$$\int_{\theta_a'}^{\theta_b'} d\theta' \int_{\theta_a}^{\theta_b} d\theta \int_{p_a'}^{p_b'} dp' \int_{p_a}^{p_b} dp\, \frac{1}{4} [A^{(\nu)} + B^{(\nu)} \cos(\theta - \theta')]$$
$$\times \left[(n_{\mathbf{q}}^{(\nu)} + 1) \delta(\varepsilon_{s'}(p') - \varepsilon_s(p) + \hbar\omega_{\mathbf{q}}^{(\nu)}) + n_{\mathbf{q}}^{(\nu)} \delta(\varepsilon_{s'}(p') - \varepsilon_s(p) - \hbar\omega_{\mathbf{q}}^{(\nu)}) \right],$$

where $A^{(\nu)}$ and $B^{(\nu)}$ are constant, depending on $\nu = ac, O, K$, and the factor $1/4$ is the product of the Jacobian of the coordinate transformations.

If we set $\xi^{(\nu)} = (\hbar\omega_{\mathbf{q}}^{(\nu)})/(\hbar v_F)$ and $r = \sqrt{p + C^2}$, we get the following expression for the previous integrals (for the details see [14])

$$\frac{n_{\mathbf{q}}^{(\nu)} + 1}{\hbar v_F} \left[\frac{1}{3} \frac{s}{s'} r^3 - \frac{1}{2} \frac{\xi^{(\nu)}}{s'} r^2 \right]_{\min \tilde{I}^-}^{\max \tilde{I}^-} + \frac{n_{\mathbf{q}}^{(\nu)}}{\hbar v_F} \left[\frac{1}{3} \frac{s}{s'} r^3 + \frac{1}{2} \frac{\xi^{(\nu)}}{s'} r^2 \right]_{\min \tilde{I}^+}^{\max \tilde{I}^+}$$

where

$$\tilde{I}^{\pm} = \left\{ r \in \mathbb{R} : \sqrt{p_a + C^2} \le r \le \sqrt{p_b + C^2},\ \sqrt{p'_a + C^2} \le \frac{s}{s'} r \pm \frac{\xi^{(\nu)}}{s'} \le \sqrt{p'_b + C^2} \right\}.$$

The same formula holds for acoustic scatterings as well and can be obtained by formally setting $\xi^{(ac)} = 0$. Observe that if $C > \xi^{(\nu)}$ then $\tilde{I}^{\pm} = \emptyset$ in the inter-band transitions.

Concerning the edge scattering, we have to evaluate the following terms

$$\begin{aligned}
w_{\alpha,\beta}^{(el-edg)} &= \int_{\theta'_a}^{\theta'_b} d\theta' \int_{\theta_a}^{\theta_b} d\theta \int_{p'_a}^{p'_b} dp' \int_{p_a}^{p_b} dp \frac{1}{4} A^{(el-edg)} \\
&\times \exp\left(-2\left(\sqrt{p}\sin\theta - \sqrt{p'}\sin\theta'\right)^2 a^2\right) \delta(\varepsilon_s(p) - \varepsilon_s(p')) \\
&\times \frac{1}{2} \exp\left(-2p\left(\sin\theta - \sin\theta'\right)^2 a^2\right) \chi_{[p'_a, p'_b]}(p) \sqrt{p + C^2} \\
&= \frac{A^{(el-edg)}}{2\hbar v_F} \int_{\theta'_a}^{\theta'_b} d\theta' \int_{\theta_a}^{\theta_b} d\theta \int_{[p_a, p_b] \cap [p'_a, p'_b]} dp \sqrt{p + C^2} \\
&\times \exp\left(-2p\left(\sin\theta - \sin\theta'\right)^2 a^2\right).
\end{aligned}$$

The discretization is completed by adopting a standard quadrature formula for the remaining integrals.

4 Numerical Results and Mobility Models

By adopting the numerical scheme outlined in the previous section, we simulate the electron transport in a suspended GNR under an external and constant electric field along the longitudinal direction.

As initial condition the equilibrium distribution functions of electrons in both bands are taken. They are Fermi-Dirac distributions

$$f_{FD}^{\pm}(\mathbf{k}) = \frac{1}{1 + \exp\left(\frac{\varepsilon_{\pm} - \varepsilon_F}{k_B T}\right)}.$$

Here ε_F represents the Fermi energy. The upper sign is for the conduction band while the lower sign is for the valence band. Therefore the initial conditions for holes reads

$$f^h_{FD}(\mathbf{k}) = 1 - f^-_{FD}(-\mathbf{k}) = 1 - \frac{1}{1 + \exp\left(\frac{\varepsilon_-(\mathbf{k}) - \varepsilon_F}{k_B T}\right)} = \frac{1}{1 + \exp\left(\frac{\varepsilon_h(\mathbf{k}) + \varepsilon_F}{k_B T}\right)},$$

where [16] $\varepsilon_h(\mathbf{k}) = -\varepsilon_-(\mathbf{k}) = \hbar v_F \sqrt{k_x^2 + k_y^2 + C^2}$.

From the distribution function the macroscopic quantities of interest are evaluated as expectation values, e.g., the electron and hole densities[1] which are given by Jacoboni [16]

$$\rho_e(t) := \frac{2}{(2\pi)^2} \int_{\mathbb{R}^2} f_e(t, \mathbf{k})\, d\mathbf{k}, \quad \rho_h(t) := \frac{2}{(2\pi)^2} \int_{\mathbb{R}^2} f_h(t, \mathbf{k})\, d\mathbf{k},$$

where $f_e(t, \mathbf{k}) = f_1(t, \mathbf{k})$ and $f_h(t, \mathbf{k}) = 1 - f_{-1}(t, -\mathbf{k})$. Note that at equilibrium there is a ono-to-one correspondence between the Fermi energy and the densities.

The Fermi energy can be controlled by applying a gate voltage [3, 4]. If $\varepsilon_F = 0\,\text{eV}$ we have a symmetrical situation between electrons and holes which is no longer valid when $\varepsilon_F \neq 0\,\text{eV}$. Positive (negative) values of ε_F imply that the majority charge constituting the electric current is that of the electrons (holes). For small W the stationary regime is reached in a few picoseconds but for high values of W one needs about 100 picoseconds, as in the case of large area graphene. This is the effect of the inter-band scatterings when the energy gap becomes comparable with the phonon energy.

Further important macroscopic quantities are the mean values of the electron and hole velocities

$$< (v_e)_x > (t) := \frac{1}{\rho_e(t)} \frac{2}{(2\pi)^2} \int_{\mathbb{R}^2} f_e(t, \mathbf{k})\, v_x(\mathbf{k})\, d\mathbf{k},$$

$$< (v_h)_x > (t) := \frac{1}{\rho_h(t)} \frac{2}{(2\pi)^2} \int_{\mathbb{R}^2} f_h(t, \mathbf{k})\, v_x(\mathbf{k})\, d\mathbf{k},$$

which are related to the currents.

The design of electron devices where the active area is made of graphene nanoribbons requires as a fundamental step devising an appropriate model for the mobilities μ_e and μ_h of electrons and holes which are defined through the relations (in the one dimensional case)

$$< (v_e)_x >= \mu_e(E, \rho_e) E, \quad < (v_h)_x >= \mu_h(E, \rho_h) E, \tag{17}$$

[1] The spin degeneracy is not included; otherwise an additional factor 2 must be added which is irrelevant to compare the mobilities.

leading to the following expression for the currents

$$J_e = -e\rho_e < (v_e)_x >= -e\mu_e(E, \rho_e)E, \quad J_h = e\rho_h < (v_h)_x >= e\mu_h(E, \rho_h)E.$$

It is customary firstly to determine the low field mobility μ_0, which is defined as

$$\mu_0(\rho) = \lim_{E \to 0} \mu(E, \rho). \tag{18}$$

The determination of the low field mobility from experimental data or Monte Carlo simulations is rather ambiguous because of the intrinsic statistical noise. We overcome such difficulties by employing an extensive simulation with the DG approach which does not suffer from any statistical noise.

Large area graphene mobilities have been already obtained from the direct solution of the Boltzmann equation by a DG approach in [3, 12, 17–20]. Here, we tackle the case of GNRs with a width of 5 nm.

The low field mobility is fitted by the following trial expression

$$\mu_0(\rho) = \frac{\tilde{\mu}}{1 + \left(\frac{\rho}{\rho_{\text{ref}}}\right)^{\alpha}}. \tag{19}$$

$\tilde{\mu}$, ρ_{ref}, and α are fitting parameters whose values, estimated by means of the least squares method, are reported in Table 2.

Then we include the electric field dependence by a modified version of the fitting described in Refs. [12, 17]

$$\mu(E, \rho) = \frac{\mu_0(\rho) + \mu_1 \left(\frac{E}{E_{\text{ref}}}\right)^{\beta_1}}{1 + \left(\frac{E}{E_{\text{ref}}}\right)^{\beta_2} + \gamma \left(\frac{E}{E_{\text{ref}}}\right)^{\beta_3}}, \tag{20}$$

where E_{ref}, $\tilde{\mu}$, β_1, β_2, β_3 and γ are fitting parameters. Their values obtained with the least square method are reported in Table 3.

In Figs. 1, 2 and 3 the low and high field mobility along with the currents versus the electric field are plotted. We observe that regarding the high field mobility, a reduction of about one order of magnitude is obtained with respect to the large area graphene (see [14]). Moreover, the hole mobility is independent on the Fermi levels when they are positive. If negative Fermi energies are considered, the role of electrons and holes is exchanged.

Table 2 Least square parameters for the low field mobility

$\tilde{\mu}$ [$\mu m^2/Vps$]	ρ_{ref} [μm^{-2}]	α [−]
1.4927410	4.2356158e+04	1.1283797

Table 3 Least square parameters for the high field mobility

ε_F [eV]	E_{ref} [V/μm]	β_1 [−]	β_2 [−]	β_3 [−]	γ [−]	μ_1 [$\mu m^2/Vps$]
0.1	1.1979e–01	2.8647e–04	1.3632	1.8102e–01	7.29058	7.4501
0.2	1.3429e–01	8.8654e–09	1.3527	1.8242e–01	6.0100	5.9609
0.3	7.1836e–02	2.8068	3.7755	2.6693	3.4984	4.4756
0.4	5.7731e–02	2.3942	3.3868	2.3360	6.4674	5.7173
0.5	2.2322e–02	1.8037	2.6963	1.7498	1.5198e+01	8.6961

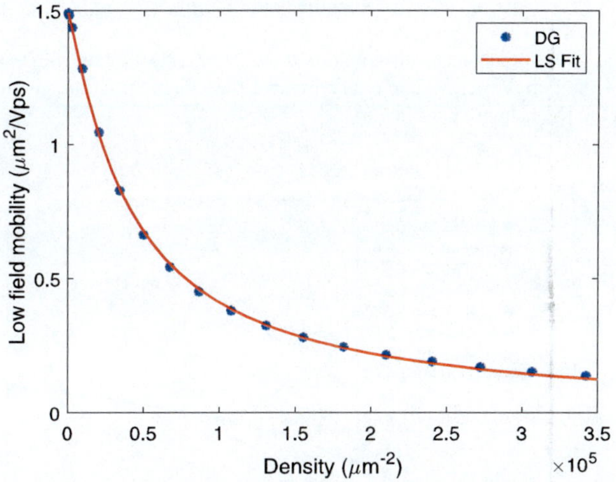

Fig. 1 Low field mobility versus density in the case of nanoribbons with a width of 5 nm. The dots represent the DG results. The continuous line is the fitting (19) with the least square parameters

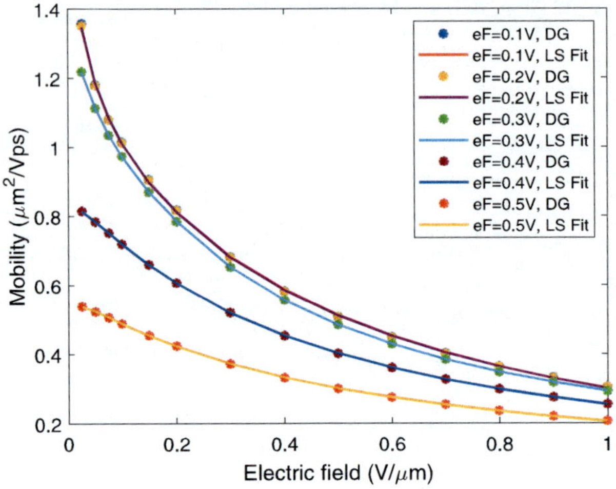

Fig. 2 High field mobility versus electric field in the case of nanoribbons with a width of 5 nm for several values of the Fermi energy. The dots represent the DG results. The continuous line is the fitting (20) with the least square parameters

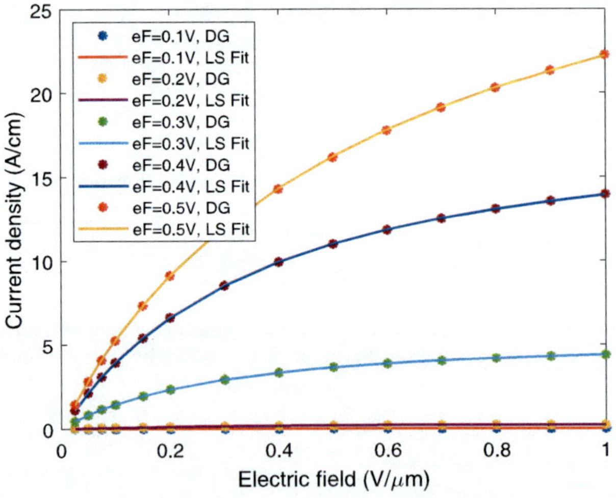

Fig. 3 Current density versus electric field in the case of nanoribbons with a width of 5 nm for several values of the Fermi energy. The dots represent the DG results. The continuous line is the fitting (20) with the least square parameters

5 Conclusions

Charge transport in graphene nanoribbons has been investigated by numerically solving the semiclassical Boltzmann equations with a discontinuous Galerkin method by taking into account all the phonon scattering mechanisms and the edge scattering. For the energy band the model in [7] has been adopted and the edge effects have been included with the approach in [13].

The numerical results have been used to get analytical models of low and high field mobilities which play a crucial role for the simulation of electron devices based on graphene nanoribbons. A degradation of the mobility is found with respect to the large area graphene.

Several open problems remain to be addressed. Just to mention a few of them: the inclusion of the crystal lattice thermal effects as in [21–24] and the transport properties at the metal-nanoribbons interface in the contacts. Moreover, quantum corrections could be included along the lines followed in [25–27].

Acknowledgments The authors acknowledge the support from INdAM (GNFM) and from Università degli Studi di Catania, *Piano della Ricerca 2020/2022 Linea di intervento 2* "QICT." G. Nastasi acknowledges the financial support from *Progetto Giovani GNFM 2020* "Trasporto di cariche e fononi in strutture a bassa dimensione."

References

1. Camiola, V.D., Mascali, G., Romano, V.: Charge Transport in Low Dimensional Semiconductor Structures. Mathematics in Industry, vol. 31. Springer International Publishing, Berlin (2020)
2. Schwierz, F.: Graphene transistors. Nat. Nanotechnol. **5**, 487–496 (2010). https://doi.org/10.1038/nnano.2010.89
3. Nastasi, G., Romano, V.: A full coupled drift-diffusion-Poisson simulation of a GFET. Commun. Nonlinear Sci. Numer. Simul. **87**, 105300 (2020). https://doi.org/10.1016/j.cnsns.2020.105300
4. Nastasi, G., Romano, V.: An efficient GFET structure. IEEE Trans. Electron. Devices. **68**, 4729–4734 (2021). https://doi.org/10.1109/TED.2021.3096492
5. Han, M.Y., Özyilmaz, B., Zhang, Y., Kim, P.: Energy band-gap engineering of graphene nanoribbons. Phys. Rev. Lett. **98**, 206805 (2007). https://doi.org/10.1103/PhysRevLett.98.206805
6. Castro Neto, A.H., Guinea, F., Peres, N.M.R., Novoselov, K.S., Geim, A.K.: The electronic properties of graphene. Rev. Mod. Phys. **81**, 109–162 (2009). https://doi.org/10.1103/RevModPhys.81.109
7. Bresciani, M., Palestri, P., Esseni, D. Selmi, L.: Simple and efficient modeling of the E-k relationship and low-field mobility in Graphene Nano-Ribbons. Solid-State Electron. **54**,1015–1021 (2010). https://doi.org/10.1016/j.sse.2010.04.038
8. Majorana, A., Nastasi, G., Romano, V.: Simulation of Bipolar charge transport in graphene by using a discontinuous Galerkin method. Commun. Comput. Phys. **26**, 114–134 (2019). https://doi.org/10.4208/cicp.OA-2018-0052
9. Coco, M., Nastasi, G.: Simulation of bipolar charge transport in graphene on h-BN. COMPEL **39**(2), 449–465 (2020). https://doi.org/10.1108/COMPEL-08-2019-0311

10. Borysenko, K.M., Mullen, J.T., Barry, E.A., Paul, S., Semenov, Y.G., Zavada, J.M., Buongiorno Nardelli, M., Kim, K.W.: First-principles analysis of electron-phonon interactions in graphene, Phys. Rev. B. **11**, 121412(R) (2010). https://doi.org/10.1103/PhysRevB.81.121412
11. Li, X., Barry, E.A., Zavada, J.M., Buongiorno Nardelli, M., Kim, K.W.: Surface polar phonon dominated electron transport in graphene. Appl. Phys. Lett. **97**, 232105 (2010). https://doi.org/10.1063/1.3525606
12. Nastasi, G., Romano, V.: Improved mobility models for charge transport in graphene. Commun. Appl. Ind. Math. **10**, 41–52 (2019). https://doi.org/10.1515/caim-2019-0011
13. Dugaev, V.K., Katsnelson, M.I.: Edge scattering of electrons in graphene: Boltzmann equation approach to the transport in graphene nanoribbons and nanodisks. Phys. Rev. B. **88**, 235432 (2013). https://doi.org/10.1103/PhysRevB.88.235432
14. Camiola, V.D., Nastasi, G., Romano, V.: Direct simulation of charge transport in graphene nanoribbons. Comm. Comp. Physics. **31**(2), 449–494 (2022). https://doi.org/10.4208/cicp.OA-2021-0032
15. Coco, M., Majorana, A., Romano, V.: Cross validation of discontinuous Galerkin method and Monte Carlo simulations of charge transport in graphene on substrate. Ricerche Mat. **66**, 201–220 (2017). https://doi.org/10.1007/s11587-016-0298-4
16. Jacoboni, C.: Theory of Electron Transport in Semiconductors, 1st edn. Springer, Berlin (2010)
17. Majorana, A., Mascali, G., Romano, V.: Charge transport and mobility in monolayer graphene. J. Math. Industry. **7**, 4 (2016). https://doi.org/10.1186/s13362-016-0027-3
18. Cockburn, B., Shu, C.-W.: Runge–Kutta discontinuous Galerkin methods for convection-dominated problems. J. Sci. Comput. **16**, 173–261 (2001). https://doi.org/10.1023/A:1012873910884
19. Cheng, Y., Gamba, I.M., Majorana, A., Shu, C.-W.: A discontinuous Galerkin solver for Boltzmann-Poisson systems in nano devices. Comput. Methods Appl. Mech. Engrg. **198**(37–40), 3130–3150 (2009). https://doi.org/10.1007/s10825-008-0247-x
20. Cheng, Y. Gamba, I.M., Majorana, A., Shu, C.-W.: A brief survey of the discontinuous Galerkin method for the Boltzmann-Poisson equations. Boletin de la Sociedad Espanola de Matematica Aplicada. **54**, 47–64 (2011). https://doi.org/10.1007/BF03322587
21. Coco, M., Romano, V.: Simulation of electron-phonon coupling and heating dynamics in suspended monolayer graphene including all the phonon branches. J. Heat Transfer. **140**, 092404 (2018). https://doi.org/10.1115/1.4040082
22. Mascali, G.: A hydrodynamic model for silicon semiconductors including crystal heating. Eur. J. Appl. Math. **26**, 447–496 (2015). https://doi.org/10.1017/S0956792515000157
23. Mascali, G., Romano, V.: Charge transport in graphene including thermal effects. SIAM J. Appl. Math. **77**, 593–613 (2017). https://doi.org/10.1137/15M1052573
24. Mascali, G., Romano, V.: Exploitation of the maximum entropy principle in mathematical modeling of charge transport in semiconductors. Entropy. **19**, 36 (2017). https://doi.org/10.3390/e19010036
25. Luca, L., Romano, V.: Quantum corrected hydrodynamic models for charge transport in graphene. Ann. Phys. **406**, 30–53 (2019). https://doi.org/10.1016/j.aop.2019.03.018
26. Barletti, L., Cintolesi, C.: Derivation of Isothermal quantum fluid equations with Fermi-Dirac and Bose-Einstein statistics. J. Stat. Phys. **148**, 353–386 (2012). https://doi.org/10.1007/s10955-012-0535-5
27. Camiola, V.D., Luca, L., Romano, V.: Equilibrium Wigner function for Fermions and Bosons in the case of a general energy dispersion relation. Entropy **22**, 1023 (2020). https://doi.org/10.3390/e22091023

A Review on a General Multi-Species BGK Model: Modelling, Theory and Numerics

Marlies Pirner and Sandra Warnecke

Abstract In this article we focus on kinetic equations for gas mixtures since in applications one often has to deal with mixtures instead of a single gas. In particular, we consider an approximation of the Boltzmann equation, the Bhatnagar–Gross–Krook (BGK) equation. This equation is used in many applications because it is very efficient in numerical simulations. In this article, we recall a general BGK equation for gas mixtures which has free parameters. Specific choices of these free parameters lead to special cases in the literature. For this model, we provide an overview concerning modelling, theoretical results and numerics.

1 Introduction

In this paper we shall concern ourselves with a kinetic description of gas mixtures. For simplicity in notation and statements, we present it here for two species, but the model can be extended to an arbitrary number of species since we only consider binary interactions. A gas of mono atomic molecules and two species is traditionally described via the Boltzmann equation for the distribution functions $f_1 = f_1(x, v, t)$, $f_2 = f_2(x, v, t)$, see, for example, [19, 20]. Here, $x \in \mathbb{R}^3$ and $v \in \mathbb{R}^3$ are the phase space variables, position and velocity of the particles, and $t \geq 0$ denotes the time. Assume that the particles of species 1 have mass m_1 and the particles of species 2 have mass m_2. The Boltzmann equation for gas mixtures is of the form

$$\partial_t f_1 + v \cdot \nabla_x f_1 = Q_{11}(f_1, f_1) + Q_{12}(f_1, f_2),$$
$$\partial_t f_2 + v \cdot \nabla_x f_2 = Q_{22}(f_2, f_2) + Q_{21}(f_2, f_1),$$

M. Pirner (✉) · S. Warnecke
Julius-Maximilians-Universität Würzburg, Würzburg, Germany
e-mail: marlies.pirner@mathematik.uni-wuerzburg.de; sandra.warnecke@mathematik.uni-wuerzburg.de

P. Barbante et al. (eds.), *From Kinetic Theory to Turbulence Modeling*, Springer INdAM Series 51, https://doi.org/10.1007/978-981-19-6462-6_17

where the intra-species collision operators $Q_{11}(f_1, f_1)$ and $Q_{22}(f_2, f_2)$ satisfy

$$\int Q_{kk}(f_k, f_k) \begin{pmatrix} 1 \\ m_k v \\ m_k|v|^2 \end{pmatrix} dv = 0, \qquad k = 1, 2 \tag{1}$$

and the inter-species collision operators $Q_{12}(f_1, f_2)$ and $Q_{21}(f_2, f_1)$ satisfy

$$\begin{aligned} \int Q_{12}(f_1, f_2) dv = \int Q_{21}(f_2, f_1) dv = 0, \\ \int \left(\begin{pmatrix} m_1 v \\ m_1|v|^2 \end{pmatrix} Q_{12}(f_1, f_2) + \begin{pmatrix} m_2 v \\ m_2|v|^2 \end{pmatrix} Q_{21}(f_2, f_1) \right) dv = 0. \end{aligned} \tag{2}$$

These properties of the collision operator ensure conservation of the number of particles, total momentum and total energy at the macroscopic level, (1) in intra-species interactions, (2) in inter-species interactions. In addition, the collision operators satisfy the inequalities

$$\begin{aligned} \int Q_{kk}(f_k, f_k) \ln f_k dv \leq 0, \qquad k = 1, 2 \\ \int Q_{12}(f_1, f_2) \ln f_1 dv + \int Q_{21}(f_2, f_1) \ln f_2 dv \leq 0. \end{aligned} \tag{3}$$

The first inequality turns into an equality if and only if f_k is a Maxwell distribution M_k given by

$$M_k = \frac{n_k}{(2\pi \frac{T_k}{m_k})^{3/2}} \exp\left(-\frac{|v - u_k|^2}{2\frac{T_k}{m_k}}\right). \tag{4}$$

Here we define for any $f_1, f_2 : \Lambda \subset \mathbb{R}^3 \times \mathbb{R}^3 \times \mathbb{R}_0^+ \to \mathbb{R}$ with $(1 + |v|^2) f_1, (1 + |v|^2) f_2 \in L^1(\mathbb{R}^3)$, $f_1, f_2 \geq 0$, the macroscopic quantities

$$\int f_k(v) \begin{pmatrix} 1 \\ v \\ m_k|v - u_k|^2 \end{pmatrix} dv =: \begin{pmatrix} n_k \\ n_k u_k \\ 3n_k T_k \end{pmatrix}, \qquad k = 1, 2, \tag{5}$$

where n_k is the number density, u_k the mean velocity and T_k the mean temperature of species k ($k = 1, 2$). For ease we write T_k instead of $k_B T_k$, where k_B is Boltzmann's constant.

In the second inequality in (3), we have equality if and only if f_1 and f_2 are Maxwell distributions M_1 and M_2 and additionally if and only if $u_1 = u_2$ and $T_1 = T_2$.

If we are close to equilibrium, the complicated interaction terms of the Boltzmann equation can be simplified by a so-called BGK approximation, consisting of a collision frequency $\nu_{kj} n_j$ multiplied by the deviation of the distributions from a local Maxwell distribution

$$\begin{aligned} \partial_t f_1 + v \cdot \nabla_x f_1 &= \nu_{11} n_1 (M_1 - f_1) + \nu_{12} n_2 (M_{12} - f_1), \\ \partial_t f_2 + v \cdot \nabla_x f_2 &= \nu_{22} n_2 (M_2 - f_2) + \nu_{21} n_1 (M_{21} - f_2). \end{aligned} \tag{6}$$

The collision frequencies per density ν_{kj} are assumed to be dependent only on x and t and not on the microscopic velocity v. For references taking into account also a dependency on the microscopic velocity v see [64] for the one-species case, [40] for the gas mixture case and [41] for the numerics of the gas mixture case.

The mixture Maxwell distributions M_{12} and M_{21} are given by

$$\begin{aligned} M_{12}(x, v, t) &= \frac{n_{12}}{(2\pi \frac{T_{12}}{m_1})^{3/2}} \exp\left(-\frac{|v - u_{12}|^2}{2\frac{T_{12}}{m_1}}\right), \\ M_{21}(x, v, t) &= \frac{n_{21}}{(2\pi \frac{T_{21}}{m_2})^{3/2}} \exp\left(-\frac{|v - u_{21}|^2}{2\frac{T_{21}}{m_2}}\right), \end{aligned} \tag{7}$$

where n_{kj}, u_{kj} and T_{kj} will shortly be defined.

This approximation should be constructed in a way such that it has the same main properties as the Boltzmann equation mentioned above.

Now, the question arises how to choose the mixture quantities n_{kj}, u_{kj} and T_{kj}. In this review article we present a general model which is published in [50] for two species. This model contains a lot of proposed models in the literature as special cases. Examples are the models of Gross and Krook [37], Hamel [42], Asinari [4], Garzo et al. [34], Sofena [62], Cercignani [19], Greene [35] and recent models by Bobylev et al. [11]; Haack et al. [38].

The second last [11] presents an additional motivation how the corresponding model can be derived formally from the Boltzmann equation, whereas the last [38] presents a derivation to macroscopic equations on the Navier-Stokes level and numerical results.

BGK models give rise to efficient numerical computations, which are asymptotic preserving, that is, they remain efficient even approaching the hydrodynamic regime [7, 9, 25, 29, 31, 58]. However, the BGK approximation is incapable of reproducing the correct Boltzmann hydrodynamic regime in the asymptotic continuum limit. Therefore, a modified version called the ES-BGK approximation was suggested by Holway for one species [44]. Then the H-Theorem of this model was shown in [57] and existence and uniqueness of mild solutions in [66]. Alternatively, the Shakov model [61] and a BGK model with velocity dependent collision frequency [64] were suggested to achieve the correct Prandtl number. For the BGK model with velocity dependent collision frequency, it is shown that a power law for the collision frequency also leads to the proper Prandtl number. The standard BGK model is extended to a velocity dependent collision frequency while it still satisfies the conservation properties. This works when replacing the Maxwell distribution by a different function, for details see [64]. For this model an H-Theorem can be proven. The existence of these modified functions is proven in [40]. However, since BGK models form the basis to build extended models as ES-BGK models, Shakov models and BGK models with velocity dependent collision frequency, we will mainly review BGK models for gas mixtures of the form (6) in this paper.

Considerations of the hydrodynamic regime for an BGK model for gas mixtures is considered, for example, in [38], a special case of the model presented in this paper. It presents a Chapman–Enskog expansion with transport coefficients in Sect. 5, a comparison with other BGK models for gas mixtures in Sect. 6 and a numerical implementation.

Additionally, we want to mention that there is also another type of BGK model for gas mixtures containing only one collision term on the right-hand side. Examples for this are Andries, Aoki and Perthame [3] and the models in [18, 36]. A derivation of the Navier-Stokes system in the compressible regime for the model in [3] and the corresponding transport coefficients can be found in section 4 of [3]. The transport coefficients of the hydrodynamic regime for the model in [18] can be found in section 5 of [18]. A comparison of these models concerning their hydrodynamic limit can be found in [15]. For gas mixtures there are also many results concerning extensions to ES-BGK models, Shakov models and BGK models with velocity dependent collision frequency [16, 36, 40, 65].

In the following, we will present theoretical and numerical results for this general BGK model for two species with two interaction terms which captures all those special cases in the literature. The outline of the paper is as follows: In Sect. 2 we will present the general multi-species BGK model for two species. For this model, we will give a review of recent theoretical results in Sect. 3. The physical meaning and possible choices of the free parameters are discussed in Sect. 4. And recent existing numerical schemes are given in Sect. 5.

2 The General BGK Model for Gas Mixtures

In this section, we will concern the question of how to choose the mixture quantities $n_{12}, n_{21}, u_{12}, u_{21}, T_{12}, T_{21}$ and the collision frequencies. The collision frequencies $\nu_{11} n_1$ and $\nu_{22} n_2$ correspond to interactions of the particles of each species with itself, while $\nu_{12} n_2$ and $\nu_{21} n_1$ are related to inter-species collisions. To be flexible in choosing the relationship between the collision frequencies, we now assume

$$\nu_{12} = \varepsilon \nu_{21}, \quad 0 < \varepsilon \leq 1. \tag{8}$$

The restriction on ε is without loss of generality. If $\varepsilon > 1$, exchange the notation 1 and 2 and choose $\frac{1}{\varepsilon}$. In addition, we assume that all collision frequencies are positive. The Maxwell distributions M_1 and M_2 in (4) are chosen to have the same density, mean velocity and temperature as f_1 and f_2, respectively. With this choice, we guarantee the conservation of mass, momentum and energy in interactions of one species with itself (1) (see section 2.2 in [50]). The remaining parameters $n_{12}, n_{21}, u_{12}, u_{21}, T_{12}$ and T_{21} will be determined using conservation of the number of particles, total momentum and energy (2), together with some symmetry considerations. If we assume that

$$n_{12} = n_1 \quad \text{and} \quad n_{21} = n_2, \tag{9}$$

we have conservation of the number of particles, see Theorem 2.1 in [50]. If we further assume that u_{12} is a linear combination of u_1 and u_2

$$u_{12} = \delta u_1 + (1-\delta) u_2, \quad \delta \in \mathbb{R}, \tag{10}$$

then we have conservation of total momentum provided that

$$u_{21} = u_2 - \frac{m_1}{m_2} \varepsilon (1-\delta)(u_2 - u_1), \tag{11}$$

see Theorem 2.2 in [50]. If we additionally assume that T_{12} is of the following form

$$T_{12} = \alpha T_1 + (1-\alpha) T_2 + \gamma |u_1 - u_2|^2, \quad 0 \leq \alpha \leq 1, \gamma \geq 0, \tag{12}$$

then we have conservation of total energy provided that

$$\begin{aligned} T_{21} = & \left[\frac{1}{3} \varepsilon m_1 (1-\delta) \left(\frac{m_1}{m_2} \varepsilon (\delta - 1) + \delta + 1 \right) - \varepsilon \gamma \right] |u_1 - u_2|^2 \\ & + \varepsilon (1-\alpha) T_1 + (1 - \varepsilon(1-\alpha)) T_2, \end{aligned} \tag{13}$$

see Theorem 2.3 in [50]. In order to ensure the positivity of all temperatures, we need to restrict δ and γ to

$$0 \leq \gamma \leq \frac{m_1}{3} (1-\delta) \left[(1 + \frac{m_1}{m_2} \varepsilon) \delta + 1 - \frac{m_1}{m_2} \varepsilon \right], \tag{14}$$

and

$$\frac{\frac{m_1}{m_2} \varepsilon - 1}{1 + \frac{m_1}{m_2} \varepsilon} \leq \delta \leq 1, \tag{15}$$

see Theorem 2.5 in [50]. For all these choices one can prove the entropy inequalities (3), see Theorem 2.7 in [50]. We observe that we have free parameters α, δ, γ. We keep the free parameters to be as general as possible. We will discuss the meaning and possible choices in Sect. 4.

3 Theoretical Results of This Model

In this section, we give an overview over recent theoretical results for the model presented in Sect. 2 concerning existence of solutions and large-time behaviour. To start with, one can prove an existence and uniqueness result of mild solutions in the periodic setting in space under certain conditions on the initial data and the collision frequencies. The proof is presented in [49]. Another existence result concerning the existence of a unique global-in-time classical solution when the initial data perturbed slightly from a global equilibrium can be found in [6].

Moreover, one can prove the following results on the large-time behaviour [25]. We denote the entropy of a function f by $H(f) = \int f \ln f dv$ and the relative entropy of f and g by $H(f|g) = \int f \ln \frac{f}{g} dv$. Then one can prove the following results on the large-time behaviour [25].

Theorem 1 *Suppose that* ν_{12} *is constant in time. Then, in the space-homogeneous case we have the following decay rate of the distribution functions* f_1 *and* f_2

$$||f_k - M_k||_{L^1(dv)} \leq 4e^{-\frac{1}{2}Ct}[H(f_1^0|M_1^0) + H(f_2^0|M_2^0)]^{\frac{1}{2}}, \quad k = 1, 2,$$

where C *is the constant given by*

$$C = \min\{\nu_{11}n_1 + \nu_{12}n_2, \nu_{22}n_2 + \nu_{21}n_1\},$$

and the index 0 denotes the value at time $t = 0$.

Theorem 2 *Suppose that* ν_{12} *is constant in time. In the space-homogeneous case, we have the following relaxation rate*

$$\partial_t(u_1 - u_2) = \nu_{12}(1-\delta)(n_2 + \frac{m_1}{m_2}n_1)(u_2 - u_1) \tag{16}$$

and a decay rate of the mean velocities

$$|u_1(t) - u_2(t)|^2 = e^{-2\nu_{12}(1-\delta)\left(n_2 + \frac{m_1}{m_2}n_1\right)t}|u_1(0) - u_2(0)|^2.$$

Theorem 3 *Suppose* ν_{12} *is constant in time. In the space-homogeneous case, we have the following relaxation rate*

$$\partial_t(T_1 - T_2) = -C_1(T_1 - T_2) + C_2|u_1 - u_2|^2 \tag{17}$$

and a decay rate of the temperatures

$$T_1(t) - T_2(t) = e^{-C_1 t}\left[T_1(0) - T_2(0) + \frac{C_2}{C_1 - C_3}(e^{(C_1-C_3)t} - 1)|u_1(0) - u_2(0)|^2\right],$$

where the constants are defined by

$$C_1 = (1-\alpha)\nu_{12}\,(n_2 + n_1)\,,$$

$$C_2 = \nu_{12}\left(n_2\left((1-\delta)^2 + \frac{\gamma}{m_1}\right) - n_1\left(1 - \delta^2 - \frac{\gamma}{m_1}\right)\right),$$

$$C_3 = 2\nu_{12}(1-\delta)\left(n_2 + \frac{m_1}{m_2}n_1\right).$$

There are also results in the space-inhomogeneous case for the linearized collision operator, see [51]. In their article, the authors study hypocoercivity for the linearized BGK model for gas mixtures in continuous phase space. By constructing an entropy functional, they prove exponential relaxation to global equilibrium with explicit rates. The strategy is based on the entropy and spectral methods adapting Lyapunov's direct method as presented in [1] for the one-species linearized BGK model.

4 Possible Choices and Meaning of the Free Parameters

In this section, we deal with the meaning and possible choices of the free parameters. One possibility is that we can choose the parameters such that we can generate special cases in the literature [4, 11, 19, 34, 35, 37, 38, 42, 62]. For instance, if we choose $\varepsilon = 1$, $\delta = \frac{m_1}{m_1+m_2}$, $\alpha = \frac{m_1^2+m_2^2}{(m_1+m_2)^2}$ and $\gamma = \frac{m_1 m_2}{(m_1+m_2)^2}\frac{m_2}{3}$, we obtain the model by Hamel in [42].

Another possibility is to choose the parameters in a way such that the macroscopic exchange terms of momentum and energy can be matched in a certain way, for example, that they coincide with the ones for the Boltzmann equation. For this, we first present the macroscopic equations with exchange terms of the BGK model (6). If we multiply the BGK model for gas mixtures by $1, m_j v, m_j \frac{|v|^2}{2}$ and integrate with respect to v, we obtain the following macroscopic conservation laws

$$\partial_t n_1 + \nabla_x \cdot (n_1 u_1) = 0,$$

$$\partial_t n_2 + \nabla_x \cdot (n_2 u_2) = 0,$$

$$\partial_t (m_1 n_1 u_1) + \nabla_x \cdot \int m_1 v \otimes v f_1(v) dv + \nabla_x \cdot (m_1 n_1 u_1 \otimes u_1) = f_{m_{1,2}},$$

$$\partial_t (m_2 n_2 u_2) + \nabla_x \cdot \mathbb{P}_2 + \nabla_x \cdot (m_2 n_2 u_2 \otimes u_2) = f_{m_{2,1}},$$

$$\partial_t \left(\frac{m_1}{2} n_1 |u_1|^2 + \frac{3}{2} n_1 T_1 \right) + \nabla_x \cdot \int m_1 |v|^2 v f(v) dv = F_{E_{1,2}},$$
$$\partial_t \left(\frac{m_2}{2} n_2 |u_2|^2 + \frac{3}{2} n_2 T_2 \right) + \nabla_x \cdot Q_2 = F_{E_{2,1}},$$

with exchange terms $f_{m_{i,j}}$ and $F_{E_{i,j}}$ given by

$$\begin{aligned} f_{m_{1,2}} &= -f_{m_{2,1}} = m_1 \nu_{12} n_1 n_2 (1-\delta)(u_2 - u_1), \\ F_{m_{1,2}} &= -F_{m_{2,1}} \\ &= \left[\nu_{12} \frac{1}{2} n_1 n_2 m_1 (\delta - 1)(u_1 + u_2 + \delta(u_1 - u_2)) + \frac{1}{2} \nu_{12} n_1 n_2 \gamma (u_1 - u_2) \right] \cdot (u_1 - u_2) \\ &+ \frac{3}{2} \varepsilon \nu_{21} n_1 n_2 (1-\alpha)(T_2 - T_1). \end{aligned}$$

Here, we can observe a physical meaning of α and δ. We see that α and δ show up in the exchange terms of momentum and energy as parameters in front of the relaxation of u_1 towards u_2 and T_1 towards T_2. So they determine, together with the collision frequencies, the speed of relaxation of the mean velocities and the temperatures to a common value. This can already be observed in Theorem 2 and Theorem 3.

Next we follow Chapter 4.1 in [38] and compare the relaxation rates in the space-homogeneous case to the relaxation rates for the space-homogeneous Boltzmann equation. In [38], they find values for ν_{kj} such that either the relaxation rate for the mean velocities (16) or the relaxation for the temperatures (17) coincides with the corresponding rate of the Boltzmann equation. But using the free parameters α, δ and γ we are able to match both of the relaxation rates at the same time. For this, we compare the coefficients of the terms $u_2 - u_1$, $T_2 - T_1$ and $|u_2 - u_1|^2$ in these Boltzmann relaxation rates and the BGK relaxation rates (16) and (17), and we derive the values of the parameters for this model:

$$(u_2 - u_1)\text{-term:} \quad \delta = 1 - \frac{\alpha_{12}}{\nu_{12}} \frac{m_1 + m_2}{2} \frac{m_1 n_1 + m_2 n_2}{m_1 n_1 m_2 n_2} \left(n_1 \frac{m_1}{m_2} + n_2 \right)^{-1},$$

$$(T_2 - T_1)\text{-term:} \quad \alpha = 1 - \frac{\alpha_{12}}{\nu_{12} n_2 n_1},$$

$$|u_2 - u_1|^2\text{-term:} \quad \gamma = \frac{1}{3} (n_1 + n_2)^{-1} \left[\frac{\alpha_{12}}{\nu_{12}} \frac{m_2 n_2 - m_1 n_1}{n_2 n_1} - m_1 n_2 (1-\delta)^2 + m_1 n_1 (1 - \delta^2) \right],$$

where α_{12} is a coefficient for energy transfer coming from Boltzmann equation, see [38] and references therein. Additionally, the constraints (12), (14) and (15) need to be satisfied. This can be verified by a corresponding choice of ν_{kj}. One possibility is

$$\nu_{kj} = \frac{1}{2} \frac{\alpha_{kj}}{n_k n_j} \frac{(m_k + m_j)^2}{m_k m_j} \tag{18}$$

and for $1 \geq \varepsilon = \frac{m_j}{m_k}$ (cf. in a plasma).

5 On Existing Numerical Schemes

In the literature, various approaches for the discretization of kinetic equations can be found, including schemes for the one-species BGK model. Contributions in numerics for multi-species BGK models have strongly increased in the last years.

To start with, we give a short overview over existing numerical methods for the one-species BGK equation. Since the contributions are very crowded, we do not claim completeness. Many ideas can be carried over to the discretization of multi-species BGK equations, and we conclude with identified publications on numerical schemes for multi-species BGK equations which can be written in the form (6).

The (one-species) Boltzmann equation captures physical phenomena very well at the kinetic level [19]. Nevertheless, numerical computation is expensive. The fastest algorithms for evaluating the Boltzmann collision operator are spectral methods with special kernels [54]. This motivates the BGK equation as an approximation of the Boltzmann equation: Even though the dimensionality is as high as for the Boltzmann equation, the BGK interaction term is better to handle and explicitly computable [31, 58]. Hence, the computational cost is much less compared to the Boltzmann equation while maintaining most of the physical properties [44, 53, 64].

The computational advantages are also useful for penalization techniques [31] where the BGK equation is solved as preconditioner for the numerical solution of the Boltzmann equation. This idea is generalized to the multi-species setting in [47]. In [27], the authors develop an improved Monte Carlo method for the BGK equation. This is supposed to be a first step towards an improved Monte Carlo simulation of the Boltzmann equation. Moreover, the BGK approach is useful when coupling different domains in which the regimes range from equilibrium to very rarefied [2].

A fully discrete scheme requires the discretization in (microscopic) velocity, space and time. First we consider the discretization in velocity before we look at the space and time variables.

Having the microscopic velocities as independent variables introduces both more degrees of freedom and more difficulties. Due to the high dimensionality, it is recommendable to use coarse grids [59] which then poses challenges regarding errors in the macroscopic quantities. This can be tackled when the conservation properties (1) are fulfilled at the discrete level. The handling of discrete moments, a discrete entropy and the corresponding discrete Maxwellians is discussed in [52].

Another approach to fulfil the conservation laws at the discrete level is given by a constrained L^2-projection in [33].

Being interested in macroscopic quantities only, the Chu reduction is a possible approach to lower the dimensionality if there are more degrees of freedom in velocity than in space [22]. Using the Chu reduction, one follows the evolution of appropriate integrals of the distribution functions, but these integrals do not correspond to macroscopic quantities yet. This method reduces the computational costs considerably.

When the mean velocities $u(x, t)$ cover a wide range and small temperatures are encountered, grid adaption becomes an important tool. This issue is tackled more and more in the last decade, e.g., in [8, 13, 17, 43].

This leads us to another advantage of multi-species BGK equations. In case of the multi-species Boltzmann equations, a large mass ratio of the species, which results in very distinct thermal speeds, requires an expensive grid resolution [55]. As particles of different species only interact through moments in the BGK model, the evolution of each species can be numerically solved on separate grids [38] which might be an important ingredient for an efficient simulation.

There can be found many different approaches for the space discretization in the literature as the BGK equation shares the same transport term with many other kinetic equations such as Boltzmann, Fokker-Planck, Vlasov, etc.

The transport term being hyperbolic, a finite volume discretization is often used. High-resolutions can be obtained by weighted essentially non-oscillatory (WENO) or discontinuous Galerkin (DG) schemes. However, for orders higher than two the corresponding formulation of the relaxation term requires additional care because it does not suffice to consider the relaxation of the cell averages, but the cell averages of the relaxation term need to be calculated. [5, 21, 45, 52, 58]

Another convenient choice is the semi-Lagrange method. The characteristics are followed exactly which requires an interpolation for the evaluation of the corresponding foot point. By conservative reconstructions or corrections, these methods can be kept conservative also for higher orders [14, 26, 29, 60, 63].

In [28], the authors present an approach for an efficient scheme based on discrete velocity models and semi-Lagrangian methods. In contrast to standard semi-Lagrangian schemes, the distribution function needs not to be reconstructed at each time step which of course accelerates numerical computations.

For the Vlasov equation, the most used method is the Particle In Cell (PIC) method [32]. But to our knowledge, it is less used for equations with interaction terms when hydrodynamic effects become more important. In [32], the authors shortly discuss different methods for the Vlasov equation and then introduce their positive and flux conservative method (PFC).

For the interaction term, a time implicit formulation is often chosen since the right-hand side becomes stiff when the collision frequencies become large (close to the hydrodynamic regime). By the implicit discretization, one can avoid tiny time steps coming from stability issues. Thanks to the special structure of the interaction term, the implication is comparably easy manageable, and the equation stays explicitly solvable [31, 58].

Usually, the transport part is evaluated explicitly. It is combined with the interaction term, e.g., by splitting methods [23, 39], implicit-explicit Runge-Kutta (IMEX RK) schemes [56, 58] or IMEX multistep methods [30].

Splitting methods must be treated with care when the right-hand side becomes stiff. In [46], the author shows that the (second-order) Strang splitting reduces actually to a first-order approximation of the equilibrium equation in the hydrodynamic limit. This leads us to the so-called asymptotic-preserving (AP) schemes which provide an adequate discretization also of the limiting equations. Using AP schemes, the correct equilibrium solutions are preserved [9, 29, 31, 45, 59]. This issue is addressed more and more in the recent years. In this context, we also want to mention the micro-macro decomposition and the parity decomposition/AP splitting. For the former approach, the distribution function is written as a sum of its equilibrium (macro) part and the remnant which represents the kinetic (micro) part. This results in one microscopic and one macroscopic equation which can be solved by individual and adequate methods [24]. For the latter approach, the distribution function is decomposed by an even and an odd parity. A new system of equations can be derived with only one time scale where splitting techniques can be applied [29, 48].

More physics is (re)introduced by multi-species models. At the discrete level, many ideas can be carried over from the single-species schemes. In the following, we give contributions of numerical schemes for multi-species BGK equations, which can be written in the form of (6).

In [25], the authors extend the work of [24] for the multi-species model. They perform a micro-macro decomposition and focus on the fluid limit. The micro part is solved by a particle method, whereas the macro part (depicting the fluid part) is solved by a standard finite volume approach. Here, an additional force term with an electric field is considered.

In [38], the authors are interested in capturing physical transport coefficient. They use the additional degrees of freedom in the multi-species setting to match relaxation rates in the space-homogeneous case equivalent to the Boltzmann ones. An extension to space-inhomogeneous simulations is done in [39] where they additionally examine the coupling to electric fields.

In [12], the authors compare numerical results for different multi-species BGK models, where one of these models is a special case of (6).

A BGK model for gas mixtures is extended to velocity dependent collision frequencies in [40]. Collision frequencies influence the relaxation process and the resulting hydrodynamic behaviour such that they also become important when calculating transport coefficients. The class of models in [40] captures a model of the form (6) as a special case, but in general the Maxwellians are replaced by more sophisticated Gaussian functions. Numerical schemes for this kind of equations have been developed in [41]. The key new ingredient is a solver based on a convex entropy minimization problem which makes possible an implicit treatment of the BGK operator.

6 Conclusions and Outlook

We recalled a general BGK model for gas mixtures with two collision terms for two species and its theoretical properties. We presented results on the H-Theorem, existence of solutions and large-time behaviour. The special feature of the presented model is the free parameters α, δ and γ. They influence the exchange of momentum and energy and can be set such that the model's behaviour matches the physics or coincides with another (more specialized) model.

An overview over existing literature on numerics for this kind of models was given.

However, BGK- type models often lack on correct parameters in the continuum limit like the Prandtl number. Therefore these models can be used as a basis for more extended models like ES-BGK models or BGK models with velocity dependent collision frequency. As a future work the Chapman–Enskog expansion of such models can be computed. Then the transport coefficients of all these models can be compared and eventually extended to match all parameters in the macroscopic equations. Here, the free parameters in the BGK model for monoatomic molecules with a sum of interaction terms might be useful.

Acknowledgement Marlies Pirner is supported by the Alexander von Humboldt foundation.

References

1. Achleitner, F., Arnold, A., Carlen, E.A.: On linear hypocoercive BGK models. In From Particle Systems to Partial Differential Equations III, Springer Proceedings in Mathematics & Statistics, vol. 162 [Cham], pp. 1–37. (2016)
2. Alaia, A., Puppo, G.: A hybrid method for hydrodynamic-kinetic flow - Part II - Coupling of hydrodynamic and kinetic models. J. Comput. Phys. **231(16)**, 5217–5242 (2012)
3. Andries, P., Aoki, K., Perthame, B.: A consistent BGK-type model for gas mixtures. J. Statist. Phys. **106**, 993–1018 (2002)
4. Asinari, P.: Asymptotic analysis of multiple-relaxation-time lattice Boltzmann schemes for mixture modeling. Comput. Math. Appl. **55**, 1392–1407 (2008)
5. Ayuso, B., Carrillo, J.A., Shu, C.-W.: Discontinuous Galerkin Methods for the one-dimensional Vlasov-Poisson system, Kinetic Related Models **4**, 955–989 (2011)
6. Bae, G., Klingenberg, C., Yun, S., Pirner, M.: Mixture BGK model near a global Maxwellian, manuscript (2021)
7. Bennoune, M., Lemou, M., Mieussens, L.: Uniformly stable numerical schemes for the Boltzmann equation preserving the compressible Navier-Stokes asymptotics. J. Comput. Phys. **227**, 3781–3803 (2008)
8. Bernard, F., Iollo, A., Puppo, G.: A local velocity grid approach for BGK equation. Commun. Comput. Phys. **16**(4), 956–982 (2014)
9. Bernard, F., Iollo, A., Puppo, G.: Accurate asymptotic preserving boundary conditions for kinetic equations on Cartesian grids. J. Sci. Comput. **65**, 735–766 (2015)
10. Bhatnagar, P.L., Gross, E.P., Krook, M.: A model for collision processes in gases. I. Small amplitude processes in charged and neutral one-component systems. Phys. Rev. **94**, 511–525 (1954)

11. Bobylev, A.V., Bisi, M., Groppi, M., Spiga, G., Potapenko, I.F.: A general consistent BGK model for gas mixtures. Kinetic Related Models **11**(6), 1377 (2018)
12. Boscarino, S., Cho, S.Y., Groppi, M., Russo, G.: BGK models for inert mixtures: comparison and applications (2021). Preprint arXiv:2102.12757 [math-ph]
13. Boscarino, S., Cho, S.Y., Russo, G.: A local velocity grid conservative semi-Lagrangian schemes for BGK model (2021). Preprint arXiv:2107.08626 [math.NA]
14. Boscarino, S., Cho, S.Y., Russo, G., Yun, S.B.: Conservative semi-Lagrangian schemes for kinetic equations Part II: applications. J. Comput. Phys. **436**, 110281 (2021)
15. Boscarino, S., Cho, S.Y., Groppi, M., Russo, G.: BGK models for inert mixtures: comparison and applications. Preprint arXiv:2102.12757
16. Brull, S.: An ellipsoidal statistical model for gas mixtures. Commun. Math. Sci. **8**, 1–13 (2015)
17. Brull, S., Mieussens, L.: Local discrete velocity grids for deterministic rarefied flow simulations. J. Comput. Phys. **266**, 22–46 (2014)
18. Brull, S., Pavan, V., Schneider, J.: Derivation of a BGK model for mixtures. Euro. J. Mech. B/Fluids **33**, 74–86 (2012)
19. Cercignani, C.: The Boltzmann Equation and Its Applications. Springer, Berlin (1988)
20. Cercignani, C.: Rarefied Gas Dynamics, From Basic Concepts to Actual Calculations. Cambridge University Press, Cambridge (2000)
21. Cheng, Y., Gamba, I.M., Proft, J.: Positivity-preserving discontinuous Galerkin schemes for linear Vlasov-Boltzmann transport equations. Math. Comput. **81**(277), 153–190 (2012)
22. Chu, C.K.: Kinetic-theoretic description of the formation of a shock wave. Phys. Fluids **8**(12), 12–22 (1965)
23. Coron, F., Perthame, B.: Numerical Passage from kinetic to fluid equations. SIAM J. Numer. Anal. **28**(1), 26–42 (1991)
24. Crestetto, A., Crouseilles, N., Lemou, M.: Kinetic/fluid micro-macro numerical schemes for Vlasov-Poisson-BGK equation using particles. Kinetic Related Models **5**, 787–816 (2013)
25. Crestetto, A., Klingenberg, C., Pirner, M.: Kinetic/fluid micro-macro numerical scheme for a two component gas mixture. SIAM Multiscale Modeling Simulation **18**(2), 970–998 (2020)
26. Crouseilles, N., Mehrenberger, M., Sonnendrücker, E.: Conservative semi-Lagrangian schemes for Vlasov equations. J. Comput. Phys. **229**, 1927–1953 (2010)
27. Degond, P., Dimarco, G., Pareschi, L.: The Moment Guided Monte Carlo method. Int. J. Numer. Methods Fluids **67**, 189–213 (2011)
28. Dimarco, G., Loubère, R.: Towards an ultra efficient kinetic scheme. Part I: Basics on the BGK equation. J. Comput. Phys. **255**, 680–698 (2012)
29. Dimarco, G., Pareschi, L.: Numerical methods for kinetic equations. Acta Numerica **23**, 369–520 (2014)
30. Dimarco, G., Pareschi, L.: Implicit-Explicit linear multistep methods for stiff kinetic equations. SIAM J. Numer. Anal. **55**(2), 664–690 (2017)
31. Filbet, F., Jin, S.: A class of asymptotic-preserving schemes for kinetic equations and related problems with stiff sources. J. Comput. Phys. **20**, 7625–7648 (2010)
32. Filbet, F., Sonnendrücker, E., Bertrand, P.: Conservative numerical schemes for the Vlasov equation. J. Comput. Phys. **172**, 166–187 (2001)
33. Gamba, I.M., Tharkabhushaman, S.H.: Spectral-Lagrangian based methods applied to computation of non-equilibrium statistical states. J. Comput. Phys. **228**, 2012–2036 (2009)
34. Garzó, V., Santos, A., Brey, J.J.: A kinetic model for a multicomponent gas. Phys. Fluids A **1**(2), 380–383 (1989)
35. Greene, J.: Improved Bhatnagar-Gross-Krook model of electron-ion collisions. Phys. Fluids **16**, 2022–2023 (1973)
36. Groppi, M., Monica, S., Spiga, G.: A kinetic ellipsoidal BGK model for a binary gas mixture. Europhys. Lett. **96**, 64002 (2011)
37. Gross, E.P., Krook, M.: Model for collision processes in gases: Small-amplitude oscillations of charged two-component systems. Phys. Rev. **102**(3), 593 (1956)
38. Haack, J.R., Hauck, C.D., Murillo, M.S.: A conservative, entropic multispecies BGK model. J. Statist. Phys. **168**, 826–856 (2017)

39. Haack, J.R., Hauck, C.D., Murillo, M.S.: Interfacial mixing in high-energy-density matter with a multiphysics kinetic model. Phys. Rev. E **96**, 063310 (2017)
40. Haack, J., Hauck, C., Klingenberg, C., Pirner, M., Warnecke, S.: A consistent BGK model with velocity-dependent collision frequency for gas mixtures. J. Statist. Phys. **184**(31), 1–17 (2021)
41. Haack, J., Hauck, C., Klingenberg, C., Pirner, M., Warnecke, S.: Numerical schemes for a multi-species BGK model with velocity-dependent collision frequency. J. Comput. Phys. **473**, 111729 (2023). https://doi.org/10.1016/j.jcp.2022.111729
42. Hamel, B.B.: Kinetic model for binary gas mixtures. Phys. Fluids **8**, 418–425 (1956)
43. Hittinger, J., Banks, J.: Block-structured adaptive mesh refinement algorithms for Vlasov simulation. J. Comput. Phys. **241**, 118–140 (2013)
44. Holway, L.H.: New statistical models for kinetic theory: methods of construction. Phys. Fluids **9**, 1658–1673 (1966)
45. Hu, J., Jin, S., Li, Q.: Asymptotic-preserving schemes for multiscale Hyperbolic and Kinetic equations. Handbook of Numer. Analy. **18**, 103–129 (2017)
46. Jin, S.: Runge-Kutta methods for hyperbolic conservation laws with stiff relaxation terms. J. Comput. Phys. **122**, 51–67 (1995)
47. Jin, S., Li, Q.: A BGK-penalization based asymptotic-preserving scheme for the multispecies Boltzmann equation. Numer. Methods Partial Differ. Eq. **29**(3), 1056–1080 (2013)
48. Jin, S., Pareschi, L.: Discretization of the multiscale semiconductor Boltzmann equation by diffusive relaxation schemes. J. Comput. Phys. **161**, 312–330 (2000)
49. Klingenberg, C., Pirner, M.: Existence, uniqueness and positivity of solutions for BGK models for mixtures. J. Differ. Equ. **264**, 702–727 (2017)
50. Klingenberg, C., Pirner, M., Puppo, G.: A consistent kinetic model for a two-component mixture with an application to plasma. Kinetic Related Models **10**, 445–465 (2017)
51. Liu, L., Pirner, M.: Hypocoercivity for a BGK model for gas mixtures. J. Differ. Equ. **267**, 119–149 (2019)
52. Mieussens, L.: Discrete velocity model and implicit scheme for the BGK equation of rarefied gas dynamics. Math. Models Methods Appl. Sci. **10**(08), 1121–1149 (2000)
53. Mieussens, L., Struchtrup, H.: Numerical comparison of Bhatnagar–Gross–Krook models with proper Prandtl number. Phys. Fluids **16**, 2797 (2004)
54. Mouhot, C., Pareschi, L.: Fast algorithms for computing the Boltzmann collision operator. Math. Comp. **75**, 1833–1852 (2006)
55. Munafo, A., Torres, E., Haack, J., Gamba, I.M., Magin, T.: A spectral-lagrangian Boltzmann solver for a multi-energy level gas. J. Comput. Phys. **264**, 152–176 (2014)
56. Pareschi, L., Russo, G.: Implicit-explicit schemes Runge-Kutta schemes and applications to hyperbolic systems with relaxation. J. Sci. Comput. **25**(1), 129–155 (2005)
57. Perthame, B., Pulvirenti, M.: Weighted L^∞ bounds and uniqueness for the Boltzmann BGK model. Arch. Rational Mech. Anal. **125**, 289–295 (1993)
58. Pieraccini, S., Puppo, G.: Implicit-explicit schemes for BGK kinetic equations. J. Sci. Comput. **32**, 1–28 (2007)
59. Puppo, G.: Kinetic models of BGK type and their numerical integration. Riv. Mat. Univ. Parma **10**(2), 299–349 (2019)
60. Qiu, J.-M., Shu, C.-W.: Positivity preserving semi-Lagrangian discontinuous Galerkin formulation: Theoretical analysis and application to the Vlasov–Poisson system, J. Comput. Phys. **230**(23), 8386–8409 (2011)
61. Shakhov, E.M.: Generalization of the Krook kinetic relaxation equation. Fluid Dyn. **3**, 95–96 (1968)
62. Sofonea, V., Sekerka, R.: BGK models for diffusion in isothermal binary fluid systems. Physica **3**, 494–520 (2001)
63. Sonndendrücker, E., Roche, J., Bertrand, P., Ghizzo, A.: The semi-Lagrangian method for the numerical resolution of Vlasov equations. J. Comput. Phys. **149**(201), 201–220 (1998)
64. Struchtrup, H.: The BGK-model with velocity-dependent collision frequency. Continuum Mech. Thermodyn. **9**(1), 23–31 (1997)

65. Todorova, B., Steijl, R.: Derivation and numerical comparison of Shakov and Ellipsoidal Statistical kinetic models for a monoatomic gas mixture. Euro. J. Mech.-B/Fluids **76**, 390-402 (2019)
66. Yun, S.-B.: Classical solutions for the ellipsoidal BGK model with fixed collision frequency. J. Differ. Equ. **259**, 6009–6037 (2015)

Gas-Kinetic Methods for Turbulent Flow

Marcello Righi

Abstract Gas-kinetic schemes are derived from the BGK-Boltzmann equation. These family of schemes for CFD are computationally more demanding than conventional upwind schemes but provide a number of advantages stemming precisely from the fact that they were not derived from the Navier–Stokes equations. We highlight three peculiarities: (i) the gas evolution is derived in space and time exactly, (ii) the viscous stress tensor is built from the collision operator and not through the diffusion operator, variations of the relaxation time are thus not simply applied linearly to the strain rate (and its moments) but input through the collision operator, (iii) the order of the Chapman–Enskog expansion is a "natural" way to further improve the physical consistence of the collision operator. In practice, gas-kinetic schemes are not only more suitable to resolve vortical structures but they also handle turbulent viscosity (or, better, a turbulent relaxation time) in a physically more relevant fashion. Whereas the first advantage is exploited mostly in approaches like Direct Numerical Simulation (DNS) and Large Eddy Simulation (LES), where the ability to resolve turbulent structures is key, the latter provides some leverage to approaches like Reynolds-Averaged Navier–Stokes (RANS) or hybrid RANS-LES where unresolved turbulence is funnelled through the collision operator. The paper aims at reviewing these advantages in the light of the results obtained by the author and those published in the recent years.

1 Introduction

Gas-kinetic schemes for hydrodynamic were developed from the 1990s from the Boltzmann-BGK equation and are still an area of active research. They are appreciated for the higher spatial and temporal accuracy as well as for the non-linear, physically consistent viscous stress tensor derived from the collision operator.

M. Righi (✉)
School of Engineering, Zurich University of Applied Sciences, Winterthur, Switzerland
Federal Institute of Technology, Zurich, Switzerland
e-mail: rigm@zhaw.ch

P. Barbante et al. (eds.), *From Kinetic Theory to Turbulence Modeling*, Springer INdAM Series 51, https://doi.org/10.1007/978-981-19-6462-6_18

Xu developed the best known gas-kinetic scheme in 2001 [49] during his PhD at Stanford University under Professor Jameson's supervision. The research work on the gas-kinetic scheme seems pretty much still concentrated around Xu and his group in Hong Kong, even though other researchers provided over time valuable contributions to the establishment of the gas-kinetic scheme [13, 23, 26, 27, 51, 52, 54].

The current research work mainly focuses on the enhancement of the numerical scheme and, in particular, on the ability to adapt it to unstructured meshes for both finite volumes and finite elements and on the addition of approximate Jacobians to allow implicit time advancement; the relevant work has been published by Xu and his group, for instance, refer to Refs. [19, 56, 58] and references therein.

A further research direction is the one focusing on the "multi-scalarity" of the scheme, in which the particles velocity are discretized allowing a more physically consistent model of rarefied flow. This area is also actively pursued by the same group [19, 50, 55].

Finally and central to this paper, gas-kinetic schemes are also increasingly applied to turbulent flow, following the RANS, LES and implicit LES, as well as hybrid RANS-LES frameworks, as recently published in Refs. [6, 8, 9] and references therein. These frameworks each have very different requirements for the numerical implementation and each may exploit different peculiarities of gas-kinetic schemes. DNS, and to a point also LES, take advantage of the exact gas-evolution provided by gas-kinetic schemes. The spatial and temporal resolution (for given spacing and time step) may be further enhanced by high-order reconstruction schemes such as Weighted Essentially Non-Oscillatory (WENO) and Discontinuous Galerkin (DG).

Turbulence modeling techniques like RANS and hybrid RANS-LES may benefit from gas-kinetic schemes through different mechanisms: these include the ability to handle the ratio of resolved-to-unresolved scales of motion (often referred to as "multi-scale" or "multi-scalarity," the high-order reconstruction also affecting the collision operator (hence the stress tensor), the Chapman–Enskog expansion order beyond the Navier–Stokes level.

Whereas the advantages provided to DNS and LES simulations mostly concern the numerical implementation, RANS and RANS-LES simulations may also benefit from the physical modeling standpoint, as shown by the author in Refs. [35–39], arguing that the ratio between unresolved and resolved scales of motion is the mechanism leading to corrections to the turbulent stress tensor.

This paper presents the implementation of a gas-kinetic scheme into a RANS solver and shows that it may be exploited to improve turbulence modeling. Moreover, it addresses the potential to further contribute to the ongoing efforts towards more accurate and physically consistent modeling of turbulence. It is structured as follows: the derivation of the gas-kinetic scheme is presented in Sect. 2 whereas the extension of the scheme for turbulent flow is in Sect. 3; Sect. 4 presents the results of numerical experiments obtained by the author, in Sect. 5 these are discussed in the light of more recent research; conclusions are finally presented in Sect. 6.

2 Gas-Kinetic Scheme

2.1 Gas Model at the Interface Between Two Computational Cells

All calculations and theoretical work is based on the Gas-Kinetic Scheme put forward by Xu in 2001 in Ref. [49]. Xu and his group have been improving and extending this scheme over the past twenty years, mostly adapting it to particular spatial schemes, enhancing the convergence rate and showing special applications. Following previous studies, the gas state is represented by a distribution function $f(x, v, t)$, whereas the conservative variables $w = [\rho \ \rho v_1 \ \rho v_2 \ \rho v_3 \ \rho E]^T$ are as customary obtained by taking moments of f:

$$w = \int \psi f d\Xi, \tag{1}$$

where $d\Xi = du_1 du_2 du_3 \, d\xi$ is the elementary volume in phase space and the vector ψ is:

$$\psi = \left[1 \quad u_1 \quad u_2 \quad u_3 \quad \frac{1}{2}\left({u_i}^2 + \xi^2\right)\right]^T, \tag{2}$$

where the internal degrees of freedom of the molecules are indicated by ξ. A time integration over the interval $0 - \Delta t$ provides the fluxes $\mathcal{F}$ through to a unit interface line normal to direction n:

$$\mathcal{F}_n = \int_0^{\Delta t} \int f \psi u_n d\Xi \, dt. \tag{3}$$

The key assumption here is to introduce the solution to the Boltzmann-BGK equation [3] for f is assumed:

$$\frac{\partial f}{\partial t} + (u \cdot \nabla) f = \frac{f^{eq} - f}{\tau}, \tag{4}$$

where τ is the usual relaxation time referred to the "average" fluctuation period which in laminar flow can be assumed to be proportional to molecular viscosity divided by static pressure:

$$\tau = \frac{\mu}{p}, \tag{5}$$

the distribution f^{eq} is a Maxwellian, providing the reference thermal equilibrium state:

$$f^{eq} = \rho \left(\frac{\lambda}{\pi}\right)^{\frac{N+2}{N}} \exp\left[-\lambda \left((u_i - v_i)^2 + \xi^2\right)\right], \tag{6}$$

the parameters $\lambda = \rho/2p = 1/2T$ and N represent the number of effective degrees of freedom of the molecules, respectively. Additional details are found in Refs. [13, 48, 49]. Equation 4 includes the transport terms and a relaxation process towards thermal equilibrium through the effects of molecular collisions. The solution to Eq. 4 is sought in terms of a series expansion of f around the Maxwellian distribution: the Chapman–Enskog expansion [10, 48] is another key step in the derivation of the scheme and is here used to model the deviations from the thermal equilibrium. In principle any expansion orders may be used in the derivation of a gas-kinetic scheme. For instance, a Chapman–Enskog expansion to the zeroth-order, i.e.:

$$f = f^{eq}, \tag{7}$$

ignores the thermal fluctuations and hence provides the Euler representation of a gas (one can insert Eq. 7 into Eq. 4 and take the moment as in Eq. 1). As the following step, we might consider a first-order expansion, which includes the linear "collisions" term:

$$f = f^{eq} - \epsilon\hat{\tau}Df^{eq} + \ldots, \tag{8}$$

where $Df^{eq} = \partial f/\partial t + u_i \partial f/\partial x_i$ (repeating the second term for all dimensions), a more formal presentation is found in Ref. [10]. We use $\hat{\tau}$ for the reference relaxation time and ϵ to measure of the deviations from thermal equilibrium. Equation 4 with a first-order Chapman–Enskog expansion leads to the derivation of the Navier–Stokes equations, as is well known [10, 48].

Further, the Chapman–Enskog expansion to the second-order:

$$f = f^{eq} - \epsilon\hat{\tau}Df^{eq} + \epsilon^2\hat{\tau}D\left(\hat{\tau}Df^{eq}\right) + \ldots. \tag{9}$$

allows recovering the Burnett equations [10]. The third truncation order, not shown here, is associated with the Super Burnett set of equations. These higher-order expansions provide a more accurate model of unresolved fluctuations; however, no convincing implementation to rarefied have been so far proposed (a discussion is found in Ref. [10]). Interestingly, higher-order expansions are being used with gas-kinetic schemes [23, 32] pursuing a higher accuracy. The schemes used by the author have used so far the first-order expansion. A further key assumption used by Xu [49] is the introduction of an upwind reconstruction, that is, a discontinuous gas state. Xu used the Monotone Upstream-centered Scheme for Conservation Laws (MUSCL). At each time step, an initial distribution f_0 is generated from the gas states in the computing cells and adopting a first-order Chapman–Enskog expansion.

A distribution function f^{BGK} is introduced as the analytical solution to the BGK equation, Eq. 4:

$$f^{BGK}(x_1, x_2, x_3, t, u_1, u_2, u_3, \xi) =$$
$$\frac{1}{\tau}\int_o^t f^{eq}(x_1', x_2', x_3', t, u_1, u_2, u_3, \xi)e^{-(t-t')/\tau}\,dt' +$$
$$+ e^{-t/\tau} f_0(x_1 - u_1 t, x_2 - u_2 t, x_3 - u_3 t), \tag{10}$$

where $x_1' = x_1 - u_1(t - t'),\ \ x_2' = x_2 - u_2(t - t'),\ \ x_3' = x_3 - u_3(t - t')$. The initial distribution f_0 depends on spatial and temporal gradients calculated from the macroscopic variables. For additional details refer to Refs. [35, 49]. Here, the resulting distribution function, solution to the BGK model and compatible with the gas initial states, f^{BGK}, is expressed in the compact form:

$$f^{BGK} = f^{NS} + e^{-t/\tau}\Delta f, \tag{11}$$

f^{NS} is a first-order Chapman–Enskog expansion around an average gas state at the interface; Δf is a function obtained as difference between f^{NS} and a Chapman–Enskog expansion obtained combining the "left" and "right" gas states. The presence of the blending function $e^{-t/\tau}$ is significant and builds the kernel of the "multi-scalar" mechanism inside the gas-kinetic scheme.

2.2 Numerical Fluxes

f^{NS} generates fluxes which might be related to those of a central conventional scheme. Inserting $f = f^{BGK}$ into Eq. 3 the fluxes are finally obtained:

$$\mathcal{F}_n = \mathcal{F}_n^{NS} + \alpha(\epsilon)\mathcal{F}_n^{\Delta}, \tag{12}$$

where $\mathcal{F}_n^{NS}$ are the fluxes from f^{NS} and $\mathcal{F}_n^{\Delta}$ the ones from Δf, in 12 $\alpha(\epsilon) = \epsilon(1 - e^{-1/\epsilon})$ and $\epsilon = \tau/\Delta t$. Equation 12 might recall an upwind scheme where central fluxes are corrected by additional terms, whose weight increases with the flow gradients. We point out that $\epsilon = \tau/\Delta t$ is the ratio between two time scales: τ relates to the relaxation process (hence to diffusion in physical space) and may be considered as a representative time scale of the unresolved scales of motion. Δt (the time step) is related to the smallest resolved time scale. ϵ is therefore the ratio of unresolved-to-resolved time scales. The "multi-scale" mechanism in gas-kinetic schemes is now evident: the larger ϵ, the larger the corrections to the Navier–stokes fluxes. However, this mechanism alone does not suffice for a gas dynamics scheme to provide a physically consistent model of rarefied gas. To this end, the relaxation time must also provide a consistent model of particle collisions. In laminar flows, some investigations have demonstrated the suitability of this gas-kinetic scheme to moderate rarefaction [53].

2.3 Artificial Dissipation

Xu in Ref. [49] introduces artificial dissipation by modifying the relaxation time, in order to provide numerical stability:

$$\tau = \frac{\mu}{p} + C\Delta t \frac{(p^l - p^r)}{(p^l + p^r)}, \tag{13}$$

where p^l and p^r are the pressure values at the interface, Δt is the time step, C is an empirical coefficient. According to the author's experience, the artificial dissipation may be necessary with coarse grids.

We recall that the BGK model implies a unity Prandtl number; the heath flux are corrected for realistic fluids, also following [49].

2.4 Boundary Conditions

A further desirable property of this type of gas-kinetic schemes is given by the straight forward application of boundary conditions. They are set on the conservative variables; in the case of the finite-volume discretization, one or more layers of "ghost" cells (depending on the order of the reconstruction) is added. The values of density and energy are set to model the desired heat flux whereas the velocities in the "ghost" computational cells must guarantee no-penetration and no-slip. For hypersonic flows, Li [22] introduced a kinetic boundary condition allowing slip and temperature difference between the flow and the wall. The hypersonic flow cases shown in this paper were computed with the kinetic boundary condition.

3 Extension of the Gas-Kinetic Scheme to Turbulent Flow

3.1 Relaxation Time Based on Eddy Viscosity

Since the effect of unresolved turbulent scales of motion can be modeled by a diffusion process, in case the eddy viscosity μ_t is available, the relaxation time to use in Eq. 4 may be trivially set to:

$$\tau = \frac{\mu + \mu_t}{p}. \tag{14}$$

A more consistent expression is (Li [44]):

$$\tau = \frac{\mu + \mu_t}{p + 2/3k}, \tag{15}$$

where k is the turbulent kinetic energy. In either case, the introduction of a turbulence model is now essential to make eddy viscosity available; this may be, for instance, a RANS model, as used by the author.

The time scale ratio ϵ is now:

$$\epsilon = \frac{\tau}{\Delta t} = \frac{\mu + \mu_t}{p\,\Delta t}. \tag{16}$$

We point out that ϵ may assume quite larger values, especially in the presence of shocks. We may recall here that the idea of eddy viscosity originates from the analogy between a distribution of particles and a distribution of eddies in a turbulent flow. The rarefaction in Eq. 16 would now imply the lack of a clear separation between the scale of motion of the eddies and of the resolved flow. As gas-kinetic scheme may handle moderate rarefaction, we may already anticipate the advantages brought by these schemes in turbulent flow. Xu [53] put forward the following approach for moderately rarefied flow: a first-order expansion is used as in Eq. 8, whereas τ is adjusted in order to locally achieve second-order accuracy, as in Eq. 9. The generalized relaxation time τ^*, now a function of local gradients, can be enforced in the Chapman–Enskog expansion, Eq. 8:

$$f = f^{eq} - \tau^* D f^{eq}, \tag{17}$$

which is then inserted into the BGK equation Eq. 4 in order to obtain the dependence of τ^* on the flow gradients:

$$\tau^* = \frac{\tau\,(1 - D\tau^*)}{1 + \tau\left(D^2 f^{eq}/D f^{eq}\right)}. \tag{18}$$

As pointed out in Ref. [53], the term $\tau D^2 f^{eq}/D f^{eq}$ can be considered as a length scale ratio, associated with the Knudsen number. To the leading order, Eq. 18 can be reduced to:

$$\tau^* \simeq \frac{\tau}{1 + \tau\left(D^2 f^{eq}/D f^{eq}\right)}. \tag{19}$$

Remarkably, a qualitatively similar result was derived in [11, 12] from different considerations:

$$\tau = \frac{\mu}{p} + \frac{\mu_t}{\rho T\left(1 + \eta^2\right)^{1/2}}, \tag{20}$$

where T is temperature and $\eta = Sk/\varepsilon = S/\omega$ is a time scales ratio, S is a measure of the local velocity gradients. k, ε, and ω are the turbulent kinetic energy, dissipation, and specific dissipation. Considering the leading order and same sign derivatives, Eqs. 19 and 20 (looking only at the turbulent part) both tune the relaxation time as a function of a scale ratio ϵ:

$$\tau^* = \tau \frac{1}{1+\epsilon}. \tag{21}$$

In this study, Eq. 20 has been preferred to Eq. 19 for practical reasons, in order to avoid the calculation of gradients. We may also recall that the unresolved-to-resolved scales ratio, a term proportional to $\eta = S/\omega$, is conventionally used to adjust or limit the expression of eddy viscosity, e.g., in the limiting value for ω proposed by Wilcox [46].

3.2 *Modification of the Timescales Ratio*

We may assume that the quantity $\epsilon = \tau/\Delta t$ may assume large values in turbulent flows, as a function of the time advancing technique and grid, leading to a lack of robustness and grid dependent results. We propose to replace it with:

$$\epsilon = \frac{\tau}{\widetilde{\tau}}, \tag{22}$$

where $\widetilde{\tau}$ is the assumed timescale of the resolved flow and may be obtained from the gradients of one of the resolved variables (i.e., $\widetilde{\tau} = \rho/D\rho$). Details are explained in [35]. We point out that the expression used in Eq. 22 is similar to the one used in rarefied gas dynamics to estimate the local Knudsen number, with the difference that in this case timescales are used instead of spatial quantities. In Sect. 4, it can be seen that this timescale ratio may assume values up to a few hundredths inside shock layers, that is, a value well above the conventional threshold for rarefaction.

3.3 *Second-Order Turbulent Stress Tensor Obtained from the Second-Order Chapman–Enskog Expansion*

As a final comment, we note that the relation between higher-order Chapman–Enskog expansion and turbulent stress models. As shown by Chen in [12], a gas-kinetic scheme built from a second-order Chapman–Enskog expansion (Eq. 8 truncated to the second-order), generates a non-linear turbulent stress tensor $R_{ij} = -\frac{2}{3}\rho k\delta_{ij} + 2\mu_t S_{ij} + \mu_t^2/(\rho k)R_{ij}^{(2)}$. The second-order term, $R_{ij}^{(2)}$, can be expressed as a function of the velocity fields u_i as:

$$R_{ij}^{(2)} = \left[C_1 \frac{\partial u_i}{\partial x_k} \frac{\partial u_j}{\partial x_k} + C_2 \left(\frac{\partial u_i}{\partial x_k} \frac{\partial u_k}{\partial x_j} + \frac{\partial u_j}{\partial x_k} \frac{\partial u_k}{\partial x_i} \right) + C_3 \frac{\partial u_k}{\partial x_i} \frac{\partial u_k}{\partial x_j} \right]. \quad (23)$$

Chen et al. [12] showed that the second-order Chapman–Enskog expansion leads to values for the numerical C_1, C_2 and C_3 similar to those selected by turbulence researchers, but based on empirical methods (refer, for instance, to Ref. [40] and references therein). This represents a further advantage provided by gas-kinetic schemes, which has not yet been implemented.

3.4 Boundary Conditions for Turbulent Flow

No particular modification to the conventional boundary conditions used in RANS computations is strictly necessary for simulations of turbulent flow with a gas-kinetic scheme.

4 Gas-Kinetic Schemes for RANS

The author presented a number of test cases calculated with a conventional RANS scheme and the gas-kinetic scheme; these are found in Refs. [35–39]. The associated RANS model was a k-ω by Wilcox [45] in all cases.

The most relevant test cases presented by the author are listed in Table 1. They include simulations of the flow around aerofoils in transonic regime, the flow in a "bump" channel also in transonic regime and several test cases in supersonic regime up to Mach 5. For convenience, two of the test cases listed in Table 1 are re-proposed in this paper, namely one airfoil case and the compression corner at Mach 5.

All cases are characterized by the interaction between a shock wave and the boundary layer leading to the separation and reattachment of the boundary layer. The turbulent structures in the boundary layer are strongly affected by the shock wave and subsequently change their spatial and temporal scales of motion dramatically in the separated flow. As is well known, these conditions are very different from those for which the evolution equations for the turbulent quantities in RANS models are traditionally defined and tuned. The book by Babinsky and the review by Bendiksen provide valuable material on the performance of RANS models in these conditions [1, 2]. The several Workshops hosted by the AIAA over the last ten years (e.g., "Drag Reduction Workshop," "Aeroelastic Prediction Workshop") also showed the clear limits of these turbulence models. As an example, the summary paper by Heeg [21] presents a striking example of the "failure" of well known models such as SA [42] and SST [29] to predict the unsteady flow characterized by shock induced separation of the boundary layer. As a side comment, it might

Table 1 Summary of flow cases presented by the author. [1] h=bump height, θ=momentum thickness, δ = displacement thickness; flow cases 2 and 4 also include a sensitivity analysis to Reynolds number. [2] Délery bump channel: freestream Mach = 0.615, the flow reaches $M \simeq 1.45$ before the shock

Case	Reynolds[1]	Mach[2]	Ref.
RAE2822 Case 9 [14]	$Re = 6{,}200{,}000$	0.72	[34]
RAE2822 Case 10 [14]	$Re = 6{,}300{,}000$	0.73	[34]
NACA0012 [20]	$Re = 9{,}000{,}000$	0.799	[34]
Délery bump channel (Case C) [15]	$Re_h = 1{,}000{,}000$	0.615	[37]
Supersonic compression corner $\alpha = 8°, 16°, 20°, 24°$ [41]	$Re_\theta = 23{,}000$	2.85	[37]
Supersonic compression corner $\alpha = 16°$ [41]	$Re_\theta = 23{,}000$	2.85	[35]
Supersonic compression corner $\alpha = 20°$ [41]	$Re_\theta = 23{,}000$	2.85	[35]
Supersonic compression corner[4] $\alpha = 24°$ [41]	$Re_\theta = 23{,}000$	2.85	[35]
Supersonic compression corner $\alpha = 24°$ [4]	$Re_\theta = 2400$	2.90	[37]
Impinging shock $\alpha = 12°$ [4]	$Re_\theta = 2400$	2.90	[37]
Impinging shock $\alpha = 7°, 8°, 8.8°, 9.5°$ [17]	$Re_\theta = 6900$	2.3	[37]
Reflected shock $\alpha = 8°$ [17]	$Re_\theta = 6900$	2.3	[35]
Reflected shock $\alpha = 8.8°$ [17]	$Re_\theta = 6900$	2.3	[35]
Reflected shock $\alpha = 9.5°$ [17]	$Re_\theta = 6900$	2.3	[35]
Supersonic compression corner $\alpha = 28°$ [16]	$Re_\delta = 877{,}000$	4.95	[37]

be worthwhile mentioning that one of the unofficial outcomes of these workshops is that the simplest model of all, the one presented by Spalart and Allmaras and known as SA, often performs better than the other ones in flow conditions considered difficult for RANS model or even "off-design". The SA model, unlike other popular models such as SST and k-ω (used by the author), only models the eddy viscosity and includes strongly non-linear functions of the strain rate to control turbulence production and destruction. Whereas there are no similarities whatsoever between the SA turbulence model and the gas-kinetic scheme, we observe that both introduce flow-dependent "corrections" into conventional models. The gas-kinetic scheme also modifies the stress tensor.

4.1 Flow Around a Supercritical Aerofoil in Transonic Regime

Case 10 is a well known test case extracted from the measurement campaign carried out by Cook [14] with the supercritical airfoil RAE2822. It is conducted at slightly higher Mach number (0.745) and similar angle of attack and Reynolds number with respect to another well known benchmark case (Case 9). The solution was obtained on a set of converging grids (ranging from 625 × 125 to 928 × 160, with a cell-clustering at the expected shock position).

The higher Mach number is sufficient to cause a proper separation. As a matter of fact, the experimental data does not include any clear indication as to the extension

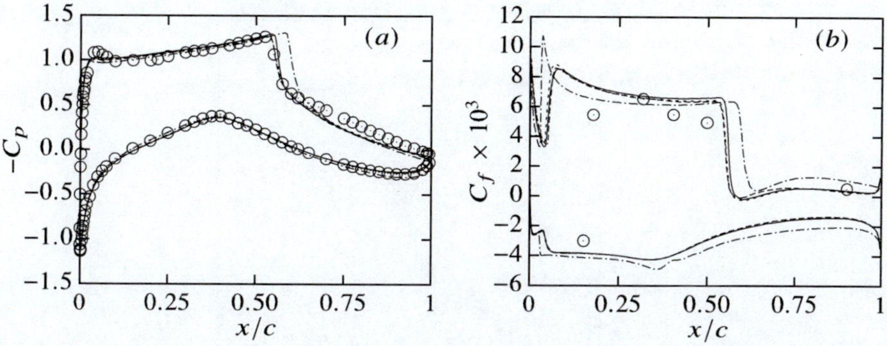

Fig. 1 Super critical airfoil RAE2822, measurements from Cook [14]. Pressure (**a**) and skin friction coefficient (**b**) for the Délery bump channel flow. (solidrule) Gas-kinetic scheme (GKS) on finest grid, (thindashedrule) GKS on medium grid, (verythindashedrule) GKS on coarsest grid, (dotdashedrule) Navier–Stokes (Roe's approximate Riemann solver) on finest grid, (open circle): experimental data

of the separated area. In the literature, results obtained with conventional schemes and two-equation models do not provide a single picture. However, most of the numerical experiments fail to predict the measured boundary layer thickening, incipient separation and consequent shift upstream of the shock (of a few percent of the chord). The results obtained in this study with the conventional RANS scheme fall within the expected range. It predicts a shock position slightly downstream of the measured point. Remarkably, the gas-kinetic scheme provides a more accurate result as can be seen in Fig. 1 Outside of the interaction region, both solvers are in good agreement with each other and with the experiments as can be observed. It is well known that more sophisticated turbulence models, such as algebraic stress model, for instance, might be able in this case to improve the performance of conventional schemes.

4.2 Supersonic Compression Corner at Mach 5

This flow case, investigated by Dolling et al. [16], has been selected to extend the Mach Number range to the borders of hypersonic flow. This flow case has been, however, calculated assuming adiabatic wall conditions. Results from RANS and hybrid simulations can be found in Edwards et al. [18]. In Fig. 2 the pressure distribution predicted by the turbulent gas-kinetic scheme is compared to the experimental values by Dolling et al. [16], highlighting an acceptable agreement. Figure 3 shows the distribution of the degree of rarefaction defined in Eq. 22. This quantity reaches values close to 0.07 in the areas where the strongest gradients appear, that is, inside shock layers. A conventional threshold for rarefaction is considered to be 0.001. It is well known that conventional RANS simulations often

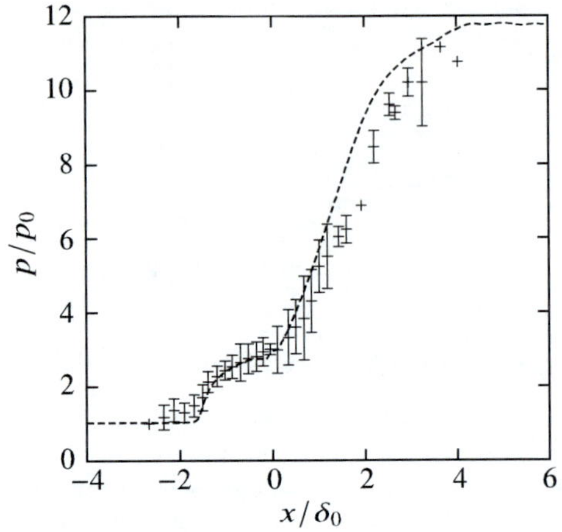

Fig. 2 Compression corner $M = 5$, Experimental values from Dolling et al. [16]. Results obtained on a grid with size: 384×160

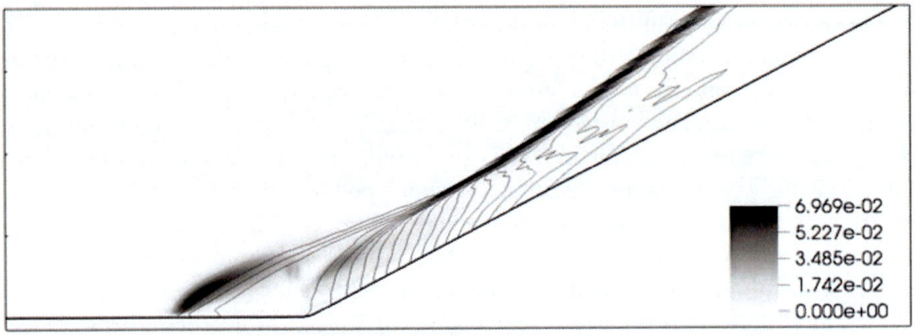

Fig. 3 Degree of rarefaction according to Eq. 22 in Flow Case 6, supersonic compression corner, $M = 4.95$ (iso-contours in grayscale). 20 pressure contour lines have been added for reference

fail to properly place the shock; this shortcoming is attributed to the fact that the flow is much different from the models "design" conditions, i.e., attached boundary layers.

5 Discussion

The conclusion from all simulations is that the gas-kinetic scheme would perform systematically better than the conventional RANS scheme. Incidentally, the ratio of unresolved-to-resolved scales of motion—being the unresolved scales of motion those modeled by the RANS model and the resolved ones those associated with the steady-state RANS solution—appears to be larger in correspondence in the interaction region. This is of course not surprising; however, it is also an indicator for the corrections introduced by the gas-kinetic scheme.

When surveying the most recent publications, two stand out: the one by Cao [6], which reports a general good agreement with experimental results obtained from RANS simulations with a gas-kinetic scheme. A RAE2822 test case is also presented. The very recent paper by Liu [25] is also interesting: it presents results of RANS simulations obtained in hypersonic turbulent flow: In this case, two innovations were introduced: (i) the Langtry [30] transitional version of the SST turbulence model was used, and (ii) remarkably the fluxes from the equation for the turbulent kinetic energy were also calculated with a gas-kinetic scheme. The test cases were chosen among the available supersonic and hypersonic ramps.

We shall also comment on the application of the gas-kinetic scheme for DNS, such as those presented in these recent papers [5, 7–9, 24, 33, 57]. The very satisfactory results are definitely due to the high spatial and temporal accuracy achieved by the scheme (for given grid and order of reconstruction). However, we might also argue that the physically exact gas-evolution contributes. Further, the gas-kinetic scheme was also applied to LES and hybrid RANS-LES simulations, mostly by Li [44]. The definition of the turbulence length scale used to switch from RANS to LES is a conventional one in [44]; however, it is theoretically possible to exploit the gas-evolution stage of the scheme to improve on it (this was the object of a private communication with Li).

We finally mention two very active areas of research, to which gas-kinetic schemes may contribute. Gas-kinetic schemes may play a role in the ongoing Machine Learning supported data-driven efforts to improve on RANS, LES, and hybrid RANS-LES models [28, 43]. The turbulent stress tensor built by gas-kinetic schemes provides more physically meaningful calibration options, such as the order of the Chapman–Enskog expansion, the estimate of the "rarefaction indicator" or the use of limiters to control the gradients of the conservative variables. For the same reasons, gas-kinetic schemes may be useful in the propagation of the epistemic uncertainties stemming from turbulence modeling. Refer to the review by Xiao [47], to the references therein and, in particular, to Mishra [31].

6 Conclusions

Gas-kinetic schemes offer a number of advantages to the simulation of turbulent flow, following both the RANS approach and working in frameworks such as LES, DNS, or hybrid RANS-LES where the numerical resolution covers a portion or the entirety of the turbulent spectrum.

In previous publications the author could show that the ratio of resolved-to-unresolved turbulent scales of motion may be exploited by gas-kinetic schemes and make the way RANS models work physically more acceptable, even and especially in the presence of particular flow patterns such as those appearing in the interaction between a shock wave and boundary layers in transonic flows.

These results appear more meaningful in the light of recent research showing (i) the growing interest in the turbulent applications of gas-kinetic schemes and (ii) the advantage of exploiting the interaction between the gas-evolution and the turbulence model. It goes without saying that much more work is necessary in order to fully understand the potential of these schemes which might go beyond correcting the behavior of RANS turbulence models.

Finally, it is worth mentioning that the emergence of data-driven analysis may provide further scope for gas-kinetic schemes. Being the turbulent stress tensor affected by fundamental choices such as the Chapman–Enskog order and the reconstruction order, data-driven machine-learning tuning of model coefficients could be carried out for different gas-kinetic schemes types and provide more robust results. Further, the propagation of epistemic uncertainties associated with turbulence modeling may be assessed by introducing perturbations of the turbulent stress tensor and measuring their effects on quantities of interest. To this end, the higher physical consistence of gas-kinetic schemes provides more options than conventional schemes.

References

1. Babinsky, H., Harvey, J.K.: Shock Wave-Boundary-Layer Interactions. Cambridge University Press, New York (2011)
2. Bendiksen, O.O.: Review of unsteady transonic aerodynamics: theory and applications. Progr. Aerosp. Sci. **47**(2), 135–167 (2011)
3. Bhatnagar, P.L., Gross, E.P., Krook, M.: A model for collision processes in gases. I. Small amplitude processes in charged and neutral one-component systems. Phys. Rev. **94**(3), 511–525 (1954)
4. Bookey, P., Wyckham, C., Smits, A.: Experimental investigations of Mach 3 shock-wave turbulent boundary layer interactions. AIAA Paper No. 2005-4899 (2005)
5. Cao, G., Pan, L., Xu, K.: Three dimensional high-order gas-kinetic scheme for supersonic isotropic turbulence i: criterion for direct numerical simulation. Comput. Fluids **192**, 104273 (2019)
6. Cao, G., Su, H., Xu, J., Xu, K.: Implicit high-order gas kinetic scheme for turbulence simulation. Aerosp. Sci. Technol. **92**, 958–971 (2019)
7. Cao, G., Pan, L., Xu, K.: Three dimensional high-order gas-kinetic scheme for supersonic isotropic turbulence II: Coarse-graining analysis of compressible Ksgs budget. J. Comput. Phys. **439**, 110402 (2021)
8. Cao, G., Pan, L., Xu, K., Wan, M., Chen, S.: Non-equilibrium time-relaxation kinetic model for compressible turbulence modeling (2021). Preprint arXiv:2112.08873
9. Cao, G., Pan, L., Xu, K.: High-order gas-kinetic scheme with parallel computation for direct numerical simulation of turbulent flows. J. Comput. Phys. **448**, 110739 (2022)
10. Cercignani, C.: The Boltzmann Equation and Its Applications. Springer, New York (1988)
11. Chen, H., Kandasamy, S., Orszag, S., Shock, R., Succi, S., Yakhot, V.: Extended Boltzmann kinetic equation for turbulent flows. Science **301**(5633), 633–636 (2003)
12. Chen, H., Orszag, S.A., Staroselsky, I., Succi, S.: Expanded analogy between Boltzmann kinetic theory of fluids and turbulence. J. Fluid Mech. **519**(1), 301–314 (2004)
13. Chou, S.Y., Baganoff, D.: Kinetic flux–vector splitting for the Navier–Stokes equations. J. Comput. Phys. **130**(2), 217–230 (1997)

14. Cook, P.H., McDonald, M.A., Firman, M.C.P.: Aerofoil RAE 2822–pressure distributions, and boundary layer and wake measurements. Experimental data base for computer program assessment. AGARD Advisory (1979)
15. Délery, J.: Experimental investigation of turbulence properties in transonic shock/boundary-layer interactions. AIAA J. **21**, 180–185 (1983)
16. Dolling, D.S., Erengil, M.E.: Unsteady wave structure near separation in a Mach 5 compression rampinteraction. AIAA J. **29**(5), 728–735 (1991)
17. Dupont, P., Haddad, C., Debiève, J.F.: Space and time organization in a shock-induced separated boundary layer. J. Fluid Mech. **559**, 255–278 (2006)
18. Edwards, J.R.: Numerical simulations of shock/boundary layer interactions using time-dependent modeling techniques: a survey of recent results. Progr. Aerosp. Sci. **44**(6), 447–465 (2008)
19. Guo, Z., Xu, K.: Progress of discrete unified gas-kinetic scheme for multiscale flows. Adv. Aerodyn. **3**(1), 1–42 (2021)
20. Harris, C.D.: Two-dimensional aerodynamic characteristics of the NACA 0012 airfoil in the Langley 8 foot transonic pressure tunnel. NASA Technical Memorandum 81-927 (1981)
21. Heeg, J., Chwalowski, P.: Investigation of the transonic flutter boundary of the benchmark supercritical wing. In: 58th AIAA/ASCE/AHS/ASC Structures, Structural Dynamics, and Materials Conference, AIAA SciTech Forum. American Institute of Aeronautics and Astronautics, AIAA–2017–0191, 9–13 January 2017 (2017)
22. Li, Q., Fu, S., Xu, K.: Application of gas-kinetic scheme with kinetic boundary conditions in hypersonic flow. AIAA J. **43**(10), 2170–2176 (2005)
23. Li, Q., Xu, K., Fu, S.: A high-order gas-kinetic Navier–Stokes flow solver. J. Comput. Phys. **229**(19), 6715–6731 (2010)
24. Liu, H., Cao, G., Chen, W., Agarwal, R.K., Zhao, W.: Gas-kinetic scheme coupled with turbulent kinetic energy equation for computing hypersonic turbulent and transitional flows. Int. J. Comput. Fluid Dyn. **35**(5), 319–330 (2021)
25. Liu, H., Agarwal, R.K., Chen, W.: Computation of hypersonic turbulent and transitional flows using an extended gas kinetic scheme. In: AIAA SCITECH 2022 Forum, p. 1050 (2022)
26. Mandal, J.C., Deshpande, S.M.: Kinetic flux vector splitting for Euler equations. Comput. Fluids **23**(2), 447–478 (1994)
27. May, G., Srinivasan, B., Jameson, A.: An improved gas-kinetic BGK finite-volume method for three-dimensional transonic flow. J. Comput. Phys. **220**(2), 856–878 (2007)
28. McConkey, R., Yee, E., Lien, F.-S.: A curated dataset for data-driven turbulence modelling. Sci. Data **8**(1), 1–14 (2021)
29. Menter, F.R.: Improved two-equation k-omega turbulence models for aerodynamic flows. NASA STI/Recon Technical Report N **93**, 22809 (1992)
30. Menter, F.R., Langtry, R.B., Likki, S.R., Suzen, Y.B., Huang, P.G., Völker, S.: A correlation-based transition model using local variables–part i: model formulation. J. Turbomach. **128**, 413–422 (2006)
31. Mishra, A.A., Mukhopadhaya, J., Iaccarino, G., Alonso, J.: Uncertainty estimation module for turbulence model predictions in su2. AIAA J. **57**(3), 1066–1077 (2019)
32. Ohwada, T., Xu, K.: The kinetic scheme for the full-Burnett equations. J. Comput. Phys. **201**(1), 315–332 (2004)
33. Pan, L., Cao, G., Xu, K.: Fourth-order gas-kinetic scheme for turbulence simulation with multi-dimensional WENO reconstruction. Comput. Fluids **221**, 104927 (2021)
34. Righi, M.: A Gas-Kinetic Scheme for Turbulent Flow. AIAA Paper No. 2014-3330 (2014)
35. Righi, M.: A modified gas-kinetic scheme for turbulent flow. Commun. Comput. Phys. **16**(1), 239–263 (2014)
36. Righi, M.: A Numerical Scheme for Hypersonic Turbulent Flow . AIAA Paper No. AIAA 2015-3341 (2015)
37. Righi, M.: A gas-kinetic scheme for turbulent flow. Flow Turbulence Combust. **97**(1), 121–139 (2016)
38. Righi, M.: Turbulence Modelling in Aeroelastic Problems. ERCOFTAC ETMM11 (2016)

39. Righi, M., Wang, R.: A gas-kinetic scheme for the simulation of turbulent flows. In: Fan, J. (ed.) Proceeding of the 29th International Symposium on Rarefied Gas Dynamics, Xi'an, pp. 1363–1370. American Institute of Physics, College Park (2014)
40. Rubinstein, R., Barton, J.M.: Nonlinear Reynolds stress models and the renormalization group. Phys. Fluids A Fluid Dyn. (1989–1993) **2**(8), 1472–1476 (1990)
41. Settles, G.S., Fitzpatrick, T.J., Bogdonoff, S.M.: Detailed study of attached and separated compression corner flowfields in high Reynolds number supersonic flow. AIAA J. **17**(6), 579–585 (1979)
42. Spalart, P., Allmaras, S.: A one-equation turbulence model for aerodynamic flows. 30th Aerospace Sciences Meeting and Exhibit (1992). https://arc.aiaa.org/doi/abs/10.2514/6.1992-439
43. Taghizadeh, S., Witherden, F.D., Girimaji, S.S.: Turbulence closure modeling with data-driven techniques: physical compatibility and consistency considerations. New J. Phys. **22**(9), 093023 (2020)
44. Tan, S., Li, Q., Xiao, Z., Fu, S.: Gas kinetic scheme for turbulence simulation. Aerosp. Sci. Technol. **78**, 214–227 (2018)
45. Wilcox, D.C.: Turbulence Modeling for CFD, 3rd edn. DCW Industries, La Canada (2006)
46. Wilcox, D.C.: Formulation of the kw turbulence model revisited. AIAA J. **46**(11), 2823–2838 (2008)
47. Xiao, H., Cinnella, P.: Quantification of model uncertainty in rans simulations: a review. Progr. Aerosp. Sci. **108**, 1–31 (2019)
48. Xu, K.: Gas-kinetic schemes for unsteady compressible flow simulations. In: VKI, Computational Fluid Dynamics, Annual Lecture Series, 29th, Rhode-Saint-Genese, Belgium (1998)
49. Xu, K.: A gas-kinetic BGK scheme for the Navier–Stokes equations and its connection with artificial dissipation and Godunov method. J. Comput. Phys. **171**(1), 289–335 (2001)
50. Xu, K.: A Unified Computational Fluid Dynamics Framework from Rarefied to Continuum Regimes. Cambridge University Press, Cambridge (2021)
51. Xu, K., Prendergast, K.H.: Numerical Navier-Stokes solutions from gas kinetic theory. J. Comput. Phys. **114**(1), 9–17 (1994)
52. Xu, K., Mao, M., Tang, L.: A multidimensional gas-kinetic BGK scheme for hypersonic viscous flow. J. Comput. Phys. **203**(2), 405–421 (2005)
53. Xu, K., He, X., Cai, C.: Multiple temperature kinetic model and gas-kinetic method for hypersonic non-equilibrium flow computations. J. Comput. Phys. **227**(14), 6779–6794 (2008)
54. Xuan, L., Xu, K.: A new gas-kinetic scheme based on analytical solutions of the BGK equation. J. Comput. Phys. **234**, 524–539 (2013). https://doi.org/10.1016/j.jcp.2012.10.007
55. Yang, X., Shyy, W., Xu, K.: Unified gas-kinetic wave-particle method for gas-particle two phase flow from dilute to dense solid-particle limit (2021). Preprint arXiv:2112.01829
56. Yang, X., Ji, X., Shyy, W., Xu, K.: Comparison of the performance of high-order schemes based on the gas-kinetic and HLLC fluxes. J. Comput. Phys. **448**, 110706 (2022)
57. Zhao, W., Wang, J., Cao, G., Xu, K.: High-order gas-kinetic scheme for large eddy simulation of turbulent channel flows. Phys. Fluids **33**(12), 125102 (2021)
58. Zhao, F., Ji, X., Shyy, W., Xu, K.: A compact high-order gas-kinetic scheme on unstructured mesh for acoustic and shock wave computations. J. Comput. Phys. **449**, 110812 (2022)

Density Functional Kinetic Theory for Soft Matter

Sauro Succi, Fabio Bonaccorso, Mihir Durve, Marco Lauricella, Andrea Montessori, and Adriano Tiribocchi

Abstract In the last decades kinetic theory has developed into a very elegant and effective framework to handle a broad spectrum of problems involving complex states of flowing matter, far beyond the original realm of rarefied gas dynamics. In this paper, we present recent applications of the lattice Boltzmann method to the computational design of soft mesoscale materials, including soft flowing crystals, dense multicore emulsions, as well as Petascale simulations of deep-sea glassy sponges. This manuscript is a tribute to the groundbreaking work of Carlo Cercignani and his undiminished impact on modern non-equilibrium statistical physics.

1 Introduction

The kinetic theory of gases in its original form as devised by Ludwig Boltzmann was restricted to binary collisions to pointlike particles hence formally limiting its application to dilute gases [1–3].

Subsequent attempts to extend it to dense gases and liquids were notoriously plagued by several problems, mostly due to infinities arising in the treatment of higher order collisions.

Several strategies have been developed over the years to cope with such problems, but the kinetic theory of dense, heterogeneous fluids remains a difficult

S. Succi (✉) · M. Durve
Center for Life Nano- & Neuro-Science, Fondazione Istituto Italiano di Tecnologia (IIT), Rome, Italy

F. Bonaccorso
Department of Physics, University of Rome "Tor Vergata" and INFN, Rome, RM, Italy

M. Lauricella · A. Tiribocchi
Istituto per le Applicazioni del Calcolo CNR, Rome, Italy

A. Montessori
Department of Engineering, Roma Tre University, Rome, Italy

P. Barbante et al. (eds.), *From Kinetic Theory to Turbulence Modeling*, Springer INdAM Series 51, https://doi.org/10.1007/978-981-19-6462-6_19

subject to this day [3–5]. A similar statement applies to complex flows with interfaces often encountered in science, engineering, soft matter, and biology. Soft droplet-based materials, namely assemblies of close-packed deformable droplets or bubbles separated by thin fluid films, include dense emulsions, foams, gels, soft granular materials, liquid crystals, and a number of biological (active) materials that are ubiquitous in industries, forming food and cosmetic products and in nature. These materials share an important common feature in that predominant physical behaviors occur at an energy scale comparable with room temperature thermal energy. A particularly interesting framework to deal with such complex flows is provided by density functional theory (DFT) [6]. Essentially the idea is that much of the physics of the complex many-body problem associated with dense fluids can be explored by investigating the dynamics of the fluid density, namely a single one-body scalar field. Of course, such dynamics is subject to self-consistent closures, typically in the form of well-educated guesses on the generating functional from which the effective one-body equation for the density can be derived via standard functional minimization of the suitable free-energy-functional (FEF). Density functional theory has met with spectacular success for the case of quantum many-body problem, leading to the development of powerful theorems and attending computational methods that still form the basis for modern computational quantum chemistry [7]. The classical version, albeit less spectacular than its classical counterpart, also provides a milestone framework to describe and simulate complex flows with interfaces, which we now proceed to illustrate in some more detail [6].

2 Density Functional Theory

The stepping stone of DFT is the free-energy functional

$$\mathcal{F}[\rho] = \int_V [\phi(\rho) + \frac{\kappa}{2}(\nabla\rho)^2]dx \tag{1}$$

where $\phi(\rho)$ is the bulk free-energy, a local function of the fluid density $\rho(x)$ and the second term represents the cost associated with the build up of interfaces within the fluid. The bulk component of the FEF determines the non-ideal equation of state via the Legendre transform $p = \rho d\phi/d\rho - \phi$, while the interface term fixes the surface tension.

For the case of a binary mixture of components A and B, it proves expedient to defined a compact order parameter

$$\phi = \frac{\rho_A - \rho_B}{\rho_A + \rho_B} \tag{2}$$

which varies between 1 in phase A and -1 in phase B. Such order parameter is conserved, hence it obeys a corresponding continuity equation of the form

$$\partial_t \rho = -\partial_x [\mu \partial_x \frac{\delta \mathcal{F}[\phi]}{\delta \phi}] \tag{3}$$

where δ denotes functional derivative.

The order parameter is convected by the barycentric velocity of the two species, $u(x, t)$, which obeys the standard Navier-Stokes equations of fluids, with an extra non-ideal pressor, known as Korteweg tensor, formally given by $K_{ab}[\rho] = \frac{\delta^2 \mathcal{F}}{\delta g_a \delta g_b}$, where $g_a \equiv \nabla_a \rho$. The divergence of the pressor determines the mechanical force acting upon the fluid interfaces, $F_a[\rho] = -\partial_b P_{ab}$ and the condition $F_a = 0$ selects the density profile realizing mechanical equilibrium of the interface.

The explicit form of the Korteweg tensor is as follows

$$K_{ab} = [p + \frac{\kappa}{2}(\partial_a \phi)^2 - \kappa \phi \Delta \phi]\delta_{ab} - \kappa \partial_a \phi \partial_b \phi \tag{4}$$

where the Latin indices a and b run over the spatial dimensions.

The continuity equation (3), jointly with the non-ideal Navier-Stokes equations provides a self-consistent mathematical framework for the quantitative description of the dynamics of the binary mixture.

What is the role of kinetic theory in the above picture?

In principle, the usual one, namely provide a bridge between the molecular description and the hydrodynamic one. And again, taking this task in a bottom-up guise, is by no means an easy enterprise. In the following, we shall take a more pragmatic top-down standpoint, namely design suitable kinetic equations capable of reproducing the non-ideal Navier-Stokes equations, without deriving them from an underlying microscale model.

3 Density Functional Kinetic Theory

DFT gives rise to a variety non-ideal hydrodynamic equations as covered in Hohenberg and Halperin [7], with no reference to kinetic theory.

In the framework of kinetic theory, the main ingredient is the non-ideal force associated with the forced-streaming term $S = F_a[\rho]\partial_{v_a} f$.

One natural question arises: why would the kinetic formulation be computationally advantageous over the hydrodynamic one?

The main point is that in the kinetic framework the term S can be brought to the right hand side and treated as a soft-collision term. Note that the partial derivative in velocity space is handled by integrated by parts, This permits to move the distribution function along unperturbed, force-free (straight) characteristics

$\Delta x_v = v\Delta t$ and include the effect of soft forces as a local correction/perturbation to this free-streaming motion.

In equations:

$$f(x + v\Delta t, v, t + \Delta t) - f(x, v, t) = (C - S)dt \tag{5}$$

where C stands for standard short-range collisions, and vector indices have been omitted for simplicity. A very popular choice is the single-time relaxation Bhatnagar-Gross-Krook model [3, 8]

$$C = (f^{eq} - f)/\tau$$

where f^{eq} is the local equilibrium and τ the relaxation time, namely the characteristic lifetime of non-equilibrium excitation [9–11]. The soft-collision term is conveniently turned into an algebraic source term

$$S = F_a[\rho] \sum_l s_l a H_l(v)$$

by integrating by parts in velocity space and exploiting recurrence relations of tensor Hermite polynomials [8, 12, 13]. In the above equation, $s_l a$ is the coefficient of the velocity gradient of f corresponding to the lth Hermite polynomial.

The advantage of the above formulation is that the streaming step at the left hand side proceeds along constant characteristics, hence it is *exact*, i.e., zero round-off error in the numerical treatment. This stands in contrast with the hydrodynamic formulation in which information moves along spacetime dependent material lines defined by the fluid velocity itself, $dx_u = u(x, t)dt$.

This simple but key advantage lies at the heart of the success of lattice kinetic techniques and most notably Lattice Boltzmann method, in which the characteristics are restricted to a suitable set of discrete velocities $\{v_i\}, i = 0, N_v$, showing sufficient symmetry to recover the correct large scale hydrodynamic limit.

Differently restated, the highly complex physics of moving interfaces is entirely absorbed within the local source S.

Clearly, such perturbative treatment is limited to sufficiently weak forces, as gauged by the so-called cell-Froude number

$$Fr = \frac{a\Delta t}{v}$$

where a is the acceleration due to the non-ideal forces. In order to preserve the stability of the numerical scheme, the time step must be chosen such that $Fr \ll 10^{-3}$, a condition which may eventually go broken in the presence of strong density gradients. This problem can be mitigated by improving the time-marching scheme, typically via locally implicit formulations, but it must be watched carefully case by case.

Lattice DFKT as discussed above is currently being used over an amazingly broad spectrum of soft-fluid problems, definitely beyond the original realm of rarefied gas dynamics.

4 Families of Lattice DFKT

In the sequel we provide a brief survey of the three major families of lattice DFKT.

4.1 Free-Energy Lattice Models

The first lattice transcription of the free-energy functional (1) has been proposed back in the mid 90s [14]. The main idea is to write the collision operator in single-relaxation form, i.e., $C = -(f - f^{eq})/\tau$, where f^{eq} is the Maxwell-Boltzmann equilibrium, and incorporate the effect of the soft term S within a generalized non-local equilibrium, reflecting the non-locality of the Korteweg tensor. Ever since its inception, it has generated a wide body of interesting results, especially in multiphase microfluidics.

4.2 Lattice Many-Body Models

An alternative and possibly more straightforward route is to connect with is to write the non-ideal pressor directly in the form of a two-body convolution:

$$P_{ab}(x) = \int r_a \psi(x - \frac{r}{2}) G(x, r) \psi(x + \frac{r}{2}) r_b dr \tag{6}$$

where $G(x, r)$ is the two-body density correlator and $\psi(\rho)$ is a local functional of the density.

Taylor expanding the correlator about $r = 0$ delivers the following series

$$P_{ab}(x) = G_0 \psi^2 \delta_{ab} + G_{2,ab} \nabla_a \psi \nabla_b \psi + \ldots. \tag{7}$$

where $G_0 = \int G(x, r)dr$, $G_{2,ab} = \int G(x, r) r_a r_b dr$ and so on to higher orders.

At variance with the free-energy approach, the correlator is designed top-down, i.e., by reverse-engineering the expression of $G(x, r)$ so as to obtain the desired physical phenomena, namely *(i) Non-ideal EoS, (ii) Tunable surface tension, (iii) Positive disjoining pressure*.

The earliest and still very popular such top-down formulation is due to Shan-Chen [15] and makes use of just a single-parameter correlator, taking the value

G_0 within the first Brillouin lattice cell and zero elsewhere Clearly, a single-parameter model does not permit to tune the three properties (i)–(iii) separately, but it nevertheless showed the way towards a very handy and flexible lattice DFKT which still takes the lion share of LB application to multiphase and multicomponent flows.

The Shan-Chen model has been subsequently extended in many directions, including the formulation of multi-range models in which the correlator extends beyond the first Brillouin region [16]. This permits to tune the surface tension independently of the EoS and also to realize a positive disjoining pressure, i.e., repulsive interaction between approaching interfaces. The multi-range method has met with significant success for the simulation of dry foams and moderately dense emulsions [17].

4.3 *Lattice Chromodynamic DFKT*

The above mentioned multi-range model has found many applications to complex states of flowing matter, particularly foams and wet emulsions. However, when it comes to dense emulsions, spurious effects have been reported on the disjoining pressure, due to lack of sufficient lattice symmetry.

To this regard, a very fruitful avenue turned out to be offered by a entirely rule-driven approach, based on the so-called *Color Gradient* technique ("color" is just a mnemonics for different chemical species or different phases of the same species, in analogy with quantum chromodynamics) [18]. Essentially the idea is to add an explicit anti-diffusive flux sending particles of each species uphill along their density gradient instead of against it. By construction, such anti-diffusive flux helps interface formation against the coalescing effect of surface tension. here too, the parameters can be adjusted to recover the properties i)-iii) above independently.

In the lattice chromodynamics models for multicomponent flows, two sets of distributions functions(let us say red and blue) are introduced to code for two different fluids.

$$f_i^k(\mathbf{x} + \mathbf{c}_i, t + 1) = f_i^k(\mathbf{x}, t) + \Omega_i^k(\mathbf{x}, t) \tag{8}$$

In the above equation, $k = R, B$ denotes the fluid and Ω_i^k is a collision operator which can be written as the combination of three sub-collisional.

$$\Omega_i^k = (\Omega_i^k)^3[(\Omega_i^k)^1 + (\Omega_i^k)^2] \tag{9}$$

where $(\Omega_i^k)^1$ is the BGK collisional, $(\Omega_i^k)^2$ is a two-phase collision operator, generating an interfacial tension between the two immiscible components and $(\Omega_i^k)^3$ is an anti-diffusive operator which favors phase segregation and keeps interfaces sharp.

Moreover, the stress-jump condition across the fluid interface can be augmented with a suitable repulsive term aiming at modelling the effect of short-range, repulsive near-contact forces induced by the presence of surfactants and colloids adsorbed at the fluid-fluid interface. The additional repulsive term can be added efficiently in the LB framework via a forcing term localized at the interface:

$$F_{rep,a} = A_h[h]n_a \tag{10}$$

where $F_{rep,a}$ is the near-contact, repulsive force acting at the fluid interface along the normal n_a and whose magnitude $A_h[h]$ depends on the distance (h) between two interacting interfaces. The interested reader is referred to [19, 20] and related literature. The main advantage in employing such a coarse-grained approach is the possibility of simulating very large droplet systems without compromising the computational efficiency of the numerical framework. Indeed, the introduction of NCIs sidesteps the explicit tracking of different phases interacting within the systems as instead occurs in immersed boundary method and multi-color volume of fluid approach [21–23]. To conclude, the extended approach still holds to a continuum description of the interface dynamics, with the governing equations modified only by the presence of a distributed body force, which can heuristically be interpreted as a coarse-grained version of the short-range molecular forces acting at the nanometer and sub-nanometer scales.

4.3.1 Near-Contact Interactions

For highly dense emulsions anti-diffusive gradient may still fail to prevent coalescence, and a more detailed description of the near-contact interactions which arise when two microscopic bodies, say droplets, come at molecular distances (say a few nanometers). The pressure associated with such forces can be estimated as

$$p_{nc} \sim A\frac{k_B T}{h^3} \tag{11}$$

where h is the film thickness between two approaching interfaces and A an amplitude coefficient scaling the near-contact forces in units of thermal ones. Such pressure withstands the coalescing action due to capillary pressure, given by

$$p_{cap} \sim \frac{\sigma}{R} \tag{12}$$

It has been shown that the ratio $N = p_{nc}/p_{cap}$ plays a major role on the multi-body configuration of the interfaces. For $N \ll 1$ the interfaces tend to coalesce, while at $N \sim 0.1$ they may eventually form an ordered pattern known as soft flowing crystals. For $N > 1$ the pattern loses order giving rise to a dense disordered emulsion.

Needless to say, these three distinct states of soft flowing matter display a very different rheology.

5 Applications: Simulating Dropland

In the following we briefly comment upon just two recent applications of density functional lattice kinetic theory.

5.1 *Dense Emulsions and Soft Granular Materials*

As a first application, we show the capability of the multicomponent model with near-contact interactions to reproduce the formation of ordered droplets clusters in microfluidic channels. In Fig. 1a we reported the formation of multilayer hexagonal droplet clusters in a channel formed by a thin inlet and an outlet chamber (see [24]). The droplets are continuously injected within the main channel by employing a recently developed internal periodic boundary condition. To note that the spontaneous ordering of the droplets into hexagonal clusters is drive by a non-trivial competition between local, short-range, repulsive interactions (i.e., the near-contact forces) and the surface tension.

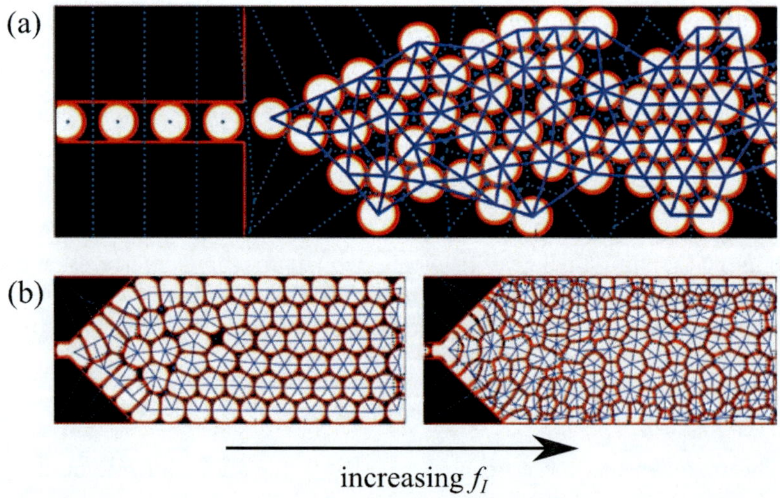

Fig. 1 (**a**) Multi-layer hexagonal droplets clusters in a microfluidic channel. Dashed lines represents Voronoi tessellation while solid lines Delauney triangulation (**b**) Droplet self-assemblies within a microfluidic channel with a divergent opening angle $\alpha = 45°$ for two different inlet channel Capillary numbers ($Ca = 0.04$ left, $Ca = 0.16$ right)

In particular, small disturbances introduced by the short-range repulsive action of the near-contact interaction forces trigger the rupture of the initial, single-file crystal symmetry, driving the droplets towards a new spatial arrangement.

It is interesting to note that the process described above is somehow similar to the instability observed in densely packed granular materials subjected to force unbalances. In panel (b) of Fig. 1 we reported the formation of dense emulsions in microfluidic devices formed by a divergent inlet channel connected to a downstream channel (see [25]). Even in this case the droplets are continuously injected within the system and let free to assemble within the outlet channel. By simply tuning the inlet capillary number it is possible to observe a spontaneous transition from a high-ordered emulsion, formed by hexagonal clusters flowing within the system (i.e., a wet emulsion) to a foam-like, dry-state, structure which results in a neat distortion of the Delauney triangulation (blue solid lines connecting the centers of neighboring droplets).

In the simulations, both the dispersed and continuous phases' discharges are kept constant and the Ca is changed via the surface tension. The observed transition is likely to be due to (i) the breakup processes downstream the injection channel, promoting the formation of liquid films and (ii) the increased deformability of the droplets interface, due to the lower values of surface tension employed. The simulations described above have been performed in two dimensions. The computational domains count, roughly, 3–500,000 nodes and the number of degrees of freedom for each time step, considering a nine speed two-dimensional lattice, is in turn, $3\text{–}5 \cdot 10^6$. The memory occupancy is of order 50–100 megabytes by considering to allocate all the arrays in double precision. Typical simulations run for several millions of cycles.

The computational framework, described in the previous section, has been capable of providing accurate predictions over a number of different dynamical modes observed in recent flow-focusing experiments of high internal phase double emulsions.

As shown in Fig. 2, the ratio between the emulsion flow rate, injected at the inlet of the main channel, and that the bulk fluid flowing through the lateral channels of the flow focuser, controls the onset of a full set of highly non trivial dynamical modes from multilayered, jetting (panel (a)) to the irregular (panel(b)) and regular dripping regime characterized by the metronomical production of roughly equally sized shells with a small number ($\sim$4–5) of cores.

It is worth noting that the simulations and experiments reported here stand as a proof of concept of the possibility of assembling discrete, soft granular, droplets based materials in microfluidic channels in a controlled manner. The simulations performed to predict the formation of SGM have been performed in three dimensions by employing an in-house HPC MPI code run on Galileo100 cluster https://www.hpc.cineca.it/hardware/galileo100. In this case the number of degrees of freedom is $\sim 200 \cdot 10^6$ nodes with a memory occupancy of order 4 GB.

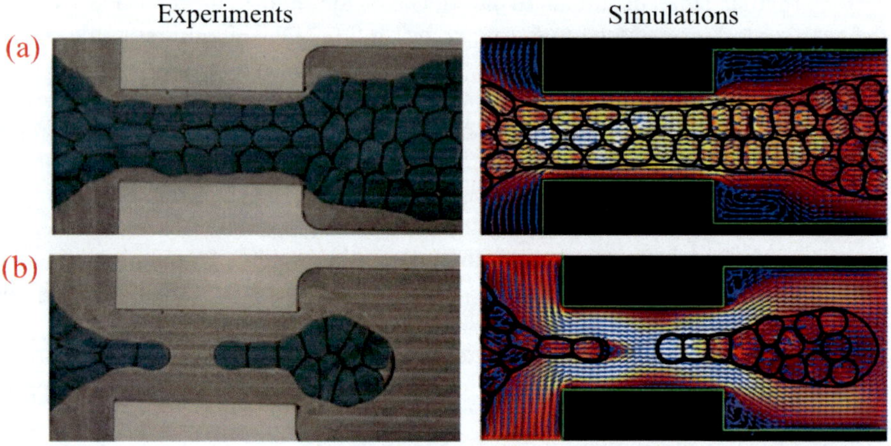

Fig. 2 Hydrodynamic simulations of soft granular materials in flow-focuser devices. For low values of $\phi \sim 1$ the low flow rates of the continuous phase favor the formation of a stable jet, with typical width of 2–3 droplet diameters (panel (**a**)) while by decreasing the inlet to continuous flow rate ratio the system transits spontaneously from a jetting to dripping regime with the formation of smaller clusters of droplets encapsulated in an external shell. The dimensions of the experimental device are $W = 1\,\text{mm}$ (width of the constriction), $L = 2\,\text{mm}$ (Length of the constriction), $H = 0.2\,\text{mm}$ (depth of the device), $W0 = 2\,\text{mm}$ (width of the outlet channel)

6 Summary and Future Outlook

Kinetic theory was originally derived in the restricted framework of dilute gases.

As of today, it has blossomed into a vigorous and ever-growing branch of non-equilibrium statistical mechanics, with countless applications to a broad spectrum of transport problems across various disciplines, such as traffic flows, finance, and soft matter, to name but a few.

In this paper, we have focused on the latter and shown how a judicious merge with density functional theory can lead to a conceptual and computational framework capable of describing soft flowing states of matter whose extreme complexity would be very hard to handle by any other method. This projects a bright future ahead of density functional (lattice) kinetic theory, a future that would have been simply unthinkable without the pioneering foundational work of Carlo Cercignani [26].

Acknowledgments The authors acknowledge funding from the European Research Council under the European Union's Horizon 2020 Framework Programme (No. FP/2014-2020) ERC Grant Agreement No.739964 (COPMAT). A.M. acknowledges the CINECA Computational Grant Iscra C "TRAHIPDE," id. HP10CU803Q under the ISCRA initiative, for the availability of high performance computing resources and support. A.T. acknowledges the CINECA Computational Grant Iscra C "WIMES," under the ISCRA initiative, for the availability of high performance computing resources and support. SS is deeply indebted to Carlo Cercignani for his profound inspirational and practical impact on his scientific trajectory.

References

1. Boltzmann, L.: Lectures on Gas Kinetic Theory, Dover Books Series. Reprint of the University of California Press, Berkeley (1964)
2. Cercignani, C.: Theory and Application of the Boltzmann Equation. Elsevier, Amsterdam (1975)
3. Cercignani, C.: The Boltzmann equation. In: The Boltzmann Equation and Its Applications, pp. 40–103. Springer, New York (1988)
4. Cercignani, C.: Rarefied Gas Dynamics: From Basic Concepts to Actual Calculations, vol. 21. Cambridge University Press, Cambridge (2000)
5. Evans, D.J., Morriss, G.P.: Statistical Mechanics of Nonequilibrium Liquids. ANU Press, Canberra (2007)
6. Bray, A.J.: Theory of phase-ordering kinetics. Adv. Phys. **51**(2), 481–587 (2002)
7. Hohenberg, P.C., Halperin, B.I.: Theory of dynamic critical phenomena. Rev. Modern Phys. **49**(3), 435 (1977)
8. Succi, S.: The Lattice Boltzmann Equation: For Fluid Dynamics and Beyond. Oxford University Press, Oxford (2001)
9. Cercignani, C., Lampis, M., Lorenzani, S.: Variational approach to gas flows in microchannels. Phys. Fluids **16**(9), 3426–3437 (2004)
10. Stefanov, S., Gospodinov, P., Cercignani, C.: Monte Carlo simulation and Navier–Stokes finite difference calculation of unsteady-state rarefied gas flows. Phys. Fluids **10**(1), 289–300 (1998)
11. Cercignani, C., Sharipov, F.: Gaseous mixture slit flow at intermediate Knudsen numbers. Phys. Fluids A Fluid Dyn. **4**(6), 1283–1289 (1992)
12. Succi, S.: The Lattice Boltzmann Equation: For Complex States of Flowing Matter. Oxford University Press, Oxford (2018)
13. Montessori, A., Prestininzi, P., La Rocca, M., Succi, S.: Lattice Boltzmann approach for complex nonequilibrium flows. Phys. Rev. E **92**(4), 043308 (2015)
14. Swift, M.R., Orlandini, E., Osborn, W.R., Yeomans, J.M.: Lattice Boltzmann simulations of liquid-gas and binary fluid systems. Phys. Rev. E **54**(5), 5041 (1996)
15. Shan, X., Chen, H.: Lattice Boltzmann model for simulating flows with multiple phases and components. Phys. Rev. E **47**(3), 1815 (1993)
16. Sbragaglia, M.R.L.S.K., Benzi, R., Biferale, L., Succi, S., Sugiyama, K., Toschi, F.: Generalized lattice Boltzmann method with multirange pseudopotential. Phys. Rev. E **75**(2), 026702 (2007)
17. Benzi, R., Sbragaglia, M., Succi, S., Bernaschi, M., Chibbaro, S.: Mesoscopic lattice Boltzmann modeling of soft-glassy systems: theory and simulations. J. Chem. Phys. **131**(10), 104903 (2009)
18. Gunstensen, A.K., Rothman, D.H., Zaleski, S., Zanetti, G.: Lattice Boltzmann model of immiscible fluids. Phys. Rev. A **43**(8), 4320 (1991)
19. Montessori, A., Lauricella, M., Tirelli, N., Succi, S.: Mesoscale modelling of near-contact interactions for complex flowing interfaces. J. Fluid Mech. **872**, 327–347 (2019)
20. Montessori, A., Lauricella, M., Tiribocchi, A., Succi, S.: Modeling pattern formation in soft flowing crystals. Phys. Rev. Fluids **4**(7), 072201 (2019)
21. Karnakov, P., Wermelinger, F., Litvinov, S., Koumoutsakos, P.: Aphros: High performance software for multiphase flows with large scale bubble and drop clusters. In: Proceedings of the Platform for Advanced Scientific Computing Conference, pp. 1–10 (2020)
22. Patel, H.V., Das, S., Kuipers, J.A.M., Padding, J.T., Peters, E.A.J.F.: A coupled volume of fluid and immersed boundary method for simulating 3D multiphase flows with contact line dynamics in complex geometries. Chem. Eng. Sci. **166**, 28–41 (2017)
23. Lörstad, D., Francois, M., Shyy, W., Fuchs, L.: Assessment of volume of fluid and immersed boundary methods for droplet computations. Int. J. Num. Methods Fluids **46**(2), 109–125 (2004)

24. Montessori, A., Tiribocchi, A., Lauricella, M., Bonaccorso, F., Succi, S.: Mesoscale modelling of droplets' self-assembly in microfluidic channels. Soft. Matt. **17**(9), 2374–2383 (2021)
25. Montessori, A., Tiribocchi, A., Lauricella, M., Bonaccorso, F., Succi, S.: Wet to dry self-transitions in dense emulsions: from order to disorder and back. Phys. Rev. Fluids **6**(2), 023606 (2021)
26. Cercignani, C.: Ludwig Boltzmann: The Man Who Trusted Atoms. Oxford University Press, Oxford (1998)

A Multi-Agent Description of Social Phenomena with Lognormal Equilibria

Giuseppe Toscani

Abstract The lognormal distribution, very common in physical and biological applications, also appears in various phenomena related to economic and social activities. In socio-economics these phenomena describe in most cases the evolution in time of the distribution of a certain attribute of agents, which aim to reach a desired target by repeated attempts. By resorting to the analogies of these problems with the classical kinetic theory of rarefied gases, we aim to illustrate the nature of the microscopic interactions which give rise to a macroscopic lognormal distribution profile.

1 Introduction

The study of random variations that occur in the data from many scientific disciplines commonly show more or less skewed probability distributions, which often closely fit the lognormal distribution [1, 8, 21].

A lognormal distribution is a continuous probability distribution of a non-negative random variable whose logarithm is normally distributed. Its density function depends on two parameters $\mu \in \mathbb{R}$ and $\sigma \in \mathbb{R}_+$, and it is expressed by

$$L(x) = \frac{1}{\sqrt{2\pi\sigma}\, x} \exp\left\{-\frac{(\log x - \mu)^2}{2\sigma}\right\}, \qquad x \in \mathbb{R}_+ \tag{1}$$

As documented by the exhaustive review by Limpert et al. [21], in natural sciences the list of phenomena which fit the distribution (1) is quite long. In addition to samples from physical and biological sciences, a relevant number of phenomena fitting distribution (1) comes from social sciences and economics, areas where it

G. Toscani (✉)
Department of Mathematics "F. Casorati", University of Pavia, and Institute for Applied Mathematics and Information Technologies "E. Magenes", Pavia, Italy
e-mail: giuseppe.toscani@unipv.it

P. Barbante et al. (eds.), *From Kinetic Theory to Turbulence Modeling*, Springer INdAM Series 51, https://doi.org/10.1007/978-981-19-6462-6_20

can be reasonably assumed that the appearance of this distribution is closely linked to the behavior of the agents.

The interest in natural phenomena described by lognormal distribution dates back to more than a century ago. Few years later the contribution by Galton [13] to demonstrate the central limit theorem through the so-called *bean machine*, a similar device for the lognormal distribution was constructed by Kapteyn [20], while studying and popularizing the statistics of the lognormal in order to help visualize it and demonstrate its plausibility. A photograph of this machine is present in [1].

Nevertheless, while the normal distribution is playing an important role in kinetic theory, where it is commonly known with the alternative name of Maxwellian distribution [3], kinetic equations leading to a macroscopic lognormal profile to our knowledge are few and recent [16, 17, 26, 27].

The kinetic modeling of social phenomena leading to lognormal profiles has been recently proposed in [16, 17] by resorting to a linear Boltzmann-type equation based on particular elementary interactions. Other phenomena have been modeled in terms of more classical interactions, similar to the collisions between molecules of rarefied gases [26, 27]. These latest kinetic models have their roots in the powerful theory of Maxwell molecules, and were inspired by the work of Carlo Cercignani.

In one of his last contributions [2], jointly written with Alexander Bobylev and Irene Gamba, Carlo Cercignani considered and studied generalized Maxwell models of the Boltzmann equation in presence of multiple interactions.

The main application of this generalized Boltzmann equation is in an economic context. Specifically, the $n \geq 1$ interacting particles are interpreted as a community of agents participating in economical trades. Denoting with $V = (v_1, v_2, \ldots, v_n)$ the n-dimensional vector of the non-negative wealth of the agents, the post-collisional state $V^* = (v_1^*, v_2^*, \ldots, v_n^*)$ gives the new wealth after a single economic trade. The n-particle interaction considered in [2] is a linear transformation of V into the post-collisional state V^*, given by

$$v_i^* = av_i + b\sum_{j=1}^{n} v_j \qquad i = 1, \ldots, n, \tag{2}$$

where the parameters a, b may be fixed or randomly distributed with a certain number of moments bounded.

Denoting by $f = f(v, t)$ the density of particles with state $v \in \mathbb{R}$ at time $t \geq 0$, and postulating the validity of molecular chaos, the evolution of any *observable quantity* φ, i.e., any quantity which may be expressed as a function of v, is given by the Boltzmann-type equation [2]

$$\frac{d}{dt}\int_{\mathbb{R}_+} \varphi(v)f(v, t)\,dv = \frac{1}{\tau n}\int_{\mathbb{R}_+^n} \sum_{i=1}^{n} \langle \varphi(v_i^*) - \varphi(v_i)\rangle \prod_{j=1}^{n} f(v_j, t)\,dv_1 \ldots dv_n, \tag{3}$$

where τ denotes a relaxation time and $\langle \cdot \rangle$ is the average with respect to the distributions of the random parameters a, b contained in (2). The collision integral depends on a constant *collision kernel*, that corresponds to consider *Maxwell-type interactions*. The right-hand side of (3) takes into account the whole set of microscopic states, and consequently it depends on the n-product of the density functions $f(v_1, t) \cdot \ldots \cdot f(v_n, t)$. Thus, if $n > 1$ the evolution of f obeys a highly non-linear Boltzmann-type equation. The interesting point remarked in [2] is that a considerable simplification occurs in presence of a large number $n \gg 1$ of participants, which results in a linearized version of Eq. (3).

The findings of [2] are an interesting example of the way kinetic theory of rarefied gases can be fruitfully employed to study multi-agents problems in economics.

The agent-based models constitute a broad class of models which have been recently introduced to describe various socio-economic phenomena of western societies [23, 24]. The mathematical modeling showed a great expansion especially in the past twenty years (cf. [4–7, 9] and the references therein). This relatively new research field borrows several methods and tools from classical statistical physics, where the macroscopic emergent behavior arises from relatively simple rules as a consequence of microscopic interactions among a huge number of agents [23, 24].

Kinetic models of Boltzmann and Fokker–Planck type are often the building block. Similarly to (3), these models can be derived by resorting to well-known tools of classical kinetic theory of gases [7, 10–12, 22], and Boltzmann-like equation for Maxwell-type molecules play the relevant rule [3, 24].

Starting from the general Boltzmann equation (3) introduced in [2], we shall here discuss several cases in which social-type interactions give rise to lognormal distributions, or in the form of self-similar solutions, or in the form of resulting equilibria. These results will take advantage of various types of asymptotics which are of common use in kinetic theory.

Section 2 refers to Eq. (3) in the simple case $n = 1$ and $v \in \mathbb{R}_+$, and the linear interaction (2) reduces to

$$v^* = (a + b)v, \quad v \geq 0, \tag{4}$$

with $a > 0$ constant and b a random variable. In order that $v_* \in \mathbb{R}_+$, in (4) the random variable b needs to satisfy some bound from below, to guarantee that $a+b \geq 0$.

Then, Sect. 3 will deal with a kinetic model based on multiple interaction of type (2), recently considered in [27] to study a jackpot game with $n \gg 1$ gamblers. Last, Sect. 4 will deal with a generalization of interaction (4) of the form

$$v^* = (a(v) + b)v, \tag{5}$$

suitable designed to obtain Boltzmann-type kinetic models deeply connected with lognormal equilibria [16, 17]. All these examples enlighten the structure of the microscopic details which give rise to a macroscopic profile fitting the lognormal distribution.

2 Gibrat's Law and Distribution of Firms

Let us consider the Boltzmann equation (3) with $n = 1$ and $v \in \mathbb{R}_+$. By further fixing $a = 1$ and $b = \eta$, where η is a centered random variable of finite variance, the elementary linear collision (2) reduces to

$$v^* = v + \eta v, \tag{6}$$

and, as discussed in [26], the Boltzmann equation is easily shown to describe the economic Gibrat's law for firm growth [14, 15], law usually known under the name of law of proportionate effect. This law states that the expected increment of a firm's size in a fixed period of time is proportional to the size of the firm at the beginning of the period. Denoting by $x(\tau)$ the size of a firm at a time $\tau \geq 0$, the postulate is expressed as

$$x(\tau + 1) = x(\tau) + \eta(\tau)x(\tau), \tag{7}$$

where $\eta(\tau)$ is a random number independent of $x(\tau)$, and $\eta(\tau)$ is independent of $\eta(\tau + k)$ for any natural number k, and there are no interactions between firms.

After a sufficiently long sequence of increments, Gibrat's law implies that

$$x(n) = x(0)(1 + \eta(1))(1 + \eta(2)) \cdots (1 + \eta(n)),$$

so that $\log x(n)$ follows a random walk. Therefore, the growth rate predicted by Gibrat's law is lognormally distributed with mean and variance linked to the mean and variance of $\eta(\cdot)$.

Resorting on classical kinetic arguments, the appearance of the lognormal distribution as a consequence of the interaction (7) has been rigorously justified in [26] by considering the *quasi-invariant limit* of the Boltzmann equation (3), namely the asymptotics as $\epsilon \to 0$ of the Boltzmann equation (3) where the relaxation time $\tau = \epsilon$ and $\eta = \sqrt{\epsilon}\mu$, with μ a centered random variable of finite variance σ, and bounded moments up to the order three. In this asymptotics (the *grazing collisions* asymptotics [28]), the solution to the Boltzmann equation (3) is well-described by the solution $f(v, t)$ of the linear diffusion equation

$$\frac{\partial f}{\partial t} = \frac{\sigma}{2} \frac{\partial^2}{\partial v^2}(v^2 f). \tag{8}$$

Equation (8) contains the main effects of Gibrat's law (7) when the random variable η produces small symmetric effects, and the relaxation time τ is small. The linear diffusion equation (8) allows to describe the evolution in time of the density $f = f(v, t)$ of the size $v \geq 0$ of firms, given their distribution $f_0(v)$ at time $t = 0$, as well as its asymptotic behavior. As shown in [26], resorting to the connection of the solution of Eq. (8) with the solution of the linear diffusion equation one shows that

Eq. (8) possesses a (unique) source-type solution given by the lognormal density

$$L_t(v) = \frac{1}{\sqrt{4\pi t}\, v} \exp\left\{ -\frac{(\log v + t)^2}{4t} \right\}, \tag{9}$$

which departs at time $t = 0$ from a Dirac delta function located in $v = 1$. The time-dependent lognormal density $L_t(v)$ has constant mean value, and its variance at time $t \geq 0$ is equal to $e^{2t} - 1$.

3 Distribution of Winnings in a Multi-Agent Jackpot Game

The behavior of a multi-agent system of online gamblers has been studied in [27] by methods of statistical physics. In particular, the analysis has been focused on a popular type of virtual-item gambling, the jackpot, i.e., a lottery-type game which occupies a big portion of the gambling market on the web. The jackpot game relies on a very simple mechanism. The gamblers participate in the game by placing a bet with a certain number of lottery tickets purchased. There is only one winning ticket in each round of the game. The winning ticket is drawn through a uniformly distributed random number with a range equal to the total number of tickets purchased in that round. The gambler who holds the winning ticket wins all the wagers after a site cut (percentage cut) has been subtracted. By assuming that the gamblers play with a part of the sum in their possess, the single game can be easily represented as a multiple collision [27]. Suppose the extracted winner is the gambler $k = 1$. Then, if ϵ is the small percentage of the sums played by gamblers in a game, and γ is the percentage cut operated by the site, the post-interactions amounts in the hands of the gamblers are

$$v_1^* = (1-\epsilon)v_1 + \epsilon(1-\gamma)\sum_{j=1}^{n} v_j; \quad v_k^* = (1-\epsilon)v_k, \quad k = 2, 3, \ldots, n. \tag{10}$$

As first noticed in [2], in presence of a large number $n \gg 1$ of gamblers, the Boltzmann equation (3) can be approximated by a linear kinetic model with interaction rule

$$v' = (1 - \epsilon\gamma)v + \sqrt{\epsilon}\, v \eta_\epsilon. \tag{11}$$

In (11), η_ϵ is a discrete random variable taking only the two values $-\sqrt{\epsilon}(1-\gamma)$, $M_\epsilon/\sqrt{\epsilon}$ with probabilities

$$P\big(\eta_\epsilon = -\sqrt{\epsilon}(1-\gamma)\big) = 1 - p_\epsilon, \qquad P\left(\eta_\epsilon = \frac{M_\epsilon}{\sqrt{\epsilon}}\right) = p_\epsilon,$$

where $p_\epsilon \in [0, 1]$ and $M_\epsilon > 0$ are two constants to be properly fixed.

The meaning of the rule (11), together with the prescribed values of η_ϵ, is the following: a gambler, who enters the game with a number of tickets (viz. an amount of money) equal to ϵv, may either win a jackpot equal to $(M_\epsilon - \epsilon\gamma)v$ (where $M_\epsilon(1+\gamma)$ is the value of the sum played in the jackpot game) with probability p_ϵ or lose the amount ϵv put into the game with probability $1 - p_\epsilon$.

In particular, p_ϵ is determined by imposing $\langle \eta_\epsilon \rangle = 0$, which guarantees that (11) reproduces the correct evolution of the mean. This gives

$$p_\epsilon = \frac{\epsilon(1-\gamma)}{M_\epsilon + \epsilon(1-\gamma)}.$$

Last, M_ϵ is determined by assuming that the second moment of η_ϵ is equal to the second moment of the system of gamblers playing according to (10). Then, if the relaxation time $\tau = \epsilon$, in the limit $\epsilon \to 0$ one shows that the limit distribution function $f = f(v, t)$ of the amount of money of the gamblers solves the Fokker–Planck equation

$$\frac{\partial f}{\partial t} f(v, t) = \frac{\sigma}{2} \frac{\partial^2}{\partial v^2}(v^2 f(x, t)) + \gamma \frac{\partial}{\partial v}(v f(v, t)). \tag{12}$$

Also in this case, one can verify that the time-dependent lognormal distribution

$$\tilde{L}_t(v) = \frac{1}{\sqrt{2\pi\sigma t}v} \exp\left(- \frac{\left(\log v + (\gamma + \sigma/2)t\right)^2}{2\sigma t} \right) \tag{13}$$

is a solution to the Fokker–Planck equation (12) corresponding to an initial value which is a Dirac delta of unit mass concentrated in the point $v = 1$. Notice that, in consequence of the percentage cut γ operated by the site, if the gamblers to not operate a refilling at the end of each game, the mean value of the lognormal density (13) decays exponentially in time at a rate γ.

4 Social Phenomena with Lognormal Equilibria

The examples of Sects. 2 and 3 underline the importance of introducing the law of proportional effect [14, 20], as random part of the elementary interactions, to produce the lognormal distribution at a macroscopic scale. On the other hand, the pure law of proportional effect requires the kinetic model to be based on elementary collisions which, in analogy with the kinetic theory of rarefied gases, are linear in the v-variable.

A different interaction, still based on properties peculiar of the lognormal distribution [1], has been first considered in [16]. This microscopic *social* interaction combines the random part of the law of proportional effect with a non-linear dependence on v of the deterministic part. This interaction is of type (5), and for any $v \in \mathbb{R}_+$ reads

$$v_* = (a(v) + b)v = v\,(1 - \Psi(v/\bar{v}_T))\,v + \eta v. \tag{14}$$

As before, the random variable η is such that $\langle \eta \rangle = 0$, while $\langle \eta^2 \rangle = \sigma$. In (14) the function Ψ is given by

$$\Psi(s) = \mu \frac{s^\delta - 1}{s^\delta + 1}, \qquad s \geq 0, \tag{15}$$

where μ, δ, and $\bar{v}_T$ are positive constants, with μ and δ satisfying $0 < \mu < 1$, $0 < \delta < 1$. This function is a bounded and concave function, positive above the reference value 1 ($v > \bar{v}_T$), while negative below ($v < \bar{v}_T$). Note that

$$-\mu \leq \Psi(s) \leq +\mu,$$

so that the value $\mu < 1$ implies that v_* is non-negative provided that the random variable η satisfies the lower bound $\eta \geq -(1 - \mu)$.

Before entering into the meaning of the various parameters, and to enlighten the social reasons behind the choice of this interaction, let us remark a key property of its deterministic part. In absence of randomness ($\eta = 0$) let us compute the interaction satisfied by the inverse variable $w = 1/v$. Since

$$\frac{1}{w_*} = \frac{1}{w}\,(1 - \Psi(\bar{w}_T/w))\,,$$

simple computations give

$$w_* = w - \Psi^*(w/\bar{w}_T)w.$$

The new function $\Psi^*(s)$ reads

$$\Psi^*(s) = \mu \frac{s^\delta - 1}{(1 - \mu)s^\delta + 1 + \mu} = \mu^* \frac{s^\delta - 1}{\alpha s^\delta + 1},$$

Thus, the functions Ψ and Ψ^* are of the same shape (with different parameters)

$$\mu^* = \frac{\mu}{1 + \mu}; \qquad \alpha = \frac{1 - \mu}{1 + \mu}.$$

A typical property of the lognormal distribution is that, if the random variable X is lognormal, the inverse $1/X$ is still a lognormal random variable. This suggests that the structure of the function $\Psi(\cdot)$ is suitable to characterize phenomena that are described by a lognormal profile.

The function Ψ plays the role of the value function in the prospect theory of Kahneman and Twersky [18, 19], and characterizes the natural asymmetry present in many aspects related to the human behavior, when trying to reach an objective characterized by a target value $\bar{v}_T$. In [16], the problem to be studied was the distribution of the service time of agents working in a call-center, which are forced to end services respecting a limit time, given by the value $\bar{v}_T$. In this case, the presence of the minus sign in front of the value function Ψ is due to the obvious fact that an agent will tend to increase the working time when $v < \bar{v}_L$, and to decrease it if $v > \bar{v}_T$. The function $\Psi(s)$ is such that, given $0 < \Delta s < 1$

$$-\Psi(1-\Delta s) > \Psi(1+\Delta s).$$

Hence, given two agents starting at the same distance from the prescribed target time $\bar{v}_T$ from below and above, it is easier for the agent starting below to move closer to the target, than for the agent starting above. Also, the prospect an agent will have of his work will be completely different depending of the sign of the value function.

In the *grazing collision limit*, corresponding to small interactions characterized by a function Ψ in which $\delta = \epsilon$, and $\eta \to \sqrt{\epsilon}\eta$, the solution to the Boltzmann equation (3) with $n = 1$ and relaxation time $\tau = \epsilon$ converges as $\epsilon \to 0$ to the solution to the Fokker–Planck equation

$$\frac{\partial f(v,t)}{\partial t} = \frac{1}{2}\left[\sigma \frac{\partial^2}{\partial v^2}\left(v^2 f(v,t)\right) + \mu \frac{\partial}{\partial v}\left(v \log \frac{v}{\bar{v}_T} f(v,t)\right)\right]. \tag{16}$$

The stationary distribution of Eq. (16) is easily found by solving the differential equation

$$\sigma \frac{d}{dv}\left(v^2 f(v)\right) + \mu v \log \frac{v}{\bar{v}_T} f(v) = 0.$$

It is immediate to conclude that the unique steady solution of unit mass is the lognormal density

$$f_\infty(v) = \frac{1}{\sqrt{2\pi\sigma^*}\, v} \exp\left\{-\frac{(\log v - \kappa)^2}{2\sigma^*}\right\}, \tag{17}$$

where

$$\sigma^* = \frac{\sigma}{\mu}, \qquad \kappa = \log \bar{v}_T - \sigma^*.$$

At difference with the examples of Sects. 2 and 3 the lognormal profile is a steady state of the underlying kinetic model. Note that the mean value of the lognormal equilibrium (17) takes the value

$$m(f_\infty) = \bar{v}_T \exp\left\{-\frac{\sigma^*}{2}\right\} < \bar{v}_T,$$

which shows that the desired target value $\bar{v}_T$ is never reached by the mean number of agents.

The situations in which an elementary interaction of type (14) holds, and consequently the social phenomena are characterized by a certain asymmetry around a target value, have been collected in a companion paper [17], which illustrates a number of examples that can be suitably described at equilibrium in terms of the lognormal distribution. The list of these phenomena is conspicuous, and include body weight distribution, behavior of drivers in traffic, distribution of agent's consumptions, distribution of city size and others. Also, in biological sciences, the Fokker–Planck equation (16) and the corresponding lognormal equilibrium profile appears when studying from a kinetic point of view the distribution of tumor size in the case of a Gompertz-type growth [25].

5 Conclusions

Recent results on kinetic modeling of social phenomena in multi-agent systems confirm that the lognormal distribution is suitable to fit them in a variety of situations, which can be described by linear diffusion equations and/or by linear Fokker–Planck equations with variable coefficients of diffusion. At the kinetic level, it is interesting to remark that the elementary collisions leading to a lognormal profile are characterized in their random part by the so-called law of proportional effect, namely by the property that the random part of the collision is proportional to the value of the pre-interaction variable. Further, the deterministic part of the pre-collision variable has to be characterized by the property that its inverse share a similar profile.

Acknowledgments This work has been written within the activities of GNFM group of INdAM (National Institute of High Mathematics), and partially supported by the Italian Ministry of Education, University, and Research (MIUR): Dipartimenti di Eccellenza Program (2018–2022) - Dept. of Mathematics "F. Casorati," University of Pavia. The editors of the volume, who gave to the author the possibility to remember the work of the notable figure of Carlo Cercignani, whose scientific results have been seminal for his research, are kindly acknowledged.

References

1. Aitchison, J., Brown, J.A.C.: The Log-Normal Distribution. Cambridge University Press, Cambridge (1957)
2. Bobylev, A.V., Cercignani, C,, Gamba I.: On the self-similar asymptotics for generalized nonlinear kinetic Maxwell models. Comm. Math. Phys. **291**(3), 599–644 (2009)
3. Cercignani, C.: The Boltzmann Equation and Its Applications. Springer Series in Applied Mathematical Sciences, vol. 67. Springer, New York (1988)
4. Chakraborti, A., Chakrabarti, B.K.: Statistical mechanics of money: effects of saving propensity. Eur. Phys. J. B **17**, 167–170 (2000)

5. Chatterjee, A., Chakrabarti, B.K., Manna, S.S.: Pareto law in a kinetic model of market with random saving propensity. Physica A **335**, 155–163 (2004)
6. Chatterjee, A., Chakrabarti, B.K., Stinchcombe, R.B.: Master equation for a kinetic model of trading market and its analytic solution. Phys. Rev. E **72**, 026126 (2005)
7. Cordier, S., Pareschi, L., Toscani, G.: On a kinetic model for a simple market economy. J. Stat. Phys. **120**, 253–277 (2005)
8. Crow, E.L., Shimizu, K. (eds.): Log-Normal Distributions: Theory and Application. Dekker, New York (1988)
9. Drăgulescu, A., Yakovenko, V.M.: Statistical mechanics of money. Eur. Phys. J. B **17**, 723–729 (2000)
10. Düring, B., Matthes, D., Toscani, G.: Kinetic equations modelling wealth redistribution: a comparison of approaches. Phys. Rev. E **78**, 056103 (2008)
11. Düring, B., Matthes, D., Toscani, G.: A Boltzmann type approach to the formation of wealth distribution curves. Riv. Mat. Univ. Parma **8**(1), 199–261 (2009)
12. Furioli, G., Pulvirenti, A., Terraneo, E., Toscani, G.: Fokker-Planck equations in the modelling of socio-economic phenomena. Math. Models Methods Appl. Sci. **27**, 115–158 (2017)
13. Galton, F.: Natural Inheritance. McMillan, London (1894)
14. Gibrat, R.: Une loi des réparations économiques: l'éffet proportionnel. Bull. Statist. Gén. Fr. **19** 469–513 (1930)
15. Gibrat, R.:Les inegalites économiques. Libraire du Recueil Sirey, Paris (1931)
16. Gualandi, S., Toscani, G.: Call center service times are lognormal. a Fokker–Planck description. Math. Models Methods Appl. Sci. **28**(8), 1513–1527 (2018)
17. Gualandi, S., Toscani, G.: Human behavior and lognormal distribution. a kinetic description. Math. Models Methods Appl. Sci. **29**(4), 717–753 (2019)
18. Kahneman, D., Tversky, A.: Prospect theory: an analysis of decision under risk. Econometrica **47**(2), 263–292 (1979)
19. Kahneman, D., Tversky, A.: Choices, Values, and Frames. Cambridge University Press, Cambridge (2000)
20. Kapteyn, J.C.: Skew Frequency Curves in Biology and Statistics. Astronomical Laboratory, Groningen. Noordhoff, The Netherlands (1903)
21. Limpert, E., Stahel, W.A., Abbt, M.: Log-normal distributions across the sciences: keys and clues. BioScience **51**(5), 341–352 (2001)
22. Matthes, D., Toscani, G.: On steady distributions of kinetic models of conservative economies. J. Stat. Phys. **130**, 1087–1117 (2008)
23. Naldi, G., Pareschi, L, Toscani, G. (eds.): Mathematical Modeling of Collective Behavior in Socio-Economic and Life Sciences. Birkhauser, Boston (2010)
24. Pareschi, L, Toscani, G.: Interacting Multiagent Systems: Kinetic Equations and Monte Carlo Methods. Oxford University Press, Oxford (2014)
25. Preziosi, L., Toscani, G., Zanella, M.: Control of tumor growth distributions through kinetic methods. J. Theor. Biol. **514**, 110579 (2021)
26. Toscani, G.: Kinetic and mean field description of Gibrat's law. Physica A **461**, 802–811 (2016)
27. Toscani, G., Tosin, A., Zanella, M.: Multiple-interaction kinetic modeling of a virtual-item gambling economy. Phys. Rev. E **100**, 012308 (2019)
28. Villani, C.: Contribution à l'étude mathématique des équations de Boltzmann et de Landau en théorie cinétique des gaz et des plasmas. PhD Thesis, Univ. Paris-Dauphine (1998)

Oscillatory Rarefied Gas Flows in Long Capillaries

Alexandros Tsimpoukis, Nikos Vasileiadis, Giorgos Tatsios, and Dimitris Valougeorgis

Abstract Oscillatory, rarefied, linear and nonlinear fully developed flows of single gases and binary gas mixtures, driven by external harmonic mechanisms with arbitrary frequency, have been recently considered by the authors in a series of works. Here, these works are reviewed by focusing on the most notable findings. More specifically, the effects of the oscillation frequency on the velocity overshooting and gas separation phenomena in gas mixture flows and of the oscillation amplitude on the flow pattern in nonlinear single gas flows are presented. Modeling is based in the former case on the McCormack kinetic model and in the latter one on the DSMC method. In general, as the flow becomes more rarefied higher frequencies are needed to trigger the overshooting phenomenon, which becomes more pronounced as the molecular mass of the gas species is increased. Notably, gas separation may be present in the whole range of gas rarefaction, provided that the flow is subject to adequate high oscillation frequency. Finally, the presence of strong external harmonic forces does not significantly affect the oscillatory macroscopic quantities, including the mass flow rate (no distortion of the amplitude-frequency curve), except of the oscillatory axial heat flux, which exhibits a non-sinusoidal pattern.

1 Introduction

Rarefied boundary-driven oscillatory flows of single gases have been extensively investigated over the last two decades [1–7]. These flows are present in various resonator structures [8, 9], while acoustic enhancement or attenuation (even cloaking) may be achieved in viscous-thermal fluids [10]. Propagation of sound waves due to mechanical and thermal excitation through binary gas mixtures has been also considered [11–13].

A. Tsimpoukis (✉) · N. Vasileiadis · G. Tatsios · D. Valougeorgis
University of Thessaly, Volos, Greece
e-mail: atsimpoukis@mie.uth.gr; nikovasi@mie.uth.gr; tatsios@mie.uth.gr; diva@mie.uth.gr

P. Barbante et al. (eds.), *From Kinetic Theory to Turbulence Modeling*, Springer INdAM Series 51, https://doi.org/10.1007/978-981-19-6462-6_21

The corresponding rarefied pressure-driven oscillatory gas flows have attracted much less attention, although there are employed in vapor deposition [14], microfluidic oscillators and pumps [15] and cryogenic pulse tubes [16]. Of course, in the hydrodynamic regime, pressure-driven oscillatory gas flows have been thoroughly examined and are encountered in numerous technological fields ranging from pneumatic lines and control systems [17], reciprocating pumps [18], combustion engines, and bioengineering to enhancement of thermal diffusion in mass and heat transfer processes, species contaminants dispersion and gas separation or mixing [19, 20]. Experimentally, oscillatory-type pressure-driven gas flows may be realized by reciprocating pistons [21] or membranes [22] or by oscillating the channel itself [23].

Although boundary and pressure gradient oscillatory flows have certain similarities, such as the traveling wave disturbance causing the flow, they also have various differences related to the involved physical phenomena and quantities of practical interest. The general mechanisms occurring in oscillatory boundary-driven flows include inertia and viscous forces, while in pressure gradient flows, in addition to the above, pressure forces are also considered. In the latter case, the difference in time scales of pressure and viscous forces may lead to unexpected results, such as the annular effect (velocity overshooting) and enhanced gas separation, which are not observed in former case. Also, in boundary-driven flows we are mainly interested in velocity and shear stresses, while in oscillatory pressure gradient flows including pulsatile flows, we are also interested in the computed flow rates.

Taking into consideration that oscillatory pressure- driven gas flows in the hydrodynamic regime are very common, along with the progress in fabrication techniques of micro devices, it is reasonable to expect that oscillatory pressure-driven rarefied flows of single gases and gas mixtures will be also widely employed, in the short future. Therefore, very recently, some theoretical studies in fully developed oscillatory gas flows in capillaries [24–27] have been reported. Here, the most notable results of the detailed analysis in [24–27] for linear and nonlinear fully developed flows of single gases and binary gas mixtures are presented.

2 Linear Oscillatory Fully Developed Binary Gas Mixture Flow

Consider the time-dependent, isothermal, rarefied flow of a binary gas mixture between two infinite long parallel plates fixed at $y' = \pm H/2$, connecting two containers, as shown in Fig. 1. The pressure in the two containers harmonically oscillates as $\tilde{P}_j\left(t'\right) = \mathbb{R}\left[P_j \exp\left(-i\omega t'\right)\right]$, $j = 1, 2$, resulting in the externally

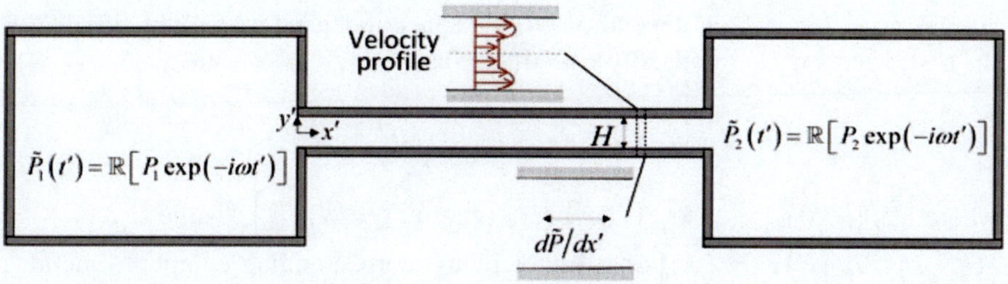

Fig. 1 Oscillatory flow configuration

imposed harmonically oscillating pressure gradient, along the parallel plates, of the form

$$\frac{d\tilde{P}}{dx'} = \mathbb{R}\left[\frac{dP}{dx'} \exp\left(-i\omega t'\right)\right]. \tag{1}$$

Here, $\tilde{P}\left(x', t'\right) = P\left(x'\right) \exp\left(-i\omega t'\right)$ is the oscillatory pressure in the $x'-$direction parallel to the plates, dP/dx' and ω refer to the amplitude and frequency, respectively, of the oscillatory pressure gradient $d\tilde{P}/dx'$ and t' is the time, while $\mathbb{R}$ denotes the real part of a complex expression $i = \sqrt{-1}$). The well-established assumption that the fluid oscillates in bulk or en mass, i.e., that all quantities oscillate with the same frequency as the pressure gradient, is applied [28]. Thus, this is an harmonically oscillating, fully developed flow (pressure and density remain constant at each cross section, while all other macroscopic distributions depend only in the $y'-$direction normal to the plates).

The binary gas mixture consists of two monatomic species of molecular masses m_α, with the index "$\alpha = 1, 2$," always referring, without loss of generality, to the light and heavy species of the mixture, respectively. The corresponding local number densities of the mixture components, defined by $\tilde{n}_\alpha\left(t'\right)$, oscillate harmonically as $\tilde{n}_\alpha\left(t'\right) = \mathbb{R}\left[n_\alpha \exp\left(-i\omega t'\right)\right]$, where n_α, $\alpha = 1, 2$, is the local amplitude of the oscillating number density of each species. The number density of the mixture is $\tilde{n}\left(t'\right) = \tilde{n}_1\left(t'\right) + \tilde{n}_2\left(t'\right)$, while the molar fraction of the mixture is defined as the ratio of the number density of the light species over the mixture number density, given by $\tilde{C}\left(t'\right) = \mathbb{R}\left[C \exp\left(-i\omega t'\right)\right]$, with $C = n_1/n = n_1/\left(n_1 + n_2\right)$, being the local amplitude of the molar fraction. The molar fraction amplitude of the heavy species is $1 - C$. The mean molecular mass of the mixture is given by $m = Cm_1 + (1 - C)\, m_2$. The number densities of the species and the mixture are related to the corresponding pressures with the equation of states as $\tilde{P}_\alpha = \tilde{n}_\alpha kT$ and $\tilde{P} = \tilde{n}kT$, respectively, where $\tilde{P}_\alpha$ are the partial pressures, $\tilde{P} = \tilde{P}_1 + \tilde{P}_2$ is the total pressure, T is the reference temperature. The mass densities of the species and the mixture are defined as $\rho_\alpha = m_\alpha n_\alpha$ and $\rho = mn$, respectively.

The deduced time-dependent flow quantities of practical interest include the bulk velocity $\tilde{U}_\alpha\left(t', y'\right)$, shear stress $\tilde{\Pi}_\alpha\left(t', y'\right)$ and heat flow $\tilde{Q}_\alpha\left(t', y'\right)$ of the two species $\alpha = 1, 2$, which depend on y', the space independent variable vertical to the plates and vary harmonically with time t' as

$$\tilde{Z}_\alpha\left(t', y'\right) = \mathbb{R}\left[Z_\alpha\left(y'\right)\exp\left(-i\omega t'\right)\right], \tag{2}$$

where $\tilde{Z}_\alpha\left(t', y'\right) = \left[\tilde{U}_\alpha\left(t', y'\right), \tilde{\Pi}_\alpha\left(t', y'\right), \tilde{Q}_\alpha\left(t', y'\right)\right]$, while $\tilde{Z}_\alpha\left(y'\right) = \left[U_\alpha\left(y'\right), \Pi_\alpha\left(y'\right), Q_\alpha\left(y'\right)\right]$ is a vector of the corresponding complex functions. In addition, the oscillatory particle flow rates of the two species are given by $\tilde{J}_\alpha\left(t'\right) = \mathbb{R}\left[J_\alpha\exp\left(-i\omega t'\right)\right]$, where $J_\alpha = n_\alpha \int\limits_{-H/2}^{H/2} U_\alpha dy'$, as well as the corresponding mixture particle flow rate $\tilde{J} = \tilde{J}_1 + \tilde{J}_2$, are complex functions.

Furthermore, the dimensionless independent space and time variables $x = x'/H$, $y = y'/H$ and $t = t'\omega$, are introduced. The dimensionless amplitude of the local oscillatory pressure gradient is

$$X = \frac{H}{P\left(x'\right)}\frac{dP\left(x'\right)}{dx'} = \frac{1}{P\left(x\right)}\frac{dP\left(x\right)}{dx} \ll 1. \tag{3}$$

The bulk velocity, shear stress and heat flow in Eq. (2) are nondimensionalized by (υX), $(2PX)$ and (υPX), respectively, with $\upsilon = \sqrt{2kT/m}$ being the characteristic speed of the mixture, to yield:

$$\tilde{\varphi}_\alpha\left(t, y\right) = \mathbb{R}\left[\varphi_\alpha\left(y\right)\exp\left(-it\right)\right] = \varphi_\alpha^{(A)}\left(y\right)\cos\left[t - \varphi_a^{(P)}\left(y\right)\right], \tag{4}$$

where $\tilde{\varphi}_\alpha\left(t, y\right) = \left[\tilde{u}_\alpha\left(t, y\right), \tilde{\varpi}_\alpha\left(t, y\right), \tilde{q}_\alpha\left(t, y\right)\right]$. In Eq. (4) the superscripts (A) and (P) refer to the amplitude and the phase angle, respectively, of each complex quantity.

Furthermore, the flow rates $\tilde{J}_\alpha\left(t'\right)$ are nondimensionalized by $(PXH/m\upsilon)$ to obtain the dimensionless oscillatory particle flow rates of each species

$$\tilde{G}_\alpha\left(t\right) = \mathbb{R}\left[G_\alpha\exp\left(-it\right)\right] = \mathbb{R}\left[G_\alpha^{(A)}\exp\left[i\left(G_\alpha^{(P)} - t\right)\right]\right] = G_\alpha^{(A)}\cos\left[t - G_\alpha^{(P)}\right], \tag{5}$$

where

$$G_\alpha = G_\alpha^{(A)}\exp\left(iG_\alpha^{(P)}\right) = 2\int_{-1/2}^{1/2} u_\alpha dy. \tag{6}$$

Also, the dimensionless oscillatory particle flow rate of the mixture is given by

$$\tilde{G}(t) = \mathbb{R}\left[G \exp(-it)\right] = \mathbb{R}\left[G^{(A)} \exp\left[i\left(G^{(P)} - t\right)\right]\right] = G^{(A)} \cos\left[t - G^{(P)}\right], \tag{7}$$

where $G = CG_1 + (1 - C)\,G_2$, with the superscripts (A) and (P), always referring to amplitudes and phase angles, respectively.

The oscillatory binary gas mixture flow between parallel plates is also characterized by the gas rarefaction and oscillation parameters, given by

$$\delta = \frac{PH}{\mu\upsilon} \quad \text{and} \quad \theta = \frac{P}{\mu\omega}, \tag{8}$$

respectively, where μ is the gas viscosity at some reference temperature T, υ is the characteristic speed of the mixture, the ratio (P/μ) is the intermolecular collision frequency. The composition of the binary gas mixture, i.e., the molecular masses m_1 and m_2, as well as the amplitude of the molar fraction C, must be also specified.

Next, the kinetic formulation, based on the McCormack model [29], is shortly presented. Due to the condition $X \ll 1$ the unknown time-dependent distribution function of each species can be linearized in a standard manner and the linearized distributions are accordingly projected to yield the following set of kinetic equations:

$$-i\frac{\delta}{\theta}\sqrt{\frac{m_\alpha}{m}}\varPhi_\alpha + c_{\alpha y}\frac{\partial \varPhi_\alpha}{\partial y} + \omega_a\gamma_\alpha\varPhi_a =$$

$$-\frac{1}{2}\sqrt{\frac{m}{m_a}} + \omega_\alpha\left\{\gamma_\alpha u_a - \upsilon_{\alpha\beta}^{(1)}\left(u_a - u_\beta\right) - \frac{1}{2}\upsilon_{\alpha\beta}^{(2)}\left(q_a - \frac{m_a}{m_\beta}q_\beta\right) + \right.$$

$$+2\sqrt{\frac{m}{m_a}}\left[\left(\gamma_\alpha - \upsilon_{\alpha\alpha}^{(3)} + \upsilon_{\alpha\alpha}^{(4)} - \upsilon_{\alpha\beta}^{(3)}\right)\varpi_a + \upsilon_{\alpha\alpha}^{(4)}\varpi_\beta\right]c_{ay} +$$

$$\left. +\frac{2}{5}\left[\left(\gamma_\alpha - \upsilon_{\alpha\alpha}^{(5)} + \upsilon_{\alpha\alpha}^{(6)} - \upsilon_{\alpha\beta}^{(5)}\right)q_a + \upsilon_{\alpha\beta}^{(6)}\sqrt{\frac{m_\beta}{m_a}}q_\beta - \frac{5}{4}\upsilon_{\alpha\beta}^{(2)}\left(u_a - u_\beta\right)\right]\left(c_{ay}^2 - \frac{1}{2}\right)\right\}, \tag{9}$$

$$-i\sqrt{\frac{m_\alpha}{m}}\frac{\delta}{\theta}\varPsi_\alpha + c_{\alpha y}\frac{\partial \varPsi_\alpha}{\partial y} + \omega_\alpha\gamma_\alpha\varPsi_a =$$

$$= \frac{4}{5}\omega_\alpha\left[\left(\gamma_\alpha - \upsilon_{\alpha\alpha}^{(5)} + \upsilon_{\alpha\alpha}^{(6)} - \upsilon_{\alpha\beta}^{(5)}\right)q_a + \upsilon_{\alpha\beta}^{(6)}\sqrt{\frac{m_\beta}{m_a}}q_\beta - \frac{5}{4}\upsilon_{\alpha\beta}^{(2)}\left(u_a - u_\beta\right)\right]. \tag{10}$$

Here, $\varPhi_a$ and $\varPsi_a$ are complex perturbed distribution functions for each species, $\omega_\alpha = \delta\left(C/\gamma_1 + (1 - C)/\gamma_2\right)\sqrt{m_a/m}$ and γ_a $(a = 1, 2)$ are the collision frequencies of each species [30]. Also, $\alpha, \beta = 1, 2$, with $\alpha \neq \beta$, while the expressions for the quantities $\nu_{\alpha\beta}^{(k)}$ are given in terms of the Chapman-Cowling integrals as in [30]. The macroscopic quantities u_α, ϖ_α and q_α at the right hand side of Eqs. (9) and (10) are defined in Eq. (4), respectively, and after applying the linearization and projection procedures, they are obtained as moments of Φ_α and Ψ_α as follows:

$$u_\alpha(y) = \frac{1}{\sqrt{\pi}} \int_{-\infty}^{\infty} \Phi_a \exp\left(-c_{ay}^2\right) dc_{ay}, \tag{11}$$

$$\varpi_\alpha(y) = \frac{1}{\sqrt{\pi}} \sqrt{\frac{m_a}{m}} \int_{-\infty}^{\infty} \Phi_a c_{ay} \exp\left(-c_{ay}^2\right) dc_{ay}, \tag{12}$$

$$q_\alpha(y) = \frac{1}{\sqrt{\pi}} \int_{-\infty}^{\infty} \left[\Psi_a + \left(c_{ay}^2 - \frac{1}{2}\right) \Phi_a\right] \exp\left(-c_{ay}^2\right) dc_{ay}. \tag{13}$$

In the present work purely diffuse reflection at the walls is assumed.

The above set of equations is computationally solved based on the discrete velocity method in the c_y space and on the second-order diamond finite difference scheme in the y space. The discretized equations are solved in an iterative manner between the kinetic equations (9) and (10) and the moment equations (11)–(13). More information about the numerical scheme may be found in [27].

Computational results are presented for the mixture flow rate amplitude and phase angle (Fig. 2), the velocity and shear stress distributions (Fig. 3) and the ratio of the flow rate amplitudes of the species (Fig. 4) in a wide range of the gas rarefaction and oscillation parameters δ and θ, as well as of the molar fraction $C \in [0, 1]$ for the He–Xe mixture with $m_2/m_1 = 32.8$.

In Fig. 2, the He–Xe flow rate amplitude $G^{(A)}$ and phase angle $G^{(P)}$ are presented in terms of δ, with $\theta = [1, 100]$ and $C = [0, 0.5, 0.9]$. The results with $C =$

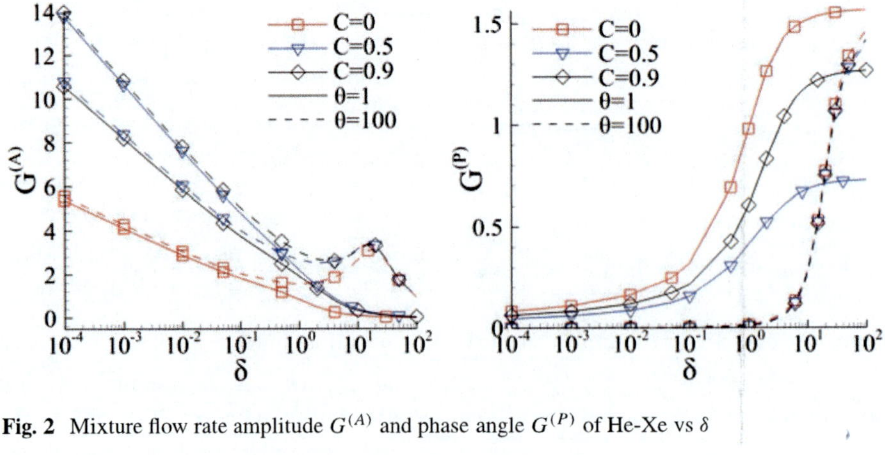

Fig. 2 Mixture flow rate amplitude $G^{(A)}$ and phase angle $G^{(P)}$ of He-Xe vs δ

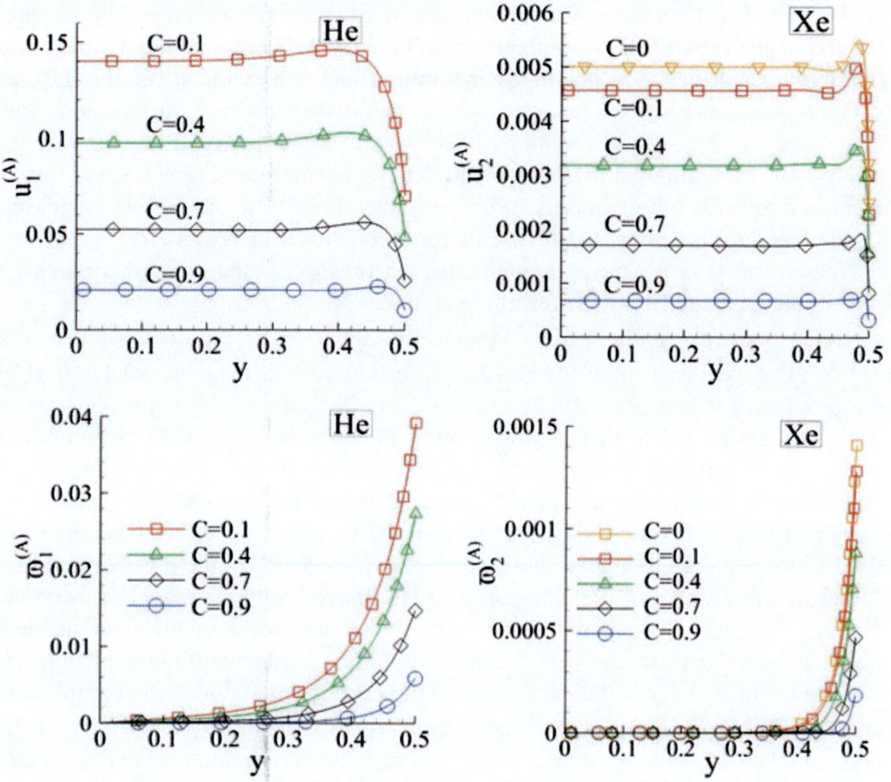

Fig. 3 Velocity $u_\alpha^{(A)}(y)$ and shear stress $\varpi_\alpha^{(A)}(y)$ amplitudes of each species of He-Xe for $\delta = 10$ and $\theta = 0.1$. Reprinted with permission from [27]. Copyright (2022) by the American Physical Society

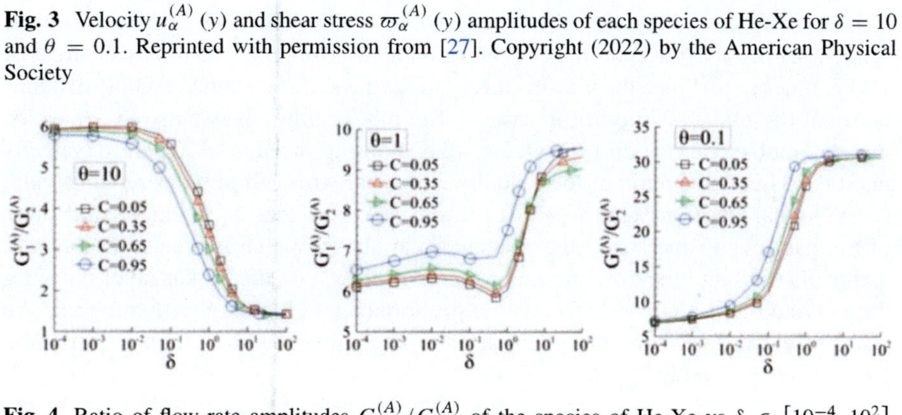

Fig. 4 Ratio of flow rate amplitudes $G_1^{(A)}/G_2^{(A)}$ of the species of He-Xe vs $\delta \in [10^{-4}, 10^2]$. Reprinted with permission from [27]. Copyright (2022) by the American Physical Society

0 correspond to the oscillatory single gas flow reported in [25]. The flow rate amplitudes and phase angles of the mixture ($C \neq 0$) depend on the flow parameters similarly to the corresponding single gas ones ($C = 0$). Always, the mixture flow

rate amplitude is larger and the phase angle is smaller than the corresponding ones of the single gas. At large θ the dependency of $G^{(A)}$ on δ, is not monotonic, indicating that there is a critical δ to obtain the maximum flow rate, while at small θ, $G^{(A)}$ is decreased monotonically. This is due to the fact that at low oscillation frequencies and as long as $\delta \ll \theta$, the variation of $G^{(A)}$ with δ has some resemblance with the steady one, including the presence of the Knudsen minimum. Then, as δ is further increased the effect of the inertia forces becomes significant and $G^{(A)}$ is decreased. In addition, as θ is decreased (the oscillation frequency is increased), $G^{(A)}$ is always decreased, while $G^{(P)}$ (the phase angle lag with respect to the pressure gradient) is always increased reaching the limiting value of $\pi/2$.

In Fig. 3, the distributions of the velocity and shear stress amplitudes $u_\alpha^{(A)}(y)$ and $\varpi_\alpha^{(A)}(y)$ of each species of the He–Xe gas mixture, with $C = [0.1, 0.4, 0.7, 0.9]$, are provided for $\delta = 10$ and $\theta = 0.1$. The specific values of δ and θ are suitable for investigating the velocity overshooting phenomenon in the light and heavy species of the mixture. Velocity overshooting is due to the fact that close to the wall, viscous and pressure gradient forces actually add to each other due to the large phase angle lag between them. As a result, the combined effect accelerates the fluid to higher velocities than those produced in the core by the pressure gradient forces acting alone. For Xe, compared to He, the velocity overshooting becomes sharper, appearing, along with its maximum value, closer to the wall inside a much thinner layer. In the core of the flow, the velocity amplitudes of both He and Xe become flat and they are close to the corresponding analytical amplitudes $u_\alpha^{(A)} = (\theta/2\delta)(m/m_\alpha)$ (see Section 3 in [27]). In parallel, $\varpi_\alpha^{(A)}(y)$ for both He and Xe take their highest values at the wall and they are monotonically decreased towards the channel center. The attenuation of the shear stress amplitude of He is smooth, diffused in the whole distance from the wall to the center, while the one of Xe is rapid in a narrow zone close to the wall and far from the wall the shear stress of Xe becomes zero. Since the viscous forces in the case of He act in the whole distance between the plates, while in the case of Xe only in thin zones close to the walls, the above observations on the velocity overshooting of He and Xe are physically justified. This description of the velocity and shear stress amplitudes remains valid for all molar fractions tested [27]. In brief, it is seen that as the molecular mass of the gas species increases, the species shear stress, which is created at the wall and is diffused into the flow, attenuates more rapidly, i.e., the Stokes layer becomes thinner and the Richardson effect more pronounced. Velocity overshooting may be also present in even lower rarefaction parameters provided that higher oscillation frequencies are applied [24].

The gas separation phenomenon for various values of δ and θ is discussed next. Gas separation in rarefied steady-state pressure-driven binary gas flows though capillaries may be analyzed by computing the ratio of the particle flow rates J_1/J_2, which is monotonically increased as δ is decreased up to its maximum value, equal to $\sqrt{m_2/m_1}(1-C)/C$, in the free molecular limit ($\delta \to 0$) [31].

In Fig. 4, the ratio of the flow rate amplitudes $G_1^{(A)}/G_2^{(A)}$ is provided in terms of δ for the He–Xe gas mixture, with $C = [0, 05, 0.35, 0.65, 0.95]$ and $\theta = [0.1, 1, 10]$.

At $\theta = 10$ the ratio $G_1^{(A)}/G_2^{(A)}$ varies qualitatively similarly as in the steady-state binary gas flow setup. It is about constant or slightly reduced in the free molecular regime (at $\delta = 0$ it is equal to the corresponding steady one) and then it is monotonically decreased asymptotically going in the slip and hydrodynamic regimes to one. In the free molecular regime, with regard to the gas rarefaction parameter, as $\delta \to 0$, with $\theta > 0$, Eqs. (9) and (10) tend to the corresponding ones for steady-state binary gas flow in the free molecular limit [30]. However, at $\theta = 1$ and $\theta = 0.1$ the behavior of $G_1^{(A)}/G_2^{(A)}$ is completely different. It remains about constant in free molecular regime, but then, it is increased in the transition regime and finally, as δ further increases, it keeps asymptotically increasing to some constant value, which is the molecular mass ratio of the heavy over the light species m_2/m_1 ($G_{He}^{(A)}/G_{Xe}^{(A)} = 32.8$). This is in accordance to the closed-from expression that as $\theta \to 0$, $G_1/G_2 = m_2/m_1$ [27]. This behavior, with the minimum and maximum values of $G_1^{(A)}/G_2^{(A)}$ appearing at the free molecular and hydrodynamic limits, respectively, and the increase in the transition regime (completely reversed compared to the steady-state behavior) becomes more pronounced as θ is decreased.

It is evident that the oscillation parameter θ has a dominant effect on the amplitude ratio of He over Xe, which is significantly increased as θ is decreased (at $\theta = 0.1$ the flow rate amplitude of He is about thirty times larger than of Xe). This behavior is due to the corresponding behavior of the velocity amplitudes and it is contributed to inertia forces, which are increased with the oscillation frequency and they influence the bulk velocity amplitude of the heavy species much more than of the light one. Therefore, as θ is decreased, the flow rate amplitude of the heavy species decreases much more significantly than the light one and although both amplitudes are decreased the velocity amplitude ratio of the light over the heavy species is increased. This effect is magnified as the flow becomes less rarefied overcoming diffusion effects due to increased intermolecular collisions and therefore, as δ increases the amplitude ratio keeps increasing. There is no contradiction to general theory, since oscillatory flows approach the hydrodynamic regime, only when both δ and θ are adequately large.

3 Nonlinear Oscillatory Fully Developed Single Gas Flow

Consider the oscillatory nonlinear fully developed flow of a monatomic rarefied gas, confined between two parallel infinite plates at temperature T_0 located at $y' = \pm H/2$, due to an external harmonic force acting on the gas per unit mass in the $x-$direction parallel to the plates [26]. The external force is defined as $\tilde{F}'(\omega, t') = F' \cos(\omega t')$, where F' is the force amplitude. The convenient complex factor $\exp(-i\omega t')$ previously used, cannot be employed since the force amplitude F' may be arbitrarily large and in nonlinear oscillatory flows the real and imaginary parts are not separable.

The oscillatory macroscopic distributions of practical interest, characterizing the flow, include the $x-$component $U_{x'}\left(y',t'\right)$ of the velocity vector, the number density $N\left(y',t'\right)$, the temperature $T\left(y',t'\right)$, and the axial and normal heat flow components $Q_{x'}\left(y',t'\right)$ and $Q_{y'}\left(y',t'\right)$, respectively, with $-H/2 \leq y' \leq H/2$ and $0 \leq t' \leq 2\pi/\omega$. The most important overall quantities are the mass flow rate and axial heat flow

$$M'\left(t'\right) = m\int_{-H/2}^{H/2} N\left(y',t'\right)U_{x'}\left(y',t'\right)dy' \text{ and } \bar{Q}_{x'}\left(t'\right) = \frac{1}{H}\int_{-H/2}^{H/2} Q_{x'}dy', \tag{14}$$

respectively, where m is the molecular mass.

The parameters defining the above dimensional flow setup include the rarefaction parameter and oscillation parameter defined in Eq. (8). Also, the external force parameter, defined as $F = F'H/\upsilon_0^2$, is needed. It is the inverse of the square of the Froude number (Fr). The effect of the external force on the flow is increased with F and nonlinear effects are becoming dominant. On the contrary, as F is decreased the corresponding linear oscillatory flow, which is linearly proportional to the force magnitude, is gradually recovered.

The following dimensionless variables are introduced:

$$x = \frac{x'}{H}, dx = \frac{dx'}{H}, y = \frac{y'}{H}, dy = \frac{dy'}{H}, t = \frac{t'}{(H/\upsilon_0)} \tag{15}$$

$$n = \frac{N}{N_0}, u_x = \frac{U_{x'}}{\upsilon_0}, \tau = \frac{T}{T_0}, p_{xy} = \frac{\Pi_{x'y'}}{2P_0}, p = \frac{P}{2P_0}, q_x = \frac{Q_{x'}}{\upsilon_0 P_0}, q_y = \frac{Q_{y'}}{\upsilon_0 P_0}. \tag{16}$$

The equation of state becomes $p = n\tau/2$.

Then, the dimensionless external force acting on the gas per unit mass becomes

$$\tilde{F}\left(\delta,\theta,\ t\right) = F\cos\left(\frac{\delta}{\theta}t\right), \tag{17}$$

while the dimensionless flow rate and axial heat flow are given by

$$M\left(t\right) = \frac{M'}{2P_0\left(H/\upsilon_0\right)} = \int_{-1/2}^{1/2} n\left(t,y\right)u\left(t,y\right)dy, \quad \bar{q}_x\left(t\right) = \int_{-1/2}^{1/2} q_x\left(y,t\right)dy. \tag{18}$$

Next, the typical DSMC approach, with the No Time Counter (NTC) scheme proposed by Bird [32], is implemented. The time evolution of the particle system within a small time interval $\Delta t'$ is split into two consecutive steps: free motion of all particles and binary collisions of particles. The time step $\Delta t'$ is nondimensionalized as $\Delta t = \Delta t' / (H/\upsilon_0)$. Purely diffuse boundary conditions are considered at the walls, while periodic boundary conditions are applied in the $x-$ and $z-$ directions. Hard sphere (HS) molecules are assumed. The external force is introduced by accordingly altering the particle velocities at each time step, during the free motion.

Numerical results of the dimensionless flow rate and axial heat flow are provided in terms of the force amplitude $F = [0.05, 0.1, 0.5]$, corresponding to small, moderate, and large force amplitudes, in a wide range of δ and θ. Since the results of the nonlinear gas flow are similar with the linear ones in terms of δ and θ, only the effect of the force amplitude is here discussed.

In Fig. 5, the flow rate amplitudes G_A are divided by the external force F in order to directly compare with the corresponding linear results (the linear solution is proportional to F) and they are presented for $\delta = [0.1, 1, 10]$, $\theta = \left[0.1, 1, 10, 20, 10^2\right]$ and $F = [0.05, 0.1, 0.5]$. The linear flow rate amplitudes obtained in [26] are also provided. It is seen that for $F = 0.05$ and $F = 0.1$ the deviation between the corresponding DSMC and linear solutions is small for $\delta \geq 1$ and for all values of θ, while for $\delta = 0.1$ and $\theta = [10, 20, 10^2]$ the deviation is increased. It is evident that nonlinear effects are becoming more pronounced in highly rarefied atmospheres (small δ) and low frequencies (large θ). For $F = 0.5$ all deviations between DSMC and linear results are further increased due to nonlinear effects. Again, the largest deviations are occurring at $\delta = 0.1$ and $\theta = [10, 20, 10^2]$ ($\delta << \theta$), while the deviations remain small for $\delta \geq 1$, even at high frequencies. Overall, it may be stated that the presence of strong external harmonic forces does not significantly affect the mass flow rate of the oscillatory flow, i.e., there is no distortion of the amplitude-frequency response curve.

The space-average axial flow $\bar{q}_x(t)$ is plotted over one cycle in Fig. 6 for $\delta = [0.1, 1, 10]$ and $\theta = \left[0.1, 1, 10, 20, 10^2\right]$ with $F = [0.05, 0, 5]$. It is readily seen that there are significant qualitative differences between the corresponding space-average heat flow for $F = 0.05$ and $F = 0.5$. For $F = 0.05$, $\bar{q}_x(t)$ for all values of δ and θ has a sinusoidal behavior over time. For $F = 0.5$, $\bar{q}_x(t)$ shows over one cycle various patterns. It is seen that for $\delta = 0.1$ with $\theta = 0.1$, $\delta = 1$ with $\theta = [0.1, 1]$ and for $\delta = 10$ and $\theta = [0.1, 1, 10]$, i.e., in all cases

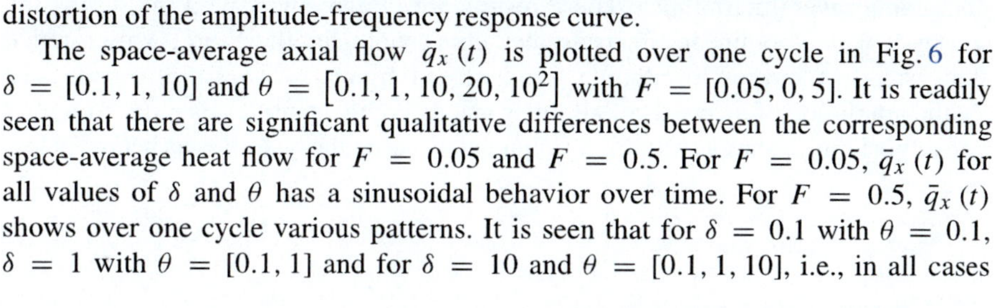

Fig. 5 Normalized oscillatory flow rate amplitude G_A/F vs $\theta \in \left[10^{-1}, 10^2\right]$

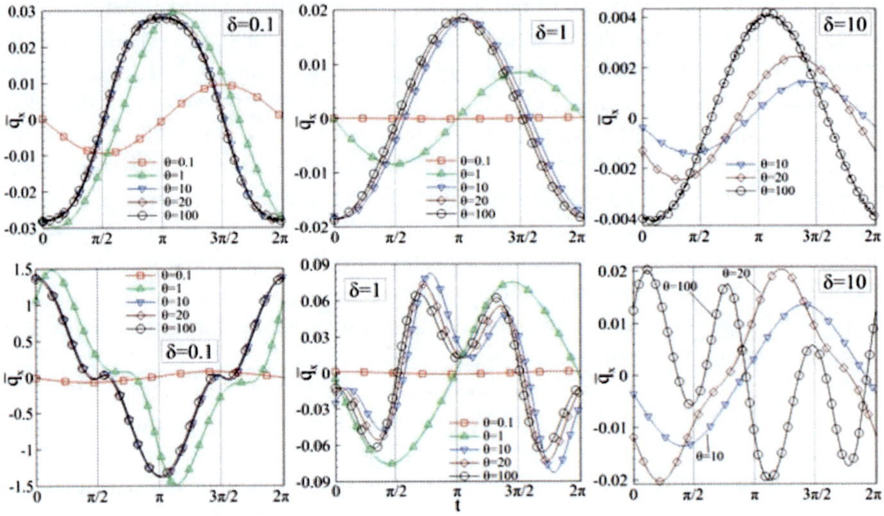

Fig. 6 Space-average axial heat flux $\bar{q}_x\,(t)$ vs t with $F = 0.05$ (up) and $F = 0.5$ (down) and $\theta = [0.1, 1, 10, 20, 100]$. Reproduced from [26], with the permission of AIP Publishing

where $\delta \geq \theta$, $\bar{q}_x\,(t)$ exhibits a sinusoidal pattern. On the contrary, in all cases where $\delta < \theta$, $\bar{q}_x\,(t)$ exhibits a rather complex non-sinusoidal pattern indicating that the introduced nonlinearities are responsible for the generation of oscillatory motion containing several harmonics. These results are in agreement with the discussion in Fig. 5, where nonlinear effects are becoming more significant in highly rarefied flow (small δ) and low oscillation frequencies (large θ). Also, for both values of F, the amplitude of $\bar{q}_x\,(t)$, as of all other macroscopic quantities, is reduced with θ and almost diminishes at very high frequencies, particularly as the gas becomes less rarefied.

4 Concluding Remarks

A brief overview of rarefied, oscillatory, pressure-driven, linear and nonlinear, fully developed flows of single gases and binary gas mixtures is provided, while the detailed analysis may be found in [24–27]. Here, the discussion is focused on the most notable findings, which include velocity overshooting, gas separation and nonlinear effects. The following concluding remarks are stated:

- Velocity overshooting (or the so-called Richardson effect) is present in oscillatory, rarefied single and binary gas mixture flows, but as the flow becomes more rarefied higher frequencies are needed to trigger this phenomenon.

- Gas separation in oscillatory binary gas mixture, may be present in the whole range of gas rarefaction provided that the flow is subject to adequate high oscillation frequency.
- Range of applicability of linear theory is much wider than expected in terms of the imposed amplitude of the oscillatory pressure gradient. The oscillatory axial heat flux is the mostly affected quantity and the only one that, due to nonlinearities, may exhibit a complex pattern.

The present results may be useful in the design of technological devices operating at moderate and high frequencies in the whole range of gas rarefaction, applicable in various technological fields.

Acknowledgments This work has been carried out within the framework of the EUROfusion Consortium and has received funding from the Euratom research and training programme 2014-2018 and 2019-2020 under grant agreement No 633053. The views and opinions expressed herein do not necessarily reflect those of the European Commission. Part of this work has been performed by using computational resources provided by the MARCONI-FUSION HPC.

References

1. Hadjiconstantinou, N.G.: Sound wave propagation in transition-regime micro- and nanochannels. Phys. Fluids **14**, 802–809 (2002)
2. Sharipov, F., Kalempa, D.: Oscillatory Couette flow at arbitrary oscillation frequency over the whole range of the Knudsen number. Microfluid. Nanofluidics **4**, 363–374 (2008)
3. Desvillettes, L., Lorenzani, S.: Sound wave resonances in micro-electro-mechanical systems devices vibrating at high frequencies according to the kinetic theory of gases. Phys. Fluids **24**, 1–24 (2012)
4. Struchtrup, H.: Resonance in rarefied gases. Contin. Mech. Thermodyn. **24**, 361–376 (2012)
5. Taguchi, S., Takata, S.: Vacuum formation behind the expansion wave in a piston motion problem. Phys. Rev. E Stat. Nonlinear Soft Matt. Phys. **86**, 016305 (2012)
6. Aoki, K., Kosuge, S., Fujiwara, T., Goudon, T.: Unsteady motion of a slightly rarefied gas caused by a plate oscillating in its normal direction. Phys. Rev. Fluids **2**, 1–33 (2017)
7. Wang, P., Zhu, L., Su, W., Wu, L., Zhang, Y.: Nonlinear oscillatory rarefied gas flow inside a rectangular cavity. Phys. Rev. E **97**, 43103 (2018)
8. Lorenzani, S., Gibelli, L., Frezzotti, A., Frangi, A., Cercignani, C.: Kinetic approach to gas flows in microchannels. Nanoscale Microscale Thermophys. Eng. **11**, 211–226 (2007)
9. Tsimpoukis, A., Valougeorgis, D.: Linear harmonic oscillatory rarefied gas flow with arbitrary frequency in comb finger blocks. Sensors Actuators A Phys. **331**, 112997 (2021)
10. Ben-Ami, Y., Manela, A.: The sound of a pulsating sphere in a rarefied gas: continuum breakdown at short length and time scales. J. Fluid Mech. **871**, 668–693 (2019)
11. Bisi, M., Lorenzani, S.: High-frequency sound wave propagation in binary gas mixtures flowing through microchannels. Phys. Fluids **28**, 052003 (2016)
12. Kalempa, D., Sharipov, F., Silva, J.C.: Sound waves in gaseous mixtures induced by vibro-thermal excitation at arbitrary rarefaction and sound frequency. Vacuum **159**, 82–98 (2019)
13. Zhang, Y., Wang, P., Guo, Z.: Oscillatory square cavity flows of binary gas mixtures. Phys. Fluids **33**, 067121 (2021)
14. Abreu, R.A., Troup, A.P., Sahm, M.K.: Causes of anomalous solid formation in the exhaust systems of low-pressure chemical vapor deposition and plasma enhanced chemical vapor deposition semiconductor processes. J. Vacuum Sci. Technol. B Microelectron. Nanometer Struct. Process. Measur. Phenomena **12**, 2763–2767 (1994)

15. Wang, S. et al.: On the modelling of the switching mechanisms of a Coanda fluidic oscillator. Sensors Actuators A Phys. **299**, 111618 (2019)
16. Jousten, K.: Applications and scope of vacuum technology. In: Jousten, K (ed.) Handbook of Vacuum Technology, pp. 19–28. Wiley-VCH Verlag GmbH & Co. KGaA, Weinheim (2016)
17. Stecki, J.S., Davis, D.C.: Fluid transmission lines–distributed parameter models part 1: a review of the state of the art. Proc. Inst. Mech. Eng. Part A Power Proc. Eng. **200**(4), 215–228 (1986)
18. Karassik, I.J., Messina, J.P., Cooper, P., Heald, C.C.: Pump Handbook. McGraw-Hill, New York (2000)
19. Kurzweg, U.H.: Enhanced heat conduction in oscillating viscous flows within parallel-plate channels. J. Fluid Mech. **156**, 291 (1985)
20. Brereton, G.J., Jalil, S.M.: Diffusive heat and mass transfer in oscillatory pipe flow. Phys. Fluids **29**, 073601 (2017)
21. Jaeger, M.J., Soepardi, T., Maddahian, A., Kurzweg, U.: Diffusional separation of gases and solutes in oscillatory flow. Sep. Sci. Technol. **26**, 503–514 (1991)
22. Jalil, S.M.: Experimental and numerical investigation of axial heat transfer enhancement by oscillatory flows. Int. J. Therm. Sci. **137**, 352–364 (2019)
23. Thomas, A.M.: Unusual effects of oscillating flows in an annulus on mass transfer and separation. Adv. Sp. Res. **32**, 279–285 (2003)
24. Tsimpoukis, A., Valougeorgis, D.: Rarefied isothermal gas flow in a long circular tube due to oscillating pressure gradient. Microfluid. Nanofluidics **22**, 5 (2018)
25. Tsimpoukis, A., Valougeorgis, D.: Pulsatile pressure driven rarefied gas flow in long rectangular ducts. Phys. Fluids **30**, 047104 (2018)
26. Tsimpoukis, A., Vasileiadis, N., Tatsios, G., Valougeorgis, D.: Nonlinear oscillatory fully-developed rarefied gas flow in plane geometry. Phys. Fluids **31**, 067108 (2019)
27. Tsimpoukis, A., Naris, S., Valougeorgis, D.: Oscillatory pressure-driven rarefied binary gas mixture flow between parallel plates. Phys. Rev. E **103**, 033103 (2021)
28. Zamir, M.: The Physics of Pulsatile Flow. Springer, New York (2000)
29. McCormack, F.J.: Construction of linearized kinetic models for gaseous mixtures and molecular gases. Phys. Fluids **16**, 2095 (1973)
30. Naris, S., Valougeorgis, D., Kalempa, D., Sharipov, F.: Gaseous mixture flow between two parallel plates in the whole range of the gas rarefaction. Phys. A Stat. Mech. its Appl. **336**, 294–318 (2004)
31. Valougeorgis, D., Vargas, M., Naris, S.: Analysis of gas separation, conductance and equivalent single gas approach for binary gas mixture flow expansion through tubes of various lengths into vacuum. Vacuum **128**, 1–8 (2016)
32. Bird, G.A.: Molecular Gas Dynamics and the Direct Simulation of Gas Flows. Clarendon Press, Oxford (1994)